UNITED STATES WATER LAW

AN INTRODUCTION

UNITED STATES WATER LAW

AN INTRODUCTION

JOHN W. JOHNSON

CRC Press
Taylor & Francis Group
Boca Raton London New York

CRC Press is an imprint of the
Taylor & Francis Group, an **informa** business

CRC Press
Taylor & Francis Group
6000 Broken Sound Parkway NW, Suite 300
Boca Raton, FL 33487-2742

First issued in paperback 2019

ISBN-13: 978-1-4200-8641-6 (hbk)
ISBN-13: 978-0-367-38608-5 (pbk)

Library of Congress Cataloging-in-Publication Data

Johnson, John W., 1946-
    United States water law : an introduction / author, John W. Johnson.
        p. cm.
    Includes bibliographical references and index.
    ISBN 978-1-4200-8641-6 (alk. paper)
        1. Water rights--United States. 2. Riparian rights--United States. 3. Water--Law and legislation--United States. 4. Water resources development--Law and legislation--United States. 5. Water-supply--United States. I. Title.

    KF645.J64 2008
    346.7304'32--dc22   2008048656

Visit the Taylor & Francis Web site at
http://www.taylorandfrancis.com

and the CRC Press Web site at
http://www.crcpress.com

Contents

Preface

Much like timber, oil, coal, and many metals, water is a natural resource. It is essential to life and its use is therefore critical. Water is a renewable natural resource since our supply is continually being renewed by precipitation and the various forms of runoff and pooling that follow.

Even though water is renewable, suitable water is limited in quantity. Demand for water often exceeds the supply. With so many individuals hoping to use the limited supply of water, it is not surprising that many are unable to obtain the rights they seek. Water is often regulated in two ways, quantity and quality. Quantity deals with the amount of water available for use and quality* deals with the amount of discharge allowed to enter the water.

There are two major systems for determining water rights in the United States. Water rights in the western United States are distinguished from those in the eastern United States, which use a system largely based on ownership of the riparian land (that which borders the body of water). The arid conditions of much of the American west, combined with the immense property owned by the federal and state governments helped to create a different system there, based on different needs.

The discussion of water rights below encompasses both of these systems. It is necessary to understand riparian rights, even in the American west, because many western states use what is known as a hybrid system, recognizing early riparian rights that pre-dated the adoption of later prior appropriation systems. Additionally, most states recognize at least some riparian rights to use the surface of water.

This work is intended only as an introductory text and those wishing for greater understanding should consult the Acknowledgments section, footnotes and other citations. There are tables of cases and statutes in the back of the text, listing the chapters and sections of each citation. There is also an index listing many of the major subjects alphabetically. The table of contents is a sectional listing of all major chapters, sections, appendices, forms, figures, and tables in the text by page number. Also included, is an extensive glossary of many of the relevant terms, used in the field.

In an effort to explain at least the basic principals and differences that exist in water law, I have included numerous references to state statutes. A list of web pages for state codes appears in the rear of this text along with a listing of western states permit agencies. I have also included a discussion of some federal case law, some important federal statutes, tables of water uses, forms, figures, etc.

This work is dedicated to my wife Alicia, whose work with the Western States Water Council, for the Army Corp. of Engineers inspired the project in the first place.

---

* Spoken another way, quality refers to the purity of the water as it is related to suitability for various purposes.

The Author

John W. Johnson earned a Bachelor of Arts in political science from Brigham Young University in 2001, where he studied comparative governments and social theory. He graduated cum laude from Seattle University School of Law, and is a member of the Washington State Bar Association.

1 The Division of Authority Between Federal and State Governments

1.1 FEDERALISM AND ITS MEANING

The rules and laws affecting water rights are heavily dependent on the division of authority between the federal and state governments. This division is often referred to as federalism.

1.2 FEDERAL AUTHORITY OVER THE WATERS GENERALLY

The federal government in the United States is one of limited power. This means "powers not delegated to the United States" belong to the people or the states.*

1.3 UNITED STATES AUTHORITY UNDER THE CONSTITUTION PART 1: THE COMMERCE CLAUSE (ART. 1 § 8 CLAUSE 3)

One of the primary claims to federal authority over the waters of the United States comes from Art. 1 § 8 clause 3 of the United States Constitution, which grants the United States power to regulate commerce "with foreign nations and among the several states, and with the Indian tribes."†

1.4 THE COMMERCE CLAUSE AND RIGHTS UNDER THE ENGLISH COMMON LAW

English Common Law recognized rights of all subjects to navigate, fish, or hunt on the surface of navigable waterways. This rule is recognized in the United States today.

---

* United States Constitution Amendment 10
† The states are not free to interfere with the federal governments exercise of its enumerated powers. State licenses, which in effect undermine federal programs, are invalid. See *First Iowa Hydro-Electric Cooperative v. Federal Power Commission*, 328 U.S. 152, 66 S.Ct. 906, 90 L.Ed. 1143 (1946). See also *California v. FERC*, 495 U.S. 490, 110 S.Ct. 2024, 109 L.Ed.2d 474 (1990). It remains possible that Congress may subject federal programs to state licensing if it chooses to, but state programs over commerce are said to be preempted when Congress chooses to act on its commerce authority. See *Patterson v. McLean Credit Union*, 491 U.S. 164, 109 S.Ct. 2363, 105 L.Ed.2d 132 (1989). Congress has on occasion deferred to state water law. In those instances, that policy can have a strong influence on the Court, which may allow the state regulation so long as any burden on federal programs is insubstantial. See *California v. United States*, 438 U.S. 645, 98 S.Ct. 2985, 57 L.Ed.2d 1018 (1978). See § 16.8. See also *Port of Seattle v. Oregon & W. R. Co.*, 255 U.S. 56, 41 S. Ct. 237, 65 L. Ed. 500 (1921) (federal authority over navigable waters in the states limited to commerce clause).

Submerged lands susceptible to these uses were owned by the Crown in trust and could not be sold or used by the monarch against such public rights. These restrictions applied to all lands affected by the ebb and flow of the tide.* In the United States the need to regulate the use of large rivers made this rule insufficient to protect public rights to such navigable bodies. The United States Supreme Court rejected the ebb and flow rule in favor of a new rule, which considers the navigable character of the river.† "Navigability" is now a factor in resolving disputes over title to submerged lands.

1.5   FEDERAL AUTHORITY UNDER THE COMMERCE CLAUSE OVER NONPUBLIC WATERS

Federal authority under the Commerce Clause includes the ability to regulate activities on nonpublic waters in order to protect visitors and wildlife.‡

1.6   THE EXTENT OF GOVERNMENT LIABILITY

The federal government is not liable for activities that raise the water level on nonpublic lands within the bed of the stream or lake but is liable for damages caused to private property beyond the bed.§

1.7   THE NEW FEDERAL TEST "NAVIGABILITY"¶

Federal authority to regulate is often based on navigability** under the Commerce Clause. The definition of navigable waters is extremely important in determining if federal regulation is appropriate under the Commerce Clause. Generally speaking, if under the test, a body of water is navigable; title to the land beneath it passed to the state upon its admission to the Union.†† This includes *noncoastal* tidelands.‡‡ The Court later held that *coastal* tidelands (meaning those lands along the ocean) were still held in trust by the federal government.§§ Congress in turn responded by turning over control of the tidelands to the states.¶¶

---

* For definition of ebb and flow test and its history see 78 Am Jur 2d, Waters § 60 with 1999 update.

† *The Propeller Chief v. Fitzhugh*, 53 U.S. 443, 12 How. 443, 13 L.Ed. 1058 (1851). See also *Port of Seattle v. Oregon & W. R. Co.*, 255 U.S. 56, 41 S. Ct. 237, 65 L. Ed. 500 (1921).

‡ *U. S. v. Brown*, 552 F.2d 817 (8th Cir. 1977).

§ *Goose Creek Hunting Club, Inc. v. U. S.*, 207 Ct. Cl. 323, 518 F.2d 579 (1975).

¶ For more on navigable waters see 78 Am Jur 2d, Waters §§ 59-81 with 1999 update; 25 Fed Proc, L Ed § 57 (Navigable Waters); 13 Fed Proc Forms 51 (Navigable Waters).

** A U.S. Army Corps of Engineers study shows a steady increase in traffic and tonnage shipped over major U.S. waterways from 1975 to 1997. For example, the Mississippi River system shipped 453.4 million tons in 1975 and increased steadily to 707.1 million tons in 1997, with a peak year in 1995 seeing 710.1 million tons shipped. See George T. Kurien, *Datapedia of the United States 1790-2005*, 340 (2001). Overall commerce across U.S. waterways remains an important use of the water.

†† *Pollard v. Hagan*, 44 U.S. 212, 3 How. 212, 11 L.Ed. 565 (1845), (also termed *Pollard*'s Lessee v. Haggin). See also § A1 (appendices).

‡‡ See *Phillips Petroleum Co. v. Mississippi*, 484 U.S. 469, 108 S.Ct. 791, 98 L.Ed.2d 877 (1988) (holding that upon admission to the Union, states take title to land beneath waters susceptible to the ebb and flow of the tide, whether navigable in fact or not) also mentioned in §§ 1.9, 3.2.

§§ *United States v. California*, 332 U.S. 19, 67 S.Ct. 1658, 91 L.Ed. 1889 (1947).

¶¶ The Submerged Lands Act, 43 U.S.C.A. §§ 1301, 1311.

The *federal test* for navigability is whether the water body was "navigable in fact" at the time the state entered the Union. The body must have been susceptible at the time to use as an avenue of commerce.*

Different definitions for "navigability" may be used in different contexts.† Federal authority sometimes extends to activities on shores and tributaries that are navigable, in fact, to protect navigation.‡ More recently, water quality statutes have maximized federal jurisdiction over navigable waters by regulating activities that might impact the quality of water in navigable waterways.§ The primary agency regulating obstruction of navigable waters is the U.S. Army Corps of Engineers.¶

The Commerce Clause grants Congress the power to authorize certain activities on and around navigable waters in the interest of assisting navigation.** These activities include flood control.††

1.8   EXTENT OF NAVIGABILITY (SEE ALSO 11.3, "NUISANCE LAW")

The Great Salt Lake was found to meet this test‡‡ because there was evidence that a few boats used the lake for trade when Utah became a state.§§ It is generally sufficient for

---

* *The Daniel Ball,* 77 U.S. 557, 19 L.Ed. 999 (1870) (saying "those rivers must be regarded as public navigable rivers in law which are navigable in fact. And they are navigable in fact when they are used, or susceptible of being used, in their ordinary condition, as highways for commerce, over which trade and travel are or may be conducted in the customary modes of trade or travel on water." The case also required capacity for interstate commerce). See *United States v. Appalachian Power Co.,* 311 U.S. 377, 61 S.Ct. 291, 85 L.Ed. 243 (1940) discussed in footnotes to § 11.4. In *Alaska v. United States,* 563 F.Supp. 1223 (D. Alaska 1983), the Court found that floatplanes were not a customary mode of trade or travel on water in Alaska for purposes of navigability or travel, even though common when Alaska was admitted to the Union in 1959.
† 13 Fed Proc Forms, Navigable Waters § 51:3, with 2005 update. See 2 Am Jur 2d, Admiralty §§ 59, 60 (navigability to determine admiralty jurisdiction); 78 Am Jur 2d, Waters § 59 (navigability to determine private ownership and right of public to use waters); 78 Am Jur 2d, Waters § 72 (navigability to determine title to the bed and stream). See Restatement (Second) of Torts § 841 for definition of "watercourse": "a stream of water of natural origin, flowing constantly or recurrently on the surface of the earth in a reasonably definite natural channel." This term "includes springs, lakes, or marshes in which a stream originates or through which it flows." See 78 Am Jur 2d, Waters § 5 (1999 update).
‡ 13 Fed Proc Forms, Navigable Waters § 51:3, with 2005 update. For what constitutes navigable in fact, see 78 Am Jur 2d, Waters § 61.
§ See 33 U.S.C.A. § 1362(7) (defining term "navigable waters"). For Army Corps of Engineers definition of navigability, see 33 C.F.R. Part 329; Coast Guard definition found at 33 C.F.R. § 2.05–25. For more on water quality see also Chapter 10.
¶ I.e., 33 U.S.C.A. § 403 (Rivers and Harbors Act). District Courts have jurisdiction to hear cases involving the removal of obstructions and issue injunctions. 33 U.S.C.A. § 406. For Corps of Engineers and federal authority to remove obstructions, see generally, 13 Fed Proc Forms, Navigable Waters, part II §§ 51:4–51:50, Proceedings for the Protection and Improvement of Navigable Waters (Corps of Engineers); 25 Fed Proc, L Ed § 57:399–57:432. For procedures promulgated by the Corps of Engineers, see 33 C.F.R. Part 329.
** *Oklahoma ex rel. Phillips v. Guy F. Atkinson Co.,* 313 U.S. 508, 85 L.Ed. 1487, 61 S.Ct. 1050 (1941).
†† 25 Fed Proc L Ed §§ 57:399–57:403, Navigable Waters: Flood Control Procedures.
‡‡ See § 1.7.
§§ *Utah v. United States,* 403 U.S. 9, 91 S.Ct. 1775, 29 L.Ed.2d 279 (1971). Ordinarily, the United States has owned navigable waters in its territories, in trust for them when they became states. Under the Equal Footing Doctrine, with few exceptions, title passed to the states upon admission to the Union. See *Pollard* § 1.5 footnotes. For rare cases where title was reserved or conveyed before statehood, see *United States v. Cherokee Nation,* 480 U.S. 700, 107 S.Ct. 1487, 94 L.Ed.2d 493 (1987) (see §§ 3.11, 9.14); *Montana v. United States,* 450 U.S. 544, 101 S.Ct. 1245, 67 L.Ed.2d 493 (1981). The extent of transfer under the Equal Footing Doctrine is addressed in *United States v. Utah,* 283 U.S. 64, 51 S.Ct. 438, 75 L.Ed. 844 (1931) (holding that interstate and international highways are passed to the states). See § 11.3.

federal purposes, that a river has been used as a highway for commerce at any time in the past and it is not necessary that it is still capable of such uses.* Privately owned waters may fall under § 10 jurisdiction if they in fact support interstate commerce.† The basic test to determine the extent of federal authority to regulate was laid out in *The Daniel Ball*.‡

More recently, the Supreme Court held that waters separated from traditional navigable waters by a nonpermeable man-made dike were not navigable solely on the basis of proximity to the more traditional waters.§

The Court found that "the phrase 'the waters of the United States' includes only those relatively permanent, standing or continuously flowing bodies of water 'forming geographic features' that are described in ordinary parlance as 'streams,' 'oceans, rivers, [and] lakes,'¶ and does not include channels through which water flows intermittently or ephemerally, or channels that periodically provide drainage for rainfall."** Courts have recognized several tests for navigability over the years, including 1) ebb and flow of the tide, 2) continuous interstate waterway, 3) navigable capacity, and 4) navigable in fact.†† Both the Corps of Engineers and the courts have interpreted the navigable waters of the United States to include some adjacent wetlands.‡‡

"Public waters" are those waters of the United States and those of the various states that are navigable under the definitions of the jurisdictions in which they are located.§§ They also include waters under state law, which are open to the public for use.¶¶ "Private waters" are those waters that are "privately owned and that do not come within the definition of public waters."***

---

* *Miami Valley Conservancy Dist. v. Alexander*, 692 F.2d 447 (1982).
† 78 Am Jur 2d, Waters § 62 with 1999 update. § 33 U.S.C.A. § 403; *United States v. Kaiser Aetna*, 408 F.Supp. 42 (D.C. Hawaii 1979) (note that use alone was not enough, but the charge of a toll for use of the canal or harbor by private persons was).
‡ *Minnehaha Creek Watershed Dist. v. Hoffman*, 597 F.2d 617 (1979). See The Daniel Ball, 77 U.S. 557, 19 L.Ed. 999 (1870). The test requires that 1) the body of water is navigable in fact, and 2) by itself or together with other bodies it forms a highway for interstate commerce. Note that intrastate highways with no interstate navigation do not fall within federal jurisdiction under the Rivers and Harbors Act. A body of water located entirely within a single state and forming no part of an interstate highway is not navigable within the meaning of the Rivers and Harbors Act. National Wildlife Federation v. Alexander, 198 App. DC 321, 613 F.2d 1054 (1979). Waters used from Utah Lake (entirely within Utah) to irrigate crops sold in interstate commerce were found within federal jurisdiction under the Federal Water Pollution Control Act, 33 U.S.C.A. §§ 1251–1376, as authorized by the Commerce Clause. *Utah Division of Parks & Recreation v. Marsh*, 740 F.2d 799 (C.A. 10 Utah 1984). Location of Jackson River entirely within Virginia exempts it from jurisdiction under 33 U.S.C.A. § 59(l). *Loving v. Alexander*, 548 F.Supp. 1079 (1982).
§ *Rapanos v. United States Army Corp. of Engineers*, 2006 U.S. Lexis 4887. (See § 10.7, Appendix A1.4).
¶ *Webster's New International Dictionary*, 2882 (2d ed.).
** *Rapanos v. United States Army Corp. of Engineers*, 547 U.S. 715 (2006). The Court has recognized that Congress has power to regulate beyond traditional notions of navigability under the Commerce Clause, but legislation must still intend the more expanded jurisdiction before it applies. See *Chemehuevi Tribe of Indians v. FPC*, 420 U.S. 395, 95 S.Ct. 1066, 43 L.Ed.2d 279 (1974). In *Kaiser Aetna v. United States*, 444 U.S. 164, 100 S.Ct. 383, 62 L.Ed.2d 332 (1979), the Court stated that Congress has authority to enact legislation regarding flood protection based on the theory that such measures protect navigable waters and facilities from harm. The court found the pond to be navigable under the commerce authority of Congress but refused to grant a public right of access. See § 3.11.
†† *Bayou Des Familles Development Corp. v. United States Corps of Engineers*, 541 F.Supp. 1025 (E.D. La. 1982).
‡‡ 78 Am Jur 2d, Waters § 60; *Jentgen v. United States*, 657 F.2d 1210 (1981), but see *Rapanos v. United States Army Corp. of Engineers*, 2006 U.S. Lexis 4887.
§§ 78 Am Jur 2d, Waters § 60 update 1999.
¶¶ *Id.*
*** 78 Am Jur 2d, Waters § 60 update 1999; Restatement (Second) of Torts § 847A.

The Court found no general right of the public to use a system of man-made canals, which were built on private property and joined artificially to navigable waters.* The Court determined that government could create such a right but must compensate the landowners for the value.†

1.9 NON-NAVIGABLE WATERS INFLUENCED BY THE TIDE

The federal government also held in trust the noncoastal, non-navigable waters influenced by the tides. These were also turned over to the states upon entry to the Union.‡ These lands include the submerged land beginning at the mean high-water line, determined by the average high-water mark over all seasons. The federal government usually turned over such lands to riparian landowners.§ A few states such as Wisconsin and Iowa took title to these lands themselves.¶

1.10 FEDERAL COMMON LAW

The definition for navigability is a matter of federal common law.** Federal common law is described as "the body of decisional law derived from federal courts adjudicating federal questions and other matters of federal concern, such as the law applying to disputes between two states, as well as federal and foreign relations law."††

1.11 UNITED STATES AUTHORITY UNDER THE CONSTITUTION PART 2: *THE ENCLAVE CLAUSE* (ART. 1 § 8 CLAUSE 17)

The Constitution grants to the federal government "Authority over all Places purchased" by the federal government "by the consent of the Legislature of the State in which the Same shall be, for Erection of Forts, Magazines, Arsenals, dock-Yards, and other needful Buildings." This combined with the clause in Art. IV § 3 clause 2 has been interpreted to grant to the federal government authority over federal lands, including National Parks. The interpretation is likely in error but federal authority in these areas still largely traces itself to interpretations of these clauses. Art. IV states that Congress has power to "dispose of and make all needful Rules and Regulations respecting the Territory or other Property belonging to the United States." This clause was almost certainly referring to lands ceded to the federal government by the original states for the purpose of selling the land to fund the new government. The authority granted by Art. IV seemingly never intended such a broad interpretation or use as we see today.

---

* *Vaughn v. Vermion Corp.*, 62 L.Ed. 2d 265, 100 S.Ct. 399 (1979 U.S.)(United States must exercise power of eminent domain to claim public right to canals created artificially and linked to navigable waters with private funds).
† 78 Am Jur 2d, Waters § 92 with 1999 update. No right under Art. I, § 8 cl. 3, the Commerce Clause. *Vaughn v. Vermion Corp.*, 62 L.Ed. 2d 265, 100 S.Ct. 399 (1979 U.S.).
‡ *Phillips Petroleum Co. v. Mississippi*, 484 U.S. 469, 108 S.Ct. 791, 98 L.Ed.2d 877 (1988) (see §§ 1.7, 3.2).
§ See Riparian § 6.2.
¶ See State Definitions of Navigability § 1.12.
** *Hughes v. Washington*, 389 U.S. 290, 88 S.Ct. 438, 19 L.Ed.2d 530 (1967).
†† *Black's* Law Dictionary, Seventh Edition, 1999.

1.12   STATE DEFINITIONS OF NAVIGABILITY

Some states have used definitions of navigability other than the federal definition, which ties navigability to the ability to sustain commercial navigation. One of the most common is the "saw log" definition, asking if a stream allows passage of logs on their way to the mill. Other states have used the "pleasure boat"* or "recreational use"† test, an expansion of the definition based on the stream's capacity for use in recreation. This definition would include a waterway useful for rowboats, canoes, or inflatable rafts. The primary significance to these definitions is in determining public rights to the navigable waters.‡

1.13   OWNERSHIP OF THE BED (SEE ALSO 6.21, "ACCRETION AND AVULSION")

The primary significance of bed ownership is in determining mineral rights.§ Tradition has it that coastal beds are the property of the federal government and noncoastal beds pass on statehood to the states. Beds under non-navigable waters are often privately owned.¶

1.14   FEDERAL AUTHORITY TO PROMOTE THE GENERAL WELFARE**

Art. 1, Section 8 authorizes Congress to collect taxes to promote the general welfare of the United States. This has been used to justify some federal projects, making the continued expansion of the commerce authority unnecessary.††

---

* *People v. Mack*, 19 Cal.App.3d 1040, 97 Cal. Rptr. 448 (Cal. App. 3 Dist. 1971).

† *Arkansas v. McIlroy*, 268 Ark. 227, 595 S.W.2d 659 (Ark. 1980).

‡ See Chapter 3 on public rights to surface waters.

§ *United States v. Louisiana*, 470 U.S. 93, 105 S.Ct. 1074, 84 L.Ed.2d 73(1985).

¶ Some states retain title to these lands themselves, though most do not. See § 1.9.

** United States Constitution, Art. I, sec. 8: "The Congress shall have Power to lay and collect Taxes, Duties, Imposts and Excises, to pay the Debts and provide for the common Defense and general Welfare of the United States; but all Duties, Imposts and Excises shall be uniform throughout the United States . . ."

†† *United States v. Butler*, 297 U.S. 1, 56 S.Ct. 312, 80 L.Ed. 477 (1936) (holding that the general welfare clause is not restricted by the enumerated powers). Note: It is unlikely that this was an intended interpretation. The General Welfare Clause appears to be used here in place of the necessary and proper clause, which, in conjunction with other grants of power in the United States Constitution, authorizes a considerable amount of federal activity but is limited to some extent by the enumerated powers. The General Welfare Clause very likely was no more than a grant to collect taxes. It was probably not intended that Congress should have no authority beyond those powers enumerated in the Constitution. While the exercise of unremunerated powers would not implicate the Supremacy Clause and would remain subject to state laws and regulations, Congress certainly had authority to do many things not specifically enumerated, such as buying and owning property. This power exists as a natural consequence of being a sovereign. It is ironic that Jefferson actually viewed the purchase of property by the federal government as unauthorized by the Constitution when he made the Louisiana Purchase. This minimalist view of the Constitution was probably more restrictive than the actual intended power of the federal government. Still, it may be somewhat closer to that intent than our present expansive reading of the Constitution's enumerated powers, at least so far as the clauses actually used to justify those powers are concerned.

1.15   PREEMPTION, FEDERAL AUTHORITY AND THE SUPREMACY CLAUSE*

Article VI cl. 2 states in part that "This Constitution, and the Laws of the United States which shall be made in Pursuance thereof; . . . shall be the supreme Law of the Land." When Congress acts within its Constitutional authority to preempt state actions, the Supremacy Clause is the source of its power to do so. Absent an expression of intent to preempt state action, states may ordinarily occupy the same field as the federal government, to the extent that the state does not interfere with the federal program or its purpose.†

1.16   EQUAL FOOTING DOCTRINE

The Equal Footing Doctrine takes its name from Northwest Ordinance of 1787.‡ That document declared it the policy of the nation to admit new states from the territory involved, on "an equal footing with the original States in all respects whatever . . ."§ With respect to waters, the courts have interpreted the doctrine as granting rights to navigable waters and the soils beneath them, within the territory, to the new state upon its admission into the Union.¶ Even with this "absolute right," the right remains subject to the power of the federal government to ensure use of the waters for interstate and international travel.**

1.17   FEDERAL COURT JURISDICTION OVER WATER RIGHTS

The jurisdiction of federal courts to hear cases over water rights is limited and must be authorized by statute. The prerequisites for federal jurisdiction are either: 1) diversity of citizenship,†† 2) federal question,‡‡ or 3) when the United States is a plaintiff.§§ Jurisdiction may also exist if the United States intervenes as a claimant to ascertain water rights.¶¶ U.S. courts also have jurisdiction to issue certain writs under the All Writs Act.*** As a note of caution, under the

---

* U.S. Const. art. VI, cl. 2.
† *Gillis v. Louisiana,* 294 F.3d 755 (5th Cir. 2002) (the state of Louisiana may regulate shipping to its ports where Congress has not acted to preempt state regulation).
‡ Northwest Ordinance, Article V.
§ David E. Engdahl. *Constitutional Federalism in a Nut Shell,* § 9.07 at 222 (1987). Note that the Equal Footing Doctrine does not limit Congress power under the Commerce Clause to regulate "navigable waters." *Wisconsin v. EPA,* 266 F.3d 741 (CA7 Wis. 2001).
¶ See *Pollard v. Hagen,* 3 How. (44 U.S.) 212 (1845); *New Orleans v. U.S.,* 10 Pet. (35 U.S.) 662 (1836).
** *United States v. Cherokee Nation of Oklahoma,* 480 U.S. 700, 94 L.Ed. 2d 704, 107 S.Ct. 1487 (1987).
†† 28 U.S.C.A. § 1332; *Mutual Chemical Co. v. Baltimore,* 33 F.Supp. 881 (D.C. Md. 1940), mod. on other grounds 122 F.2d 385 (C.A.4 Md.). 25 Fed Proc L Ed § 56:1933.
‡‡ 28 U.S.C.A. § 1331; *Weiland v. Pioneer Irrig. Co.,* 259 U.S. 498, 66 L.Ed. 1027, 42 S.Ct. 568 (1922). 25 Fed Proc L Ed § 56:1933.
§§ 28 U.S.C.A. § 1345; *Cappaert v. United States,* 426 U.S. 128, 48 L.Ed. 2d 523, 96 S. Ct. 2062 (1976).
¶¶ 28 U.S.C.A. § 1345; *New Mexico ex rel Reynolds v. Molybdenum Corp. of America,* 570 F.2d 1364 (C.A.10 NM 1978). 25 Fed Proc L Ed § 56:1933.
*** 43 U.S.C.A. § 1651; *United States v. United States Dist. Court,* 206 F.2d 303 (C.A.9 1953). 25 Fed Proc L.Ed § 56:1933.

McCarran Amendment,* the United States may be joined under certain conditions as a defendant in state court regarding the administration of federal water rights. Federal courts may determine ancillary nonfederal claims once its jurisdiction has been properly invoked.†

---

* 43 U.S.C.S. § 666. 25 Fed Proc L Ed § 56:1933.
† *New Mexico ex rel Reynolds v. Molybdenum Corp. of America*, 570 F.2d 1364 (C.A.10 NM 1978). 25 Fed Proc L Ed § 56:1933.

2 Allocation of Interstate Water Rights*†

2.1 GENERALLY

Because water is a limited resource that often crosses state boundaries, it can lead to competition among the states‡ and may sometimes lead to collective action problems (see Chapter 12). Use of both ground and surface waters by one state may have a profound impact on the water resources of another. An upstream state may choose to appropriate an entire stream so as to deny a downstream state any use of the resource. Likewise, use of an aquifer by one state may lower the water table in another. As one may well imagine, this has led to a fair amount of squabbling among states, particularly in the West. As a general rule, the authority of a state to issue and regulate water in its borders must yield to the vested rights of other states.§

2.2 EQUITABLE APPORTIONMENT

The United States Supreme Court has jurisdiction in disputes between states under Article III § 2 of the United States Constitution. The Court has stated that there must be

---

* For discussion of game theory, see Chapter 12.
† For more on water rights as between states, see 78 Am Jur 2d, Waters §§ 309–315.
‡ For recent cases, see *North Dakota v. U.S. Army Corps of Engineers*, 05-628 (2006); *Arizona v. California*, 547 U.S. __ (2006). See §§ 2.2, 9.2. Another conflict is currently brewing between the states of Florida and Georgia regarding the release of water from an Army reservoir in Georgia for use in Florida. Georgia, like much of the United States, is in the midst of a drought, and the release to Florida threatens municipal supplies in Georgia.

Negotiations are ongoing between the states of Florida, Alabama, and Georgia as the two downstream states (Alabama and Florida) complain that Georgia consumption is threatening the downstream supply. Georgia has sued the Army Corps of Engineers to limit further releases. Florida ranked third in total water consumption (20,100 million gallons per day) among the states in 2000, behind only California (51,200 million gallons per day) and Texas (29,600 million gallons per day). In comparison, Georgia consumed 6,500 million gallons per day. For complete comparison among states, see Table 1 in Appendix 3. The 2005 data is being compiled but as of the date of this entry had not yet been released by USGS.

§ *State of Wyo. v. State of Colo.*, 298 U.S. 573, 56 S. Ct. 912, 80 L. Ed. 1339 (1936); *State of Arizona v. State of California*, 298 U.S. 558, 56 S. Ct. 848, 80 L. Ed. 1331 (1936). Note that downstream states are not entitled to the natural flow of a stream. *State of Colo. v. State of Kan.*, 320 U.S. 383, 64 S. Ct. 176, 88 L. Ed. 116 (1943).

an equitable distribution of water between the states.* The doctrine of equitable apportionment will not help private individuals where the Supreme Court has not apportioned
a stream between the states.† In a case between New York and New Jersey, Justice
Holmes stated, "New York has the physical power to cut off all the water within its jurisdiction. But clearly the exercise of such a power to the destruction of the interest of lower
States could not be tolerated."‡ The Court weighs all of the factors that create equities,
including costs and benefits and does not strictly apply the priority rule.§ In order for
the Court to intercede, there must be a clear showing of harm by one state against
another.¶ Ordinarily, a special water master is appointed to mediate the disputes before
the Court reviews it. Many of these disputes are reoccurring and have been reviewed
intermittently, over the years, by the Court. The Court recently issued further rulings in
a dispute between California and Arizona, which has been going on for generations.**

2.3   TRANSPORT OUTSIDE THE WATERSHED

The United States Supreme Court does not allocate between states based on the location of the use.†† Movement of water outside the watershed is irrelevant with regard
to interstate disputes and rights.‡‡

---

* *Kansas v. Colorado*, 206 U.S. 46, 27 S.Ct. 655, 51 L.Ed. 956 (1907) (the court found against Kansas on
this occasion but suggested if increased use in Colorado continued, it would eventually warrant relief
for Kansas by throwing out of balance the equitable division of benefits); *State of Connecticut v. Com.
of Mass.*, 282 U.S. 660, 51 S. Ct. 286, 75 L. Ed. 602 (1931) (this does not mean an equal distribution
and diversion for domestic supply will not be enjoined absent show of significant harm downstream);
State of New York v. State of Illinois, 274 U.S. 488, 47 S. Ct. 661, 71 L. Ed. 1164 (1927) (a state may
not get injunction to protect water for power where no plans are in place and no use is currently being
made). See § 16.5. Injunctions are allowed if a state can prove imminent danger of substantial injury.
State of Washington v. State of Oregon, 297 U.S. 517, 56 S. Ct. 540, 80 L.Ed. 837 (1936).
† *Vineyard Land & Stock Co. v. Twin Falls Salmon River Land & Water Co.*, 245 F. 9 (9th Cir. 1917).
The courts have adjudicated according to priority as between competing interests across state lines
when both states follow the appropriations doctrine. See *Bean v. Morris*, 221 U.S. 485, 31 S.Ct. 703, 55
L. Ed. 821 (1911). The problems are greater when the dispute is between individuals from states with
different doctrines.
‡ *New Jersey v. New York*, 283 U.S. 336, 51 S.Ct. 478, 75 L. Ed. 1104 (1931). See § 2.4.
§ *Colorado v. Kansas*, 320 U.S. 394, 64 S.Ct. 181, 88 L.Ed. 116 (1943); *Colorado v. New Mexico*, 459 U.S.
176, 103 S.Ct. 539, 74 L.Ed.2d 348 (1982) (addressing the factors of just apportionment and finding in
favor of Colorado) (see § 7.9); 467 U.S. 310, 104 S.Ct. 2433, 81 L.Ed.2d 247 (1984) (addressing the evidentiary standards for equitable apportionment and finding that Colorado had not met these requirements).
¶ This showing was not met in the case *Colorado v. Kansas*, therefore relief was denied in that case.
It should be noted that this is a high hurdle to overcome.
** *Arizona v. California*, 547 U.S. __ (2006); 283 U.S. 423, 51 S.Ct. 522, 75 L.Ed. 1154 (1931); 292 U.S.
341 (1934); 298 U.S. 558, 56 S.Ct. 848, 80 L.Ed. 1331 (1936); 344 U.S. 919, 73 S. Ct. 385, 97 L.Ed.
708 (1953); 373 U.S. 546, 83 S.Ct. 1468, 10 L.Ed.2d 572 (1963); 439 U.S. 419, 99 S.Ct. 995, 58 L.Ed.2d
627 (1979); 460 U.S. 605, 103 S.Ct. 1382, 75 L.Ed.2d 318 (1983). See §§ 2.1, 9.2. This is just a partial list of the major disputes to reach the court between these parties. Other recent cases involved a
dispute between Arkansas and Florida that the Court refused to hear, regarding river pollution from
the Arkansas poultry industry. The Court's refusal to hear the case indicates that it did not meet the
evidentiary standard required by the Court, of clear harm.
†† *Wyoming v. Colorado*, 259 U.S. 419, 42 S.Ct. 552, 66 L.Ed. 999 (1922) (ultimately using appropriation
as the doctrine to settle disputes between these two states).
‡‡ See *Stratton v. Mt. Hermon Boys' School*, 216 Mass. 83, 103 N.E. 87 (1913) (state of Conn. relying on
strict common law rule that all uses outside the watershed are per se unreasonable, to argue unsuccessfully against outside uses).

2.4 CONFLICT BETWEEN RIPARIAN STATES

The Supreme Court has allocated rights in these states but makes it clear that such allocations are not appropriations and give no superior right to one state, over the other.* The Court has required a minimum flow be delivered to the downstream state.

2.5 EQUITABLE APPORTIONMENT IN FISH AND OTHER NATURAL RESOURCES

The Supreme Court has held that the doctrine of equitable apportionment applies to fish as well as water.†

2.6 SPECIAL MASTER

A special master is an individual, usually with some expertise in the field, who is appointed by the Court to assist in a matter or case.

2.7 INTERSTATE COMPACTS‡

Interstate compacts have become more popular in resolving disputes over water rights between states, than litigation in court. These compacts or agreements between the states are considered binding by the courts, the same as if the court had created an apportionment itself.§ Water rights, granted by the state before the compact was entered into, may still be affected by the agreement. The Court has made clear that no state can grant rights to water that the state does not first have to give.¶ The Court has been willing to hear disputes over compact performance but is not ordinarily willing to rewrite a compact once Congress has approved it. The compact is limited to its terms.** The Court considers compacts as contracts, and violations of

---

* *New Jersey v. New York*, 283 U.S. 336, 51 S.Ct. 478, 75 L. Ed. 1104 (1931). See §, 2.2.

† *Idaho v. Oregon*, 462 U.S. 1017, 103 S.Ct. 2817, 77L.Ed.2d 387 (1983) (finding for Idaho in its suit to have an apportionment of the chinook salmon and steelhead trout migrating up the Columbia and Snake Rivers).

‡ For a list of compacts, see Table 15 in the rear of this volume. Compacts must be approved by the United States Congress in order to be binding. In 1969 and 1971, California and Nevada ratified a compact that Congress refused to ratify. The states honored it voluntarily for a number of years. Congress based its refusal on objection from the Departments of Justice and the Interior, which believed the compact failed to protect federal and Indian interests. A recent agreement (December 2007) among states of the Colorado River Basin, including California, Nevada, Arizona, Utah, Colorado, New Mexico, and Wyoming, has been reached regarding division of those water resources in light of eight years of recent drought in the region. The previous compact was agreed to in 1922. The basis of the new agreement is to allow states to use a high percentage of water they save now, when the drought ends.
See http://seattletimes.nwsource.com/html/nationworld/2004071258_water14.html, available December 14, 2007.

§ See *Hindlider v. La Plata River & Cherry Creek Ditch Co.*, 304 U.S. 92, 58 S.Ct. 803, 82 L.Ed. 1202 (1938) (overruling the finding of a lower court that prior appropriation would dictate rights above the compact and finding that the equitable apportionment is binding even if the compact was made after the granting of the water rights affected).

¶ *Id.*

** See *Texas v. New Mexico*, 462 U.S. 554, 103 S.Ct. 2558, 77 L.Ed.2d 1 (1983); 467 U.S. 1238, 104 S.Ct. 3505, 82 L.Ed.2d 816 (1984).

contractual duties are resolved accordingly. The Court has also treated compacts as statutes, preferring the law of statutory interpretation to contract law when reviewing evidence of the meaning of ambiguous terms.*

2.8   FEDERAL-INTERSTATE COMPACTS†

These are compacts in which the federal government is a party. There are no major legal differences in how these compacts are viewed or treated in the courts.

2.9   JURISDICTION OF THE UNITED STATES SUPREME COURT OVER INTERSTATE DISPUTES

The United States Supreme Court usually enjoys exclusive jurisdiction in interstate disputes.‡ When one state sues another, the Supreme Court will usually oversee the matter.§

2.10   CONGRESSIONAL APPROVAL AND THE COMPACT CLAUSE

The Compact Clause of the United States Constitution¶ requires congressional approval of any state compact or agreement with other states. This approval is treated as statutory law** by the courts, and interstate compacts become federal law once approved by Congress.

---

* See *Oklahoma v. New Mexico*, 501 U.S. 221, 111 S.Ct. 2281, 115 L.Ed.2d 207 (1991) (preferring the law of statutory interpretation over contract law to allow extrinsic evidence not permitted in contract law). This was a closely divided 5–4 decision. See § 2.10.
† Delaware River Basin Compact, among Delaware, New Jersey, New York, Pennsylvania, and the United States; Susquehanna River Basin compact, among Maryland, New York, Pennsylvania, and the United States. The power of the federal government of the United States to cede federal works and powers to the states was first recognized in *Seawright v. Stokes,* 44 U.S. (3 How.) 150 (1845).
‡ United States Constitution, Art. III, Sec. 2. See §, 2.2.
§ Note that a suit against a federal agency is not necessarily an interstate conflict within the exclusive jurisdiction of the Supreme Court. See *Alabama v. U.S. Army Corps of Engineers*, 424 F.3d 1117 (11th Cir. 2005).
¶ U.S. Const. art. I § 10 cl. 3.
** This requires presidential consent the same as other legislation. See H.R. Doc. No. 690, 77th Cong., Sess. (1942) (veto message). Sax, Joseph L., Legal Control of Water Resources 725-726 (3rd Ed, 2000). See *Oklahoma and Texas v. New Mexico*, 501 U.S. 221 (1991) (treating compact under principles of statutory interpretation and not contract law, which would involve the parol evidence rule). See also *Texas v. New Mexico*, 462 U.S. 554 (1983) (illustrating potential problems of ambiguity in interstate compacts). Compacts are federal law and can defeat state laws inconsistent with the compact. See *Hinderlider v. La Plata River & Cherry Creek Ditch Co.*, 304 U.S. 92 (1938).

3 Public Rights to Use Surface Waters

3.1 LIMITS ON PRIVATE PROPERTY RIGHTS

Even in the absence of legislation forbidding such activity, the Commerce Clause forbids any private property owner from obstructing navigability of the waters. Most navigable waterways are state- or privately owned and are dealt with primarily under state law. As such, many states have defined a broad range of permissible public uses of the waterways.* Title to the underlying land is subordinate to state authority to secure the rights of its inhabitants to full navigation of the water.†

---

* For a discussion of state ownership rights in water, see *Kansas v. Meek*, 246 Kan. 99, 785 P.2d 1356 (1990) (holding that the state cannot force a private riparian land owner to open the non-navigable water, under the federal definition, over his property to the public). Compare with *Day v. Armstrong* mentioned in footnotes to 3.4. For the federal question of navigability, see *United States v. Holt Bank*, 270 U.S. 49, 46 S.Ct. 197, 70 L.Ed. 465 (1926) (establishing ownership of the beds of navigable streams and lakes as matter of federal law and defining navigable by the following test: 1) the body of water was used or susceptible for use for commerce, 2) such use for commerce was possible under natural conditions, 3) commerce was or could be conducted in customary modes of trade and travel on the water, and 4) all these conditions were satisfied at the time of statehood) (see § 11.3).

† See *Diana Shooting Club v. Husting*, 156 Wis. 261, 145 N.W. 816 (Wis. 1914). In a WA case, *Wilbour v. Gallagher*, 462 P.2d 232 (Wash. 1969), the court ruled that an individual who filled part of his land to avoid seasonal flooding caused by a hydroelectric dam had no right to do so, since the public had seasonal rights to use the water. This case may be an anomaly, due in part to the failure of the individual to acquire permits. Even so, it is a troubling case for property owners who wish to maintain exclusive use of their property. A WI court held that a landowner who purchased the banks around a lake suitable only for duck hunting was not entitled to the diminution of the property value for loss of exclusive use when the state condemned a right-of-way. *Branch v. Oconto County*, 109 N.W.2d 105 (Wis. 1961) (see § 3.4). In *State v. Bollenbach*, 63 N.W.2d 278 (Minn. 1954), the MN court held that a similar statute to that involved in *Branch* applied only to navigable waters under the federal definition. See also *Head v. Amoskeag Mfg. Co.*, 113 U.S. 9, 5 S. Ct. 441, 28 L. Ed. 889 (1885) (use of water subject to reasonable regulation); *City of Trenton v. State of New Jersey*, 262 U.S. 182, 43 S. Ct. 534, 67 L. Ed. 937, 29 A.L.R. 1471 (1923) (state has dominion over waters in its boundaries, subject to the Constitution); *State of Ark. v. State of Tenn.*, 246 U.S. 158, 38 S. Ct. 301, 62 L. Ed. 638 (1918); *U.S. v. Cress*, 243 U.S. 316, 37 S. Ct. 380, 61 L. Ed. 746 (1917); *In re Opinion of the Justices*, 103 Me. 506, 69 A. 627 (1908) (states may pass laws to conserve water resources). The state may delegate its authority to regulate to an administrative agency. See *Ouachita Power Co. v. Donaghey*, 106 Ark. 48, 152 S.W. 1012 (1912); *In re Board of Water Com'rs of City of Hartford*, 87 Conn. 193, 87 A. 870 (1913), aff'd, 241 U.S. 649, 36 S. Ct. 552, 60 L. Ed. 1221 (1916) (state may delegate to a municipal corporation); *McGuire v. City of Rapid City*, 43 N.W. 706 (Dakota 1889) (municipal corp. authority within adjoining territory).

3.2 THE PUBLIC TRUST DOCTRINE (SEE ALSO 3.8, 5.4 ON IN-STREAM VALUES)

Even though the state takes title to the lands beneath the waters, it holds such lands in trust for the public and as such cannot convey them, unless it would convey some public purpose.* The United States Supreme Court has found a grant to a railroad invalid where the Court reasoned that the state held title to the land beneath the water in public trust and could not convey the lands inconsistent with that trust. At least thirty-nine states have considered the question and all have concluded the same thing, that the state holds the lands beneath the water in trust for the public.† Most have overruled legislative attempts to circumvent the trust.‡ Private uses are scrutinized under the doctrine to determine if they are consistent with the trust.§ The Public Trust Doctrine has been interpreted to protect public uses including navigation, commerce, fishing, hunting, bathing, swimming, and recreation.¶ The doctrine is often applied to retain free and unobstructed recreational use, even when title of the soil is in private hands.**

The public has a right to use a navigable river and the bed up to the high-water mark, for navigational, fishing, recreational, and other authorized purposes.††

3.3 NON-NAVIGABLE TRIBUTARIES OF NAVIGABLE WATERWAYS

States have sometimes expanded the Public Trust Doctrine to include non-navigable tributaries of navigable waterways.‡‡ The modern approach is that ownership of the

---

* *Shively v. Bowlby*, 152 U.S. 1, 14 S. Ct. 548, 38 L. Ed. 331 (1894); *Illinois Cent. R. Co. v. State of Illinois*, 146 U.S. 387, 13 S. Ct. 110, 36 L. Ed. 1018 (1892). But see *Appleby v. City of New York*, 271 U.S. 364, 46 S.Ct. 569, 70 L.Ed. 992 (1926) (stating *Illinois Central Railroad* was a statement of state law for the State of Illinois). It is worth noting that the later court was probably incorrect in its description of Illinois Railroad as applying state law. It is difficult to imagine this being the case. The Court was likely looking for a way to overrule the case while keeping up the appearance of precedence. *Appleby* was seemingly reaffirmed, however, by *Phillips Petroleum Co. v. Mississippi*, 484 U.S. 469, 108 S.Ct.791, 98 L.Ed.2d 877 (1988) (see §§ 1.7, 1.9). In recent years the Public Trust Doctrine has been used as a tool to protect against negative environmental impacts. See *National Audubon Society v. Superior Court of Alpine County*, 33 Cal.3d 419, 189 Cal Rptr. 346, 658 P.2d 709 (1983). The Public Trust Doctrine creates a presumption that the legislature does not intend to impair public rights under the trust. See *Gwathmey v. State Through Dept. of Environment, Health, and Natural Resources Through Cobey*, 342 N.C. 287, 464 S.E.2d 674 (1995).
† *Chatfield East Well Co., Ltd. v. Chatfield East Property Owners Ass'n*, 956 P.2d 1260 (Colo. 1998).
‡ See *Arizona Center For Law in the Public Interest v. Hassell*, 172 Ariz. 356, 837 P.2d 158 (Ariz. App. Div. 1 1991).
§ *Kootenai Environmental Alliance v. Panhandle Yacht Club, Inc.*, 105 Idaho 622, 671 P.2d 1085 (Idaho 1983).
¶ *Marks v. Whitney*, 98 Cal.App.3d 1040, 491 P.2d 374 (Cal. 1971). Some states restrict the right of the state to dispose of lands in such a way as to interfere significantly with the public's right to access water. See *Rettkowski v. Department of Ecology*, 122 Wash. 2d 219, 858 P.2d 232 (1993); *Lewis Blue Point Oyster Cultivation Co. v. Briggs*, 229 U.S. 82, 33 S. Ct. 679, 57 L. Ed. 1083 (1913); *Johnson v. Seifert*, 257 Minn. 159, 100 N.W.2d 689 (1960); *Cordovana v. Vipond*, 198 Va. 353, 94 S.E.2d 295, 65 A.L.R.2d 138 (1956).
** 78 Am Jur 2d, Waters § 118. *Brown v. Newport Concrete Co.*, 44 Ohio App. 2d 121, 336 N.E.2d 453 app. dismd.
†† *Bess v. County of Humboldt*, 3 Cal. App. 4th 1544, 5 Cal Rptr 2d 399 (1992). 78 Am Jur 2d, Waters § 87 with 1999 update.
‡‡ *National Audubon Society v. Superior Court*, 189 Cal. Rptr. 346, 658 P.2d 709 (Cal. 1983) (see footnotes to § 3.2).

land beneath the waters is irrelevant to the rights of the public to use the waterway for public purposes. It is the ability of the waterway to sustain recreational uses that determines the availability of the water to the public for use.* Many states, including Montana, Wyoming, and New Mexico hold that all waters in the state are held in public trust even on private land. This view is not universal, though. A Colorado court determined that waters entirely on private land were not for public use.† Colorado is among a small minority of states, including Kansas,‡ Alabama, Pennsylvania, and Indiana, which deny public use. Louisiana, Missouri, and Virginia grant no public hunting or fishing rights in non-navigable waters. Some older case law in Connecticut, Illinois, Maine, Massachusetts, New York, North Carolina, and Tennessee hold that privately owned beds yield only to an easement for public passage over them. Most recent decisions favor at least some public recreational use.

3.4   PUBLIC ACCESS TO WATERS

Public use of waterways often requires crossing private lands.§ States may take land by eminent domain to provide access to the public.¶ Such condemnations often lead to disputes as to how much compensation is due but it is clear the state may seize an easement right for the public. Several theories have been applied to claim that property, bordering waterways, is burdened with an easement for public access.** Many states may allow public use of the surface of waters but do not guarantee a right to cross private lands to access the waters.†† Some states follow the minority view and do not allow public use of waters bounded by private lands.‡‡

3.5   CUSTOM

Oregon courts found that the public has a right to use of the dry sand areas of ocean beaches based on custom.§§ Florida and Hawaii also cite custom as the source of

---

* *Montana Coalition for Stream Access v. Curran*, 210 Mont. 38, 682 P.2d 163 (Mont. 1984) (holding that the public may use the bed and banks up to the ordinary high-water mark and may go above the mark if necessary to navigate around obstacles); compare with *Galt v. State Dept. of Fish, Wildlife, and Parks*, 731 P.2d 912 (Mont. 1987) (stating that use of the beds and banks must be related to use of the water itself). See also Cal. Fish & Game Code § 5943, providing access to captured waters stored by dams for fishing, during the open season.

† *People v. Emmert*, 198 Colo. 137, 597 P.2d 1025 (Colo. 1979).

‡ *Meek v. Hays*, 246 Kan. 99, 785 P.2d 1356 (Kan. 1990) (a riparian owning land on both sides of stream may fence the stream to prevent trespass).

§ See *Day v. Armstrong*, 362 P.2d 137 (Wyo. 1961) (determining that ownership of waters is in the state and uses made of water by the state were not limited by law such that the legislature had power to allow people to float in streams); compare to *Kansas v. Meek* described in footnotes to § 3.1; See Minn Stat. Ann. § 103G.005(15)(b) declaring all water in Minnesota capable of substantial public use to be public; Ind. Code Ann. §§ 13-2-14-1, 13-2-14-2, claiming state control of all lakes "used by the public with the acquiescence of any or all riparian owners," for use fishing, boating, swimming, and storage of water; Iowa Code Ann. § 107.24, providing for the condemnation of access ways for water suitable for hunting, fishing, and trapping. See also Cal. Const. Art. I, Sec. 25.

¶ *Branch v. Oconto County*, 13 Wis.2d 595, 109 N.W.2d 105 (Wis. 1961) (see footnote to § 3.1).

** *Walbridge v. Robinson*, 22 Idaho 236, 125 P. 812 (1912) (title to public waters vested in the state).

†† See *Shellow v. Hagen*, 101 N.W.2d 694 (Wis. 1960).

‡‡ I.e., Colorado and Kansas.

§§ *Thornton v. Hay*, 254 Or. 584, 462 P.2d 671 (Or. 1969).

beach access.* Some New England laws that grant access only to unimproved or unenclosed lands are also rooted in custom.

3.6 IMPLIED DEDICATION

Dedication requires both intent to create an easement by the landowner and acceptance of the easement by the public. In California this may be implied when members of the public use private land adversely for five years and the landowner never gives permission but offers no protest though fully aware of the use.†

3.7 PRESCRIPTION (SEE ALSO 6.20, 7.10, AND 13.7)

Prescription occurs when the public has used an access way openly, continuously, and adversely for the required lawful time, similar to adverse possession.‡ Wisconsin recognizes the right to moor boats may be acquired this way. It is uncommon for western states to recognize prescriptive water rights.§ Many states now require an application with the state to obtain a water right and any adverse use is considered a use against the state and the community as a whole.¶ Some states have allowed prescription against appropriated water but not against the unappropriated waters of the state.** It may be possible to acquire an easement for overflow on the land of another by prescription.††

---

* *Public Access Shoreline Hawaii v. Hawaii County Planning Comm'n*, 903 P.2d 1246 (Hawaii 1995).

† *Gion v. Santa Cruz*, 84 Cal. Rptr. 162, 465 P.2d 50 (Cal. 1970).

‡ See *Mathewson v. Hoffman*, 77 Mich. 420, 43 N.W. 879 (1889).

§ *Thompson v. Bingham*, 78 Idaho 305, 302 P.2d 948 (1956) (adjoining landowner cannot acquire prescriptive right to waste or seepage); see also *Burgett v. Calentine*, 56 N.M. 194, 242 P.2d 276 (1951); *Tongue Creek Orchard Co. v. Town of Orchard City*, 131 Colo. 177, 280 P.2d 426 (1955). For a contrary view, see *Lomas v. Webster*, 109 Colo. 107, 122 P.2d 248 (1942) (holding a prescriptive right in seepage valid).

¶ *Cambell v. Wyoming Dev. Co.*, 100 P.2d 124 (Wyo. 1940); see also Utah Code Ann. 73-3-1 (for an example of a state statute forbidding adverse rights in water). Utah courts previously allowed adverse possession of water rights though such rights were very difficult to prove. See *Hammond v. Johnson*, 94 Utah 20, 66 P.2d 894 (1937); see Wash Rev. Code Ann. 90.14.220 (forbidding any prescriptive rights to water in state); other states not allowing prescriptive water rights by statute include Alaska, Idaho, Nevada, Arizona, Montana (see Mont. Code Ann. § 85-2-301(3)), and Kansas. Still others, such as California, Colorado and Wyoming, disallow prescriptive rights by judicial decree, finding the state permit systems in appropriation to be exclusive. Texas recently joined this group as well. See *In re Adjudication of the Brazos River Basin*, 746 S.W.2d 207 (Tex. 1988).

** *Mountain Meadow Ditch & Irrigation Co. v. Park Ditch & Reservoir*, 277 P.2d 527 (Colo. 1954); *Douville v. Pembina County Water Resource Dist.*, 2000 ND 124, 612 N.W.2d 270 (N.D. 2000) (prescription not allowed against state); *Kray v. Muggli*, 84 Minn. 90, 86 N.W. 882 (1901) (recognizing prescriptive rights to artificial waters); *Smith v. Youmans*, 96 Wis. 103, 70 N.W. 1115 (1897) (rights to maintain artificial waters acquired by prescriptive use); but see *Greisinger v. Klinhart*, 282 S.W. 473 (Mo. Ct. App. 1926), quashed, 292 S.W. 75 (Mo. 1926), for want of jurisdiction; *Mitchell Drainage Dist. v. Farmers' Irr. Dist.*, 127 Neb. 484, 256 N.W. 15 (1934); *Lake Drummond Canal & Water Co. v. Burnham*, 147 N.C. 41, 60 S.E. 650 (1908) (rights may depend on whether change is permanent or temporary). An artificial condition may be treated as natural under some circumstances. *Tapoco, Inc. v. Peterson*, 213 Tenn. 335, 373 S.W.2d 605 (1963); *Diversion Lake Club v. Heath*, 126 Tex. 129, 86 S.W.2d 441 (1935); *Tapoco, Inc. v. Peterson*, 213 Tenn. 335, 373 S.W.2d 605 (1963) (requiring abandonment for public to acquire rights in artificial waters).

†† 78 Am Jur 2d, Waters § 208; *Snowden v. Wilas*, 19 Ind. 10; *Ireland v. Bowman & Cockrell*, 130 Ky 153, 112 SW 56, *Swan v. Munch*, 65 Minn. 500, 67 NW 1022; *Alcorn v. Sadler*, 71 Miss. 634, 14 So. 444; *Hood v. Slefkin*, 88 RI 549; *Diking Dist. of Pend Oreille County v. Calispel Duck Club*, 11 Wash 2d 131, 118 P.2d 780; *Charnley v. Shawano Water Power & River Improv. Co.*, 109 Wis. 563, 85 NW 507. Prescriptive rights may not be recognized when the use constitutes a nuisance, though. See *Charlotte v. Pembroke Iron-Works*, 82 Me 391, 19 A. 902 (regarding flooding of highway creating danger to public).

Adverse use against a private party is generally considered an implied grant of an easement.* As such, it must be in a manner as to raise the presumption of such an easement by the other party's failure to complain. This is difficult in water law, since it is often difficult to prove knowledge on the part of the aggrieved party, equally difficult to prove usage, and given the flow of the water it is difficult to know when an adverse use is being made. As a result, many states do not allow prescriptive rights at all. In those, which do, they are rare. The use ordinarily must be open and notorious, hostile (meaning against the rights of another), adverse (without permission), and for the statutory period of time required.

Where prescription is allowed, the general rule is that it only runs against downstream riparian users.† Some jurisdictions require actual harm to the downstream user before a right may be recognized. The amount of water an adverse user may be entitled to varies among jurisdictions. Some states restrict prescriptive rights to a specific amount of water and others use a percentage of the total flow.‡ It is worth noting that the modern trend appears to be moving away from prescriptive rights, especially where water is said to belong to the state and permits are required to appropriate it. Prior-appropriation states almost never allow prescription of water rights. A few states have allowed a junior user to take a senior's priority by adverse possession.§ Ordinarily, the junior may argue for abandonment and claim a new appropriation as of the new filing date and is not entitled to the priority date of the senior.

In California, one of the rules allowing a riparian rights holder to enjoin appropriation uses has worked to undermine the riparian significantly. The rule makes it easier for appropriators to acquire prescriptive rights because any appropriative use starts the statute running against the riparian. A later California amendment¶ requires all uses to be reasonable, which further weakens riparian rights.**

3.8   PUBLIC TRUST AND EASEMENTS (SEE ALSO 3.2 "THE PUBLIC TRUST DOCTRINE" AND 5.4 "IN-STREAM VALUES")

Many courts have applied the Public Trust Doctrine as the reason for granting a public right to cross private land,†† including public use of the dry sands near the ocean.‡‡

---

* *Pabst v. Finmand*, 190 Cal. 124, 211 P. 11 (1922) (granting a prescriptive right for adverse use over the five-year statutory period). Note California law has changed to deny this right under the circumstances of this case. See Cal. Const. Article X, Section 2 (making such uses of unused water no longer actionable).
† *Wells Water Dist. v. Maine Turnpike Auth.*, 84 A.2d 433 (Me. 1951); but see *Donanello v. Gust*, 150 P. 420 (Wash 1915).
‡ *Mally v. Weidensteiner*, 153 P. 342 (Wash. 1915) (using both limits of total water and no more than one-third of the total flow); *Akin v. Spencer*, 69 P.2d 430 (Cal. Ct. App. 1937) (adverse user is entitled to full quantity of water used as against all downstream riparians).
§ E.g., Idaho, Montana, and Utah.
¶ Cal. Const., Art. X, § 2.
** The amendment was a reaction to *Herminghaus v. S. California Edison Co.*, 252 P. 607 (Cal. 1926).
†† *Hudson County Water Co. v. McCarter*, 209 U.S. 349, 28 S. Ct. 529, 52 L. Ed. 828 (1908).
‡‡ *Matthews v. Bay Head Improvement Association*, 95 N.J. 306, 471 A.2d 355 (N.J. 1984); see footnotes to § 3.4 for state statutes on this point. It is worth noting that cases such as *Curran* and *Day* expressly or implicitly hold that the public has no right to cross private lands. See footnotes to § 3.3 and 3.4. Some states also apply the "wild river" concept that the water should flow in its natural flow and banks. See Me. Rev. Stat. Ann. Tit. 12, § 661; Wis. Stat. Ann. § 30.26; Mich. Comp. Laws. Ann. § 281.761. These laws ordinarily deal with lands considered public lands and are for the purpose of preserving the land through regulations on land use and zoning of the banks.

3.9   POLICE POWER IN CALIFORNIA AND OTHER COASTAL STATES

California imposes extensive regulations on property owners for the benefit of the public. Other states may impose regulations to a lesser degree. Such conditions often include easements for view and for public access.* The California Coastal Act was ruled largely constitutional to the extent it served the legislative purpose. A provision requiring lateral access along the shoreline of an objecting property owner seeking a building permit was found by the Court not specifically tailored enough to satisfy the stated purpose of ensuring views of the ocean for back lots.† The Supreme Court ruled that there must be a reasonable relation between the restriction and the public need. Where a city sought a public right-of-way on land along a creek as a condition for a commercial building permit, the court ruled that the city must prove that the exaction is "roughly proportional" to the impact of the construction.‡

3.10   EMINENT DOMAIN (SEE ALSO 7.9 "WASTEWATER")

The Fifth Amendment§ of the United States Constitution states, "nor shall private property be taken for public use, without just compensation."¶ This includes water rights.** The greatest controversy of eminent domain is not the government's right to seize lands†† and waters but what amount of compensation is required.‡‡ The Court often applied what is known as the navigation servitude, referring to the unique right of

---

* See *Ranch Ass'n v. California Coastal Comm'n*, 454 U.S. 1070, 102 S.Ct. 622, 70 L.Ed.2d 606 (1981).
† *Nollan v. California Coastal Comm'n*, 483 U.S. 825, 107 S.Ct. 3141, 97 L.Ed.2d 677 (1987).
‡ *Dolan v. City of Tigard*, 512 U.S. 374, 114 S.Ct. 2309, 129 L.Ed.2d 304 (1994).
§ The Fifth Amendment is considered the exclusive remedy for landowners whose property is damaged by the public in Montana, where a Stream Access Law guarantees the public access to all streams in the state and forbids landowners from restricting access. *Madison v. Graham*, 316 F.3d 867 (9th Cir. 2002).
¶ Constitution of the United States of America, Amendment 5.
** *Franco-American Charolaise, Ltd. v. Oklahoma Water Resources Bd.*, 1990 OK 44, 855 P.2d 568 (Okla. 1990), reissued (Apr. 13, 1993) (taking of riparian rights requires compensation); *Mississippi State Highway Com'n v. Gilich*, 609 So. 2d 367 (Miss. 1992) (riparian rights are revocable privilege); *Hillebrand v. Knapp*, 65 S.D. 414, 274 N.W. 821, 112 A.L.R. 1104 (1937).
†† Some states declare the beneficial use of water a public use and allow any person so using water the right of eminent domain when necessary to continue the beneficial use. See RCWA 90.03.040.
‡‡ See *United States v. Rands*, 389 U.S. 121, 88 S.Ct. 265, 19 L.Ed.2d 329 (1967) (declaring that the federal government need not compensate for special potential uses, such as suitability as a port). It should be noted that the cases do not deal with consumptive rights (see § 2.10.1, Appendix 1). *Rands* dealt with a taking of property on the Columbia River being leased to the State of Oregon with an option to purchase. The option was not exercised, because the United States condemned the property and conveyed it to the state at one-fifth the price of the option. The Court found that since the United States could alter the course of the river without the need to compensate, they did not need to give special compensation for potential uses on the river. The court reiterated that navigable waters are the property of the United States, conferring on the United States a "dominant servitude," extending the entire stream and bed. Since the riparians are always subject to federal power over the stream and bed, compensation for losses is not necessary. Criticism of this decision eventually led to § 111 of the Rivers and Harbors Act, 33 U.S.C.A. § 595a, which effectively reversed the *Rands* decision (see below). See also *United States v. Grand River Dam Authority*, 363 U.S. 229, 80 S.Ct. 1134, 4 L.Ed.2d 1186 (1960) (finding that no compensation was due when the U.S. built a dam on the site an authority, created by state agency, intended to use to generate power). The authority claimed losses and filed suit. The unauthorized construction of bridges or other obstructions to passage are in many ways considered public nuisances and not compensable.

the federal government to regulate navigable waters (see Chapter 1 on the Commerce Clause). This doctrine allows certain takings under the commerce clause without compensation.* Congress overruled the doctrine in section 111 of the 1970 Rivers and Harbors Flood Control Act.† Today by act of Congress compensation is given for land above the normal high-water mark considering the lands "highest and best use."

Refusal of the Army Corps of Engineers to grant a permit, even if it is a necessary permit, does not constitute a taking, simply because it denies the owner of property the "highest and best use" of his land.‡

3.11 TAKING OF WATERS AND NAVIGATION SERVITUDE

During the Great Depression the federal government took over a state project in California to rearrange the water system there. When water was directed away from certain lands, which had previously enjoyed natural irrigation, the landowners sued. The United States Supreme Court found compensation due, saying, "No reason appears why those who get the waters should be spared from making whole those from whom they are taken. Public interest requires appropriation; it does not require expropriation."§

Another case holds that while Congress has the right to guarantee a public right of access to a pond in Hawaii if it so chooses, it does not follow that Congress's powers under the Commerce Clause provide an automatic right of public access. The Court went on to say that compensation for such a taking to provide public access was required. The Court stated that while the Congress could have refused to allow dredging that created the pond on the grounds that it impaired navigation, now that the pond existed, petitioners have a body of water that is private property under Hawaiian law.¶

While the Commerce Clause does grant Congress power to regulate interstate commerce, it also limits the rights of the states to interfere with national interests.** The navigational servitude is still alive and used in certain cases where navigation interests are clearly at stake and takings necessary.†† For the most part, these cases involve alterations to the waterway in conjunction with interstate commerce and not seizure

---

* *Id.*

† The act says, "In all cases where real property shall be taken by the United States for public use in connection with any improvement of rivers, harbors, canals, or waterways of the United States, and in all condemnation proceedings by the United States to acquire lands or easements for such improvements, the compensation to be paid for real property taken by the United States above the normal high water mark of navigable waters of the United States shall be the fair market value of such real property based upon all uses to which such real property may reasonably be put, including its highest and best use, any of which uses may be dependent upon access to or utilization of such navigable waters." 33 U.S.C.A. § 595a.

‡ *Delton Corp. v. United States*, 657 F.2d 1184 (1981).

§ *United States v. Gerlach Live Stock Co.*, 339 U. 725, 70 S.Ct. 955, 94 L.Ed. 1231 (1950). See §§ 6.13 and 16.5.

¶ *Kaiser Aetna v. United States*, 444 U.S. 164, 100 S.Ct. 383, 62 L.Ed.2d 332 (1979) (refusing to grant a public right-of-access to a dredged pond that became navigable under the Constitution by private action). See § 1.8.

** See *Sporhase v. Nebraska*, 458 U.S. 941, 102 S.Ct. 3456, 73 L.Ed.2d 1254 (1982) (finding part of a Nebraska statute unconstitutional for interference with national interests in interstate commerce). See § 3.12.

†† See *United States v. Cherokee Nation of Oklahoma*, 480 U.S. 700, 107 S.Ct. 1487, 94 L.Ed.2d 794 (1987) (holding that when the United States acts pursuant to the commerce authority regarding navigable waters, it does not matter that the rights interfered with [here gravel and minerals] were unrelated to navigation, no compensation need be paid). See §§ 1.8, 9.14. *Black*'s Law Dictionary Seventh Edition (1999) defines a navigation servitude as "1. An easement allowing the federal government to regulate commerce on navigable water without having to pay compensation for interference with private ownership rights."

of riparian lands. In a sense, the government is seizing the water and affecting the land as a result. The limit of the government's navigational servitude is the ordinary high-water line, which is not readily susceptible to uniform and precise definition.*

3.12   FACIAL DISCRIMINATION AGAINST OUT-OF-STATE INTERESTS (SEE ALSO 5.9 "GEOGRAPHIC RESTRICTIONS ON USE")

The district court in New Mexico struck down a New Mexico law that discriminated against the out-of-state interests of the city of El Paso, Texas, in the issuance of rights to groundwater in New Mexico.† The dormant Commerce Clause is readily understood to forbid such discrimination against out-of-state interests without some strong reason for doing so. Reasons might include serious threats to public health, for example. Pending appeal, New Mexico changed its statute to require the state engineer to determine that any out-of-state uses would not impede existing rights before issuing an out-of-state permit. A balancing test weighing the statewide supply and demand for water against the supply available and the out-of-state demand. The Court rejected a constitutional challenge to the new requirements.‡ Ordinarily, a state may restrict the issuance of unallocated rights in such a way as to insure primary benefits to prior users within the state,§ as a market participant.¶ The state may not restrict the sale of allocated water to out-of-state interests, though.**

---

* 78 Am Jur 2d, Waters § 96 with 1999 update. *United States v. Cameron*, 466 F.Supp 1099 (1978).
† *City of El Paso v. Reynolds*, 563 F. Supp.379 (D.N.M. 1983).
‡ *City of El Paso II*, 597 F. Supp. 694 (D.N.M. 1984).
§ *City of Trenton v. State of New Jersey*, 262 U.S. 182, 43 S. Ct. 534, 67 L.Ed. 937, 29 A.L.R. 1471 (1923). Note that states have a duty to conserve resources to make certain they are available to all of its inhabitants according to their need. The exercise of this responsibility is not a violation of the law.
¶ See *Hughes v. Alexandria Scrap Corp.*, 426 U.S. 794, 96 S.Ct. 2488, 49 L.Ed.2d 220 (1976); see also *Sporhase v. Nebraska*, 458 U.S. 941, 102 S.Ct. 3456, 73 L.Ed.2d 1254 (1982) (dealing with market participation in water allocation and finding that while Nebraska may favor in state users in making allocations, they may not prevent the sale of allocated water out of state). But see *City of El Paso v. Reynolds*, 563 F.Supp. 379 (D.N.M. 1983) (Interpreting *Sporhase* to forbid out-of-state discrimination unless it is essential to human survival). The Court in *El Paso* seems to have ignored the market participant exception to the dormant Commerce Clause, mentioned by the *Sporhase* court. Nonetheless, the statute was modified before the case could be appealed. The new statute was challenged in *City of El Paso II*, 597 F.Supp. 694 (D.N.M. 1984) (distinguishing between economic protectionism and other noneconomic benefits, but still seemingly ignoring the market participant exception). See N.M. Stat. Ann. § 72-5-23. The Court in *Sporhase* stated, "Finally, given appellee's conservation efforts, the continuing availability of ground water in Nebraska is no simply happenstance; the natural resource has some indicia of a good publicly produced and owned in which a State may favor its own citizens in times of shortage." *Sporhase*, 458 U.S. 941. See also *Intake Water Co. v. Yellowstone River Compact Comm'n*, 769 F.2d 568 (9th Cir. 1985) (finding that since Congress ratified the compact, the provision requiring unanimous consent of the agreeing states for out-of-state diversion was valid and outside the negative commerce clause). It appears that while the Court in *Intake Water* reached what is likely the proper conclusion, they, too, may have misread *Sporhase*. As mentioned above, the states are market participants in unallocated water rights and can discriminate in the initial allocation of water in their boundaries, even for economic purposes. While some lower courts have struggled to grasp this concept, the Supreme Court has been relatively clear in its opinions, and nothing in those decisions seems to support the notion that states cannot favor their own interests in allocating unallocated water. The confusion is likely the result of attorneys not properly reading and briefing the case law for the Court. The lower court decisions in *El Paso* above are almost certain to be overturned if challenged.
** *Id.* See § 3.11.

4 Water Ownership

4.1 OWNERSHIP OF NAVIGABLE WATERS

Generally, ownership of navigable waters, in their natural state, is vested in the public or the state in trust for the public.* Water in its natural state is often treated as real property† but can be considered personal property when captured.‡ *Real property* refers to land, where as *personal property* includes nearly all that is not intimately connected or secured to the land.

4.2 STATE OWNERSHIP RIGHTS AND FEDERAL RESERVED RIGHTS (SEE ALSO 9.2 "INDIAN RIGHT")

The Desert Land Act of 1877§ authorizes western states to appropriate waters within their boundaries.¶ In most jurisdictions, non-navigable waters may be privately owned.** The states' rights over waters are generally considered to be regulatory and not proprietary in nature.†† The federal government has the power to reserve water

---

* *State of Oklahoma v. State of Texas*, 258 U.S. 574, 42 S. Ct. 406, 66 L. Ed. 771 (1922), modification denied, 260 U.S. 711, 43 S. Ct. 251, 67 L. Ed. 476 (1923); *U.S. v. Twin City Power Co.*, 350 U.S. 222, 76 S. Ct. 259, 100 L. Ed. 240 (1956) (navigable waters not subject to private ownership); *Northport Irr. Dist. v. Jess*, 215 Neb. 152, 337 N.W.2d 733 (1983) (water in running stream not subject to private ownership). See also 78 Am Jur 2s, Waters §§ 229-230 with 1999 update.

† *Smith v. Municipal Court*, 202 Cal. App. 3d 685, 245 Cal. Rptr. 300 (1st Dist. 1988); *Bayou Land Co. v. Talley*, 924 P.2d 136 (Colo. 1996); *Foster v. Sunnyside Valley Irr. Dist.,* 102 Wash. 2d 395, 687 P.2d 841 (1984). See RCWA 90.03.005 declaring all waters in the state as belonging to the public, subject to existing rights.

‡ *McCarter v. Hudson County Water Co.*, 70 N.J. Eq. 695, 65 A. 489 (Ct. Err. & App. 1906); but see *Stanislaus Water Co. v. Bachman*, 152 Cal. 716, 93 P. 858 (1908) (holding that water in a canal or conduit for irrigation remains real property).

§ 43 U.S.C.A. §§ 321-25

¶ *Idaho Dept. of Water Resources v. U.S.*, 122 Idaho 116, 832 P.2d 289 (1992), cert. granted, 506 U.S. 939, 113 S.Ct. 373, 121 L. Ed. 2d 285 (1992) and judgment rev'd on other grounds, 508 U.S. 1, 113 S. Ct. 1893, 123 L. Ed. 2d 563 (1993). See also *California Oregon Power Co. v. Beaver Portland Cement Co.*, 295 U.S. 142, 55 S. Ct. 725, 79 L. Ed. 1356 (1935); *Port of Seattle v. Oregon & W. R. Co.*, 255 U.S. 56, 41 S. Ct. 237, 65 L. Ed. 500 (1921) (law of State of Washington). Note that under existing federal law land grants carry no water rights but are subject to prior appropriations. Grantees take subject to changed conditions that occurred while government owned land. *Isaacs v. Barber*, 10 Wash. 124, 38 P. 871 (1894).

** *Cinque Bambini Partnership v. State*, 491 So. 2d 508 (Miss. 1986), cert. granted, 479 U.S. 1084, 107 S. Ct. 1284, 94 L. Ed. 2d 142 (1987) and judgment aff'd, 484 U.S. 469, 108 S. Ct. 791, 98 L. Ed. 2d 877 (1988); *Adirondack League Club, Inc. v. Sierra Club*, 92 N.Y.2d 591, 684 N.Y.S.2d 168, 706 N.E.2d 1192 (1998). Appropriation states are much more likely not to recognize a private right of ownership of water. The following states have statutory or constitutional provisions declaring water in some form as belonging to the public or the state: Arizona, Nevada, New Mexico, Oregon, Idaho, Montana, North Dakota, Texas, Wyoming, California, Colorado, South Dakota, Nebraska, and Utah. See Getches at 85.

†† *City of Barstow v. Mojave Water Agency*, 23 Cal. 4th 1224, 99 Cal. Rptr. 2d 294, 5 P.3d 853 (2000).

rights,* but such reservation must be clearly stated or implied.† Reserved rights are exempted from the appropriations system.‡

States may be estopped from asserting ownership claims over lakes in order to enforce new ecological standards where ownership vested in private individuals by virtue of state and federal patents and longstanding deeds, when no reservation was made.§ Generally speaking, states cannot abandon ownership of waters held in trust for the public.¶

4.3   RIPARIAN OWNERSHIP RIGHTS OF WATER BED

In riparian states (discussed in Chapter 6), the owner of land bordering a water body is said to own the bed to the center of the river, stream, lake, or pond.** The traditional right to exclude persists only in a handful of states (see 2.3). There is a common right among all riparian landowners to the reasonable use of the surface of the water. Under some state laws landowners may be subject to the public's use of surface waters as well.†† Riparian rights apply to land between the water level and ordinary high-water mark.‡‡

---

* The earliest mention of federal reserved rights was in dicta in *United States v. Rio Grande Irrigation Co.*, 174 U.S. 690 (1899). According to Frank J. Trelease, actual application of this to non-Indian matters did not occur until 1955, *FPC v. Oregon*, 349 U.S. 435, 75 S.Ct. 832, 99 L.Ed. 1215 (1955). See Trelease, Frank J., Federal Reserved Water Rights Since PLLRC, 54 Denver L.J. 473 (1977) (cited in Gould, George A. and Douglas L. Grant, Cases and Materials on Water Law, Fifth Ed., p. 639 [1995]). See also 78 Am Jur 2d, Waters § 230 with 1999 update. See also USCS Constitution, Article 1, § 8, cl 3, n 157. Cappaert v. United States, 426 US 128, 48 L. Ed. 2d 523, 96 S.Ct. 2062 (1976) (federal government has authority to reserve water rights via Commerce Clause).

† *State of Alaska v. Babbitt*, 72 F.3d 698 (9th Cir. 1995), adhered to, 247 F.3d 1032 (9th Cir. 2001); *Shoshone-Bannock Tribes v. Reno*, 56 F.3d 1476 (D.C. Cir. 1995); *Totemoff v. State*, 905 P.2d 954 (Alaska 1995) (reserved right limited to what is needed to fulfill purpose of reservation); *In re General Adjudication of All Rights to Use Water in Gila River System and Source*, 195 Ariz. 411, 989 P.2d 739 (1999), cert. denied, 530 U.S. 1250, 120 S. Ct. 2705, 147 L. Ed. 2d 974 (2000) and cert. denied, 530 U.S. 1250, 120 S. Ct. 2705, 147 L. Ed. 2d 974 (2000) (groundwater may be reserved where other waters are not sufficient).

‡ *Idaho Dept. of Water Resources v. U.S.*, 122 Idaho 116, 832 P.2d 289 (1992), cert. granted, 506 U.S. 939, 113 S. Ct. 373, 121 L. Ed. 2d 285 (1992) and judgment rev'd on other grounds, 508 U.S. 1, 113 S. Ct. 1893, 123 L. Ed. 2d 563 (1993); *U.S. v. Jesse*, 744 P.2d 491 (Colo. 1987). Reserved rights exist from the time of the creation of the enclave to which the right attaches. *Navajo Development Co., Inc. v. Sanderson*, 655 P.2d 1374 (Colo. 1982). Reserved rights are owned by the federal government and can be leased. *State ex rel. Greely v. Confederated Salish and Kootenai Tribes of Flathead Reservation*, 219 Mont. 76, 712 P.2d 754 (1985). If the reservation is only for a secondary purpose then U.S. must acquire rights some as others. *U.S. v. State*, 135 Idaho 655, 23 P.3d 117 (2001), as amended (May 1, 2001).

§ *Odom v. Deltona Corp.*, 341 So.2d 977 (Fla 1976) (public officials had long acquiesced under color of law). See 78 Am Jur 2d, Waters § 77, with 1999 update.

¶ 78 Am Jur 2d, Waters § 231 with 1999 update. But see *State v. Sunapee Dam Co.*, 70 NH 458, 50 A. 108. This appears to be an unusual holding, but often states may (and do) authorize private use of waters for beneficial purpose without violating the trust. Also when there is said to be some public purpose served by abandoning the trust courts have sometimes allowed it. See *Shively v. Bowlby*, 152 US 1, 38 L.Ed. 331, 14 S.Ct. 548.

** *Wheeler v. United States*, 770 F.Supp. 1205 (W.D.Mich.1991) (riparian takes title to unsurveyed island from shore to center of stream).

†† *Johnson v. Seifert*, 257 Minn. 159, 100 N.W.2d 689 (Minn. 1960); *Snively v. Jaber*, 48 Wash.2d 815, 296 P.2d 1015 (Wash 1956). The state does not relinquish ownership rights when granting permits for beneficial use. See *Stockman v. Leddy*, 55 Colo. 24, 129 P. 220 (1912) (overruled on other grounds by *Denver Ass'n for Retarded Children, Inc. v. School Dist. No. 1 in City and County of Denver*, 188 Colo. 310, 535 P.2d 200 [1975]).

‡‡ *State v. McFarren*, 62 Wis. 2d 492, 215 N.W.2d 459 (1974). Temporary unusual changes do not change the rights of affected parties. *Rice v. Naimish*, 8 Mich. App. 698, 155 N.W.2d 370 (1967).

4.4 RIGHT TO ARTIFICIAL WATERS*

The common right of owners of land bordering water may not apply to artificially created waters such as dikes and dams.† Landowners adjacent to a pit mine in Florida were not allowed to assert rights against the mine owner after the mine filled with water.‡ The Florida courts reached a different conclusion where the lake was artificially created as part of a development project with covenants in the deeds.§ Courts are divided whether a riparian who alters a body of water in a manner that benefits downstream riparians is required to maintain the benefit.¶ Because all owners have rights to use the whole surface of the water, some courts hold that an owner may not build out into the lake on submerged land.** It is sometimes suggested that this result may be different when the project is water-related, such as a dock.

4.5 REASONABLE USE RULE (SEE ALSO 7.9 ON WASTEWATER, 6.2 "RIPARIAN RIGHTS," 8.19 "STREAM FLOW")

A resort owner may allow guests to use the water surface but must limit the number of guests and their activities to prevent unreasonable interference with other landowners.†† The same rule applies to landowners, who wish to invite others to use the surface or the water, and to the state when it condemns a piece of land and opens it

---

* *Village of Pewaukee v. Savoy*, 103 Wis. 271, 79 N.W. 436 (1899) (rights to artificial waters acquired same as real property); *Case v. Hoffman*, 100 Wis. 314, 72 N.W. 390 (1897), reh'g granted, 100 Wis. 314, 74 N.W. 220 (1898) and modified on other grounds on reh'g, 100 Wis. 314, 75 N.W. 945 (1898) (implied contract for artificial water rights). Developed waters not previously part of any stream or body are ordinarily not part of the appropriation system and not subject to rules of priority. *Smith v. Duff*, 39 Mont. 382, 102 P. 984 (1909); *Wiggins v. Muscupiabe Land & Water Co.*, 113 Cal. 182, 45 P. 160 (1896) (landowner who captures water otherwise lost in some manner entitled to amount made available). See § 6.12. One who artificially diverts or stores water is not ordinarily required to maintain the diversion for the benefit of others. *Mitchell Drainage Dist. v. Farmers' Irr. Dist.*, 127 Neb. 484, 256 N.W. 15 (1934). Contract law may create an obligation under certain circumstances. Also see the rules of equity when one acts in reasonable reliance on the promise of another and it would seem unfair not to enforce the promise. *Kray v. Muggli*, 84 Minn. 90, 86 N.W. 882 (1901). See also 78 Am Jur 2d, Waters §§ 195–99 with 1999 update.
† *Anderson v. Bell*, 433 So.2d 1202, (Fla. 1983); *Lee v. Cercoa, Inc.*, 433 So. 2d 1 (Fla. Dist. Ct. App. 4th Dist. 1983).
‡ *Publix Super Markets, Inc. v. Pearson*, 315 So.2d 98 (Fla. App. 2 Dist. 1975).
§ *Silver Blue Lake Apartments, Inc. v. Silver Blue Lake Home Owners Ass'n, Inc.*, 245 So.2d 609 (Fla. 1971); *Johnson v. Board of County Com'rs of Pratt County*, 259 Kan. 305, 913 P.2d 119 (1996) (under certain conditions riparian rights may apply to artificially altered waters).
¶ Some courts find no obligation to maintain the benefit. *Goodrich v. McMillan*, 217 Mich. 630, 187 N.W. 368, 26 A.L.R. 801 (1922) (distinguishing *Mathewson v. Hoffman*, 77 Mich. 420, 43 N.W. 879[1889]); *Drainage Dist. No. 2 of Snohomish County v. City of Everett*, 171 Wash. 471, 18 P.2d 53, 88 A.L.R. 123 (1933). Others find an obligation. *Marshall Ice Co. v. La Plant*, 136 Iowa 621, 111 N.W. 1016 (1907); *Kray v. Muggli*, 84 Minn. 90, 86 N.W. 882 (1901); *Hammond v. Antwerp Light & Power Co.*, 132 Misc. 786, 230 N.Y.S. 621 (Sup. 1928) (decision based on prescription); *Lake Drummond Canal & Water Co. v. Burnham*, 147 N.C. 41, 60 S.E. 650 (1908) (decision based on description of condition as temporary or permanent).
** *Bach v. Sarich*, 74 Wash.2d 575, 445 P.2d 648 (Wash. 1968). See § 6.9.
†† *Thompson v. Enz*, 379 Mich. 667, 154 N.W.2d 579 (Mich. 1971); *Garbarino v. Noce*, 181 Cal. 125, 183 P. 532, 6 A.L.R. 1433 (1919); *Nolte v. Michels Pipeline Const., Inc.*, 83 Wis. 2d 171, 265 N.W.2d 482 (1978) (actor not liable for interference with others riparian rights unless causes unreasonable harm). See § 6.6.

up for the benefit of the public.* Reasonable use is an attempt to allow more generous use of the water than for mere domestic purposes.† The general rule of riparian water use is that each riparian has a right to use the water passing through his land,‡ but no riparian has the right to use the water to the injury of another.§

4.6 REASONABLE USE, THE STATE AS A RIPARIAN OWNER, AND PUBLIC ACCESS

When the state opens riparian land for public access, it is expected to police and maintain the site so that there is no unreasonable interference with other riparian users.¶ This usually includes preventive measures to protect the rights of neighboring landowners. Some states have held that swimming or boating over the land of another constitutes a trespass.** The "common use rule" allows all common riparian owners the right to enjoy use of the surface of the whole water.††

---

* *Botton v. State*, 69 Wash.2d 751, 420 P.2d 352 (Wash. 1966). See § 4.6.

† *Harris v. Brooks*, 225 Ark. 436, 283 S.W.2d 129 (1955). The Harris case established several general rules: (a) the right to use water for strictly domestic purposes is superior to many other uses of water; (b) other than the use mentioned above, all other lawful uses of water are equal; (c) when one lawful use of water is destroyed by another lawful use, the latter must yield, or it may be enjoined; (d) when one lawful use of water interferes with or detracts from another lawful use, then a question arises as to whether, under all the facts and circumstances of that particular case, the interfering use shall be declared unreasonable and as such enjoined, or whether a reasonable and equitable adjustment should be made, having due regard to the reasonable rights of each. A holder of an easement may assign to another under certain circumstances, though reasonable use may still apply. See *Miller v. Lutheran Conference & Camp Ass'n*, 331 Pa. 241, 200 A. 646, 130 A.L.R. 1245 (1938); *Hard v. Boise City Irrigation & Land Co.*, 9 Idaho 589, 76 P. 331 (1904). Easements often held not transferable. See *Davis v. Briggs*, 117 Me. 536, 105 A. 128 (1918); *Ross v. McGee*, 98 Md. 389, 56 A. 1128 (1904); *Williamson v. Yingling*, 80 Ind. 379, 1881 WL 7160 (1881). But see *U.S. v. Ahtanum Irr. Dist.*, 236 F.2d 321 (1956) (an individual's water right, no matter how measured or described, may not exceed his needs), cert. Denied 77 S.Ct. 386m 352 U.S. 988, 1 L.Ed. 2d 367.

‡ *Jones v. Oz-Ark-Val Poultry Co.*, 228 Ark. 76, 306 S.W.2d 111 (1957); *Town of Purcellville v. Potts*, 179 Va. 514, 19 S.E.2d 700, 141 A.L.R. 633 (1942); *Pyle v. Gilbert*, 245 Ga. 403, 265 S.E.2d 584 (1980) (rights of lower riparian subject to reasonable irrigation use by upper riparian); *Strobel v. Kerr Salt Co.*, 164 N.Y. 303, 58 N.E. 142 (1900); *U.S. v. Fallbrook Public Utility Dist.*, 165 F. Supp. 806 (S.D. Cal. 1958) (applying law of California); *Bollinger v. Henry*, 375 S.W.2d 161 (Mo. 1964); *Martin v. British Am. Oil Producing Co.*, 1940 OK 218, 187 Okla. 193, 102 P.2d 124 (1940).

§ Restatement (Second) of Torts, Chapter 41. For factors used by the Restatement in determining "reasonable use," see Restatement (Second) of Torts § 850A. See also §§ 6.12, 11.5. *Brown v. Tomlinson*, 246 Ga. 513, 272 S.E.2d 258 (1980) (right of lower riparian to flow); *Maddocks v. Giles*, 1999 ME 63, 728 A.2d 150 (Me. 1999); *Wisniewski v. Gemmill*, 123 N.H. 701, 465 A.2d 875 (1983); *Whorton v. Malone*, 209 W. Va. 384, 549 S.E.2d 57 (2001); *Indian Refining Co. v. Ambraw River Drainage Dist.*, 1 F. Supp. 937 (E.D. Ill. 1932); *Scranton Gas & Water Co. v. Delaware, L. & W. R. Co.*, 240 Pa. 604, 88 A. 24 (1913); *Seneca Consol. Gold Mines Co. v. Great Western Power Co. of California*, 209 Cal. 206, 287 P. 93, 70 A.L.R. 210 (1930); *Fall River Valley Irr. Dist. v. Mt. Shasta Power Corp.*, 202 Cal. 56, 259 P. 444, 56 A.L.R. 264 (1927).

¶ *Botton v. State*, 69 Wash.2d 751, 420 P.2d 352 (1966) (upholding a lower court injunction against the state). See § 4.5.

** *Lembeck v. Nye*, 24 N.E. 686 (Ohio 1890); *Decker v. Baylor*, 19 A. 351 (Pa. 1890); *Shandalee Camp v. Rosenthal*, 233 N.Y.S. 11 (1929); *Sanders v. De Rose*, 191 N.E. 331 (Ind. 1934).

†† *Beach v. Hayner*, 173 N.W. 487 (Mich. 1919). This rule was also adopted in the *Botton* case. See also *Florio v. State ex rel. Epperson*, 119 So. 2d 305, 80 A.L.R.2d 1117 (Fla. Dist. Ct. App. 2d Dist. 1960) (one may not enjoy to the exclusion of other riparians).

4.7   THE NATURAL FLOW THEORY (SEE ALSO 6.19 "NATURAL FLOW")

Generally, under this theory a riparian landowner may take water only for domestic uses. This theory has been largely displaced by "reasonable use." The natural-flow language still exists in some jurisdictions, even though "reasonable use" is nearly universally recognized in American courts today.

4.8   THE RIGHT TO TRANSFER WATER*

Part of owning property is the right to transfer it to others. In most jurisdictions, water rights may be transferred,† but not in excess of what the original right holder had.‡ This means that any use by the purchaser in excess of the use of the original holder may be invalid as against other rights holders. Reasonable use is still generally applied in order to protect the rights of other owners. Ordinarily water rights

---

* See Form 2.
† *Duckworth v. Watsonville Water & Light Co.*, 158 Cal. 206, 110 P. 927 (1910); *Pyle v. Gilbert*, 245 Ga. 403, 265 S.E.2d 584 (1980); *Dallas Creek Water Co. v. Huey*, 933 P.2d 27 (Colo. 1997); *McMillan v. Etter*, 229 Mich. 366, 201 N.W. 499 (1924) (construction of right depends on terms used); *Cornish Town v. Koller*, 758 P.2d 919 (Utah 1988) (rules of construction same as for deeds); *Town of Gordonsville v. Zinn*, 129 Va. 542, 106 S.E. 508, 14 A.L.R. 318 (1921) (language of instrument determines if exclusive). Water rights are transferred by deed. *Salt Lake City v. Silver Fork Pipeline Corp.*, 2000 UT 3, 5 P.3d 1206 (Utah 2000); *Department of State Lands v. Pettibone*, 216 Mont. 361, 702 P.2d 948, 26 Ed. Law Rep. 823 (1985) (rights appurtenant to land pass with land unless expressly reserved); *Lindenmuth v. Safe Harbor Water Power Corporation*, 309 Pa. 58, 163 A. 159, 89 A.L.R. 1180 (1932) (transferability of right depends on nature of right and mode of transfer); *Eastern Pennsylvania Power Co. v. Lehigh Coal & Navigation Co.*, 246 Pa. 72, 92 A. 47 (1914) (not a grant in proprietary sense); *Johnston v. Little Horse Creek Irr. Co.*, 13 Wyo. 208, 79 P. 22 (1904) (transfer not affected by failure of recording device in state); *State of Wyo. v. State of Colo.*, 298 U.S. 573, 56 S. Ct. 912, 80 L. Ed. 1339 (1936) (holding that water rights acquired by appropriation are transferable, in whole or in part, permanently or temporarily). State constitutions may prohibit transfers of municipal water rights. *Nephi City v. Hansen*, 779 P.2d 673 (Utah 1989); *Ficklen v. Fredericksburg Power Co.*, 133 Va. 571, 112 S.E. 775 (1922) (courts prefer an interpretation that does not restrict water use or that least encumbers the grantee, but restrictions are possible); *Eastern Pennsylvania Power Co. v. Lehigh Coal & Navigation Co.*, 246 Pa. 72, 92 A. 47 (1914) (where language is not ambiguous and clearly intends to restrict use it will be given effect).
‡ *Smith v. Stanolind Oil & Gas Co.*, 172 P.2d 1002 (Okla. 1946); *State v. Apfelbacher*, 167 Wis. 233, 167 N.W. 244 (1918). There is no right to transfer based on an application for water right. See *Catherland Reclamation Dist. v. Lower Platte North Natural Resources Dist.*, 230 Neb. 580, 433 N.W.2d 161 (1988). A right may be impertinent or in gross. See *Chase v. Cram*, 39 R.I. 83, 97 A. 481 (1916), modified, 97 A. 802 (R.I. 1916); *Cadwalader v. Bailey*, 17 R.I. 495, 23 A. 20 (1891); *Bradley v. Jackson County*, 347 S.W.2d 683 (Mo. 1961) (disfavoring easement in gross); *Maclay v. Missoula Irr. Dist.*, 90 Mont. 344, 3 P.2d 286 (1931) (finding right severed from land assumes nature of easement in gross). Water is often appurtenant to land on which beneficial use is made. *Axtell v. M.S. Consulting*, 1998 MT 64, 288 Mont. 150, 955 P.2d 1362 (1998); *Salt Lake City v. Silver Fork Pipeline Corp.*, 2000 UT 3, 5 P.3d 1206 (Utah 2000); *Dermody v. City of Reno*, 113 Nev. 207, 931 P.2d 1354 (1997); *Loosle v. First Federal Sav. & Loan Ass'n of Logan*, 858 P.2d 999 (Utah 1993) (inchoate right not appurtenant to the land); *Department of State Lands v. Pettibone*, 216 Mont. 361, 702 P.2d 948, 26 Ed. Law Rep. 823 (1985) (determination of appurtenance is one of fact); *Sun Vineyards, Inc. v. Luna County Wine Development Corp.*, 107 N.M. 524, 760 P.2d 1290 (1988) (sale of land pending application for extending water rights does not vest pending rights in sold land).

may be severed from the land, whether riparian or otherwise.* Water rights may also be leased on occasion.†

4.9 PRIVATE WATER OWNERSHIP

Many jurisdictions treat all water as belonging to the public.‡ Private ownership of water in these jurisdictions is impossible, since all water belongs to the state. A private individual has only the right to use water as authorized by the state for beneficial use.§

In other jurisdictions¶ water may become personal property when captured.** The appropriator's rights in the molecules of the water may not end when the water is used.†† Washington State uses a combination of two tests to determine ownership of water, the "geographical" and "control and possession" tests.‡‡ The "geographical" test uses the location of the water on the owner's property as the basis for ownership, while the "control and possession" test concerns itself with whether the appropriator intends to recapture the water for further use.§§

Beneficial use is often defined by statute, specifying uses the state recognizes as beneficial.

---

* *Thurston v. City of Portsmouth*, 205 Va. 909, 140 S.E.2d 678 (1965); *Adams v. Chilcott*, 182 Mont. 511, 597 P.2d 1140 (1979) (appurtenant rights severable from land they benefit). Absent clear severance, the courts will presume water rights were transferred with the land. *Bayou Land Co. v. Talley*, 924 P.2d 136 (Colo. 1996). Note that the ordinary rules of contract law generally apply to water contracts. *Johnson v. Armour & Co.*, 69 N.D. 769, 291 N.W. 113, 127 A.L.R. 828 (1940) (contract; easement); *Painter v. Alexandria Water Co.*, 202 Va. 431, 117 S.E.2d 674 (1961) (contracts can run with the land to future owners).

† *Bergeron v. Forger*, 125 Vt. 207, 214 A.2d 85 (1965) (holding that instrument was a lease rather than a license); *White v. Board of Land Com'rs*, 595 P.2d 76 (Wyo. 1979) (examining rights of the lessees); *Dallas Creek Water Co. v. Huey*, 933 P.2d 27 (Colo. 1997).

‡ I.e., Nevada, Montana, Wyoming, and New Mexico. See § 3.3. For more complete list of jurisdictions see § 4.2.

§ A. Dan Tarlock, Water Resource Management, 4th ed. Ch. 3 at 364 (1993).

¶ I.e., Idaho.

** *Department of Ecology v. Bureau of Reclamation*, 118 Wash.2d 761, 827 P.2d 275 (1992).

†† See *Ide v. United States*, 263 U.S. 497, 44 S.Ct. 182, 68 L.Ed. 407 (1924).

‡‡ Barker and Scharf, 1B Wash Prac 3rd ed. § 58.22.5, *Methods of Practice, Real Estate: Scope of Water Right* with 1993 update.

§§ *Department of Ecology v. Bureau of Reclamation*, 118 Wash.2d 761, 827 P.2d 275 (1992).

5 Water Supply and Uses*

5.1 EVAPORATION† AND PRECIPITATION‡ (FOR AVERAGE DISTRIBUTION OF RAINFALL IN THE UNITED STATES, SEE FIGURE 12 AND FIGURE 13)

About 80,000 cubic miles of water evaporates from the ocean annually, and about 15,000 cubic miles is evaporated from lakes and land surfaces.§ This water goes into the atmosphere and later falls as precipitation in amounts equal to the total evaporation. About 24,000 cubic miles of water fall as rain each year, over land.¶

5.2 USES GENERALLY**

Generally, a water right is a right to use water in some amount for a beneficial purpose.†† In the United States, total withdrawals equaled 408 billion gallons per day in 2000.‡‡ Freshwater withdrawals were 85 percent of the total. Seventy-nine percent of total withdrawals were from surface water.§§ Thermoelectric-power plants withdrew an estimated 195 billion gallons per day.¶¶ (See chart in Appendix 3)

---

* See Tables 1–12 for jurisdictional breakdown of water withdrawals and uses; Figures 5 Water Cycle, 6 Water Cycle, 8 Aquifers and Wells, 11 Average pH of precipitation, 12 Average Rainfall in United States, 13 Average Rainfall of World, 25 Total Water Use in United States, 26 Groundwater Withdrawals in United States, 27 Surface Withdrawals in United States, 28 Bureau of Reclamation Water Operations, 30 Distribution of Earth's Water, 31 Global Annual Components of Precipitation.

† See http://ga.water.usgs.gov/edu/watercycleevaporation.html. See also Figures 5 Hydrologic Cycle, 6 Water Cycle.

‡ See Figures 5 Hydrologic Cycle, 6 Water Cycle, 11 pH of rainfall, 12 Annual Rainfall in United States, 13 World annual precipitation, 31 Global annual Components of Precipitation.

§ George A. Gould and Douglas L. Grant, *Cases and Materials on Water Law*, 71 (1995).

¶ George A. Gould and Douglas L. Grant, *Cases and Materials on Water Law*, 71 (1995).

** For a discussion of major water users in the United States see 25 Am Jur Proof of Facts, 249–52, Water Pollution § 11.

†† *Smith v. Denniff*, 23 Mont. 65, 57 P. 557 (1899), rev'd on other grounds, 24 Mont. 20, 60 P. 398 (1900).

‡‡ http://ga.water.usgs.gov/edu/wateruse2000.html (accessed November 8, 2006).

§§ *Id.*

¶¶ *Id.*

5.3   ABANDONMENT* OF RIGHTSt (SEE ALSO 6.3 "RIPARIAN," 7.6 "APPROPRIATION")

The United States Supreme Court found that a state may treat water rights, which have not been used for twenty years and have not been filed with the state, as abandoned.‡ In such cases there is no taking and no compensation is necessary. Virtually all appropriation states have some statutory provision or common law rule§ dealing with abandonment.¶ Some states permit leases as a means of avoiding abandonment.** There is a division when abandonment is followed by renewed use whether the right is revived.†† It is a little surprising that courts have been rather reluctant to strictly enforce abandonment statutes‡‡

---

* *Crow v. Carlson*, 107 Idaho 461, 690 P.2d 916 (1984); *Massee v. Schiller*, 237 Ark. 809, 376 S.W.2d 558 (1964) (abandonment of well easement); *City of Anson v. Arnett*, 250 S.W.2d 450 (Tex. Civ. App. Eastland 1952), writ refused n.r.e.; *Axtell v. M.S. Consulting*, 1998 MT 64, 288 Mont. 150, 955 P.2d 1362 (1998); *Davenport v. Town of Danvers*, 336 Mass. 106, 142 N.E.2d 753 (1957) (abandonment a question of intention); *Fruit Growers' Ditch & Reservoir Co. v. Donald*, 96 Colo. 264, 41 P.2d 516, 98 A.L.R. 1288 (1935)(mere nonuse not enough). Many statutes list defenses to forfeiture. See Ariz. Rev. Stat. Ann. § 45–189; N.M. Stat. Ann. § 72-5-28; Idaho Code § 42-222(2); Cal. Water Code § 1241 (using term reversion). Even in the absence of such a statute most states will accept some reasonable excuses. See also *In re Drainage Area of Bear River in Rich County*, 12 Utah 2d 1, 361 P.2d 407 (1961) (statute providing for forfeiture of water right by nonuser for five years); *Steele v. Pfeifer*, 310 N.W.2d 782 (S.D. 1981) (use of a water well). See also 78 Am Jur 2d, Waters § 240 with 1999 update.
† See Form 5 Forfeiture, Form 14 notice of abandonment.
‡ *Texico, Inc. v. Short*, 454 U.S. 516, 102 S.Ct. 781, 70 L.Ed.2d 738 (1982). For an example of termination of unused riparian rights, see *In the Matter of Deadman Creek drainage Basin*, 694 P.2d 1071 (Wash. 1985) (the court determined that fifteen years was a reasonable time and held that riparian rights not exercised by 1932 were relinquished).
§ I.e., Colorado; see Cororado Rev. Stat. § 37-92-402 (11). Washington and Montana also use rebuttable presumptions like Colorado. See also *Okanogan Wilderness League v. Town of Twisp*, 133 Wash. 2d 769, 947 P.2d 732 (1997); *79 Ranch, Inc. v. Pitsch*, 204 Mont. 426, 666 P.2d 215 (1983). Note Washington permits holding water for use in low flow periods as an exception to abandonment. See Wash. Rev. Code § 90.14.140(2)(b).
¶ I.e., Idaho Code sec. 42.222(2). Note courts tend to be reluctant to find abandonment unless the evidence is quite strong. See *Beaver Park Water, Inc. v. Victor*, 649 P.2d 300. (Colo. 1982).
** N.M. Stat. Ann. § 72-6-3; Idaho Code § 42-108A; S.D. Codified Laws §§ 46-5-30.3. At least one case held that leasing a water right worked an immediate abandonment. See *Slosser v. Salt River Valley Canal Co.*, 7 Ariz. 376, 65 P.332 (1901). It is doubtful whether this decision would hold up today though.
†† *Town of Eureka v. Office of State Engineer*, 826 P.2d 948 (Nev. 1992); *Sheep Mountain Cattle Co. v. State*, 45 Wash. App. 427, 726 P.2d 55 (1986) (holding that automatic forfeiture violates due process rights of holder). Oregon found a forfeiture when water was used for some other purpose than it was acquired for. See *Hennings v. Water Reservoir Dept.*, 50 Or. App. 121, 622 P.2d 333 (1981). Rights transferred with the land are only good to the extent of beneficial use at time of the transfer. See *Little v. Greene & Weed Inv.*, 839 P.2d 791 (Utah 1992). Many states authorize extensions of time under certain circumstances to avoid abandonment. See Utah Code Ann. § 73-1-4-3(a) (authorizing successive 5-year periods); Idaho Code § 42-222(2) (authorizing single 5 year exemption); Wash Rev. Code Ann. § 90.14.140(2)(c) (rights may be unused for future development within 15 years of most recent beneficial use). See also *R.D. Merrill Co. v. State*, 137 Wash.2d 118, 969 P.2d 458 (1999) (irrigation rights not used during time permission sought for resort).
‡‡ *Beaver Park Water, Inc., v. Victor*, 649 P.2d 300 (Colo. 1982) (finding that water rights not used in 30 years was not sufficient to show abandonment); *Gilbert v. Smith*, 97 Idaho 735, 552 P.2d 1220 (1976) (refusing to recognize prescriptive right because no finding of intent to abandon and insufficient evidence of adverse, hostile open use). It should be noted that Colorado does not have a strict statutory forfeiture law as many other states do. See *Southeastern Colorado Water Conservancy Dist. v. Twin Lakes Associates, Inc.*, 770 P.2d 1231 (Colo. 1989). Instead Colorado law states that 10 years non-use creates a rebuttable presumption of forfeiture. Colo. Rev. Stat. § 37-92-402(11).

and enforcement at all is uncommon in most jurisdictions. It should be noted that there is a difference between abandonment and forfeiture of water rights.* While often used interchangeably, abandonment is a common law principle requiring intent and forfeiture† is a statutory loss of rights set out by the legislature. Riparian rights are seldom impacted by disuse.‡

5.4 IN-STREAM VALUES (SEE ALSO 3.8 "PUBLIC TRUST DOCTRINE")

These values would include such activities as recreation, aesthetics, fish and wildlife maintenance, and general environmental preservation. The earliest effort included an Oregon law almost a century ago, where lawmakers withdrew from appropriation, waters entering the Columbia River.§

5.5 MINIMUM FLOWS¶ (SEE ALSO 7.1 "PRIOR APPROPRIATION")

A variation of the withdrawal is to limit withdrawals in order to preserve certain minimum flow levels in the river or waterbed. This policy does not restrict all use or appropriation of water from the body, but limits use to amounts above the minimum flow level.** Many states refuse to make further appropriations in a river or lake, in order to maintain minimum flows for fish and wildlife.

5.6 IN-STREAM APPROPRIATION†† (SEE ALSO 5.13 AND 7.5)

These appropriations are for uses that keep the water in the stream. The essential elements of this technique are the elimination of diversion as a requirement for an appropriation and the recognition of the value of preserving in stream uses.‡‡

5.7 WATER AS A NATURAL RESOURCE

Water has important economic value as a natural resource. Water is considered a renewable natural resource because as new precipitation falls, stores are replenished. Even though water is renewable, it is in limited supply. Demand often exceeds supply. Water law does not treat the water as real property in the same sense land

---

* See *Jenkins v. State Department of Water Resources*, 103 Idaho 384, 647 P.2d 1256 (1982).
† See Idaho Code § 42-222(2).
‡ See § 6.3.
§ Or. Rev. Stat. § 538.200
¶ For a discussion of the water quality needs of fish, see 25 Am Jur Proof of Facts p. 247 Water Pollution § 9.
** Wash. Rev. Code Ann. §§ 90.22 and 90.54; see also Mont. Code Ann. § 85-2-316; Cal. Water Code § 1257.5. See *Hubbard v. State,* 86 Wash. App. 119, 936 P.2d 27 (Wash. App. Div. 3 1997) (requiring a dam to allow certain minimum flows to pass).
†† See Table 20 for list of western state in-stream flow uses. See also § 7.5 on diversion.
‡‡ Cal. Pub. Res. Code §§ 5093.50 to 5093.69 (establishing state policy to preserve natural flow of certain rivers with extraordinary recreational and scenic value).

is considered. It grants rights to use the water for a particular purpose. In this way, several individuals may use the same water for different purposes.

5.8   IRRIGATION* (SEE ALSO APPENDIX 3, TABLE 7)

Irrigation† consumed nearly 137,000 million gallons of water per day in the year 2000. Of that, about 80,000 million gallons were surface water and 56,900 million gallons were from groundwater sources.‡ Irrigation runoff creates pollution problems as fertilizer and sediments that would ordinarily take thousands of years to build up, enter the water system at unnatural rates. (See 7.2 on Clean Water Act) Pesticides are another large problem from irrigation runoff. In some states water appropriated for irrigation is not severable from the land it benefits.§

Water consumption for irrigation varies widely among the several states. California consumes the greatest amount of irrigation water (30,500 million gallons per day).¶ The least amount of irrigation water was consumed in the state of West Virginia (0.04 million gallons per day). Irrigation tends to be more common out west where Idaho ranks second in consumption with 17,100 million gallons per day.**

5.9   GEOGRAPHIC RESTRICTIONS ON USE (SEE ALSO 3.12 "FACIAL DISCRIMINATION AGAINST OUT-OF-STATE INTERESTS" 7.30 "PROTECTING WATER BASINS OF ORIGIN")

Some jurisdictions have laws that forbid or make it difficult to transport water outside of one water basin and into another.†† The intention of these restrictions may vary. They may be used to make exporters accountable for the costs they impose on the place of origin. There is a possible Dormant Commerce Clause issue, where

---

* For more information on irrigation methods see http://ga.water.usgs.gov/edu/irmethods.html. For a discussion of the harmful effects of overfertilization on water, see 25 Am Jur Proof of Facts p. 254–256, Water Pollution § 13.

† *Union Falls Power Co. v. Marinette County,* 238 Wis. 134, 298 N.W. 598, 134 A.L.R. 958 (1941) (discussing local significance of irrigation water rights).

‡ http://ga.water.usgs.gov/edu/summary95.html.

§ *State ex rel. Blome v. Bridgeport Irrigation Dist.,* 205 Neb. 97, 286 N.W.2d 426 (1979).

¶ http://pubs.usgs.gov/circ/2004/circ1268/htdocs/table02.html.

** *Id.* (See Chart 2 in Appendix 3.)

†† See *Coffin v. Left Hand Ditch Co.,* 6 Colo. 443 (1982); *Osterman v. Central Nebraska Pub. Power and Irrigation Dist.,* 268 N.W. 334 (Neb. 1936) (overruled by *Little Blue Natural Resources Dist. v. Lower North Platte Natural Resources Dist.,* 294 N.W.2d 598 (Neb. 1980); *U. S. v. Fallbrook Public Utility Dist.,* 165 F. Supp. 806 (S.D. Cal. 1958) (applying law of California); See Texas Water Code Ann. § 11.085; Okl. Stat. Tit. 82, § 105.12(4); Alaska Stat. § 46.15.035; Kan. Stat. Ann. § 82a-1502; and Mont. Code Ann. § 85-2-301. See also Or. Rev. Stat. §§ 537.801; Neb. Rev. Stat. § 46-289; Wy. Stat § 41-2-121(a)(ii)(E)(VIII). Note that in most jurisdictions, riparian rights extend only to tracts of land that touch and border the water. States vary some on how much land in this tract may be benefited by riparian rights. Ordinarily land outside the water basin is not considered riparian and water used on such land would be termed an appropriation. Many eastern states also have laws designed to protect the original basins. See Minn. Stat. § 103G.265(2); Conn. Gen. Stat. § 22a-369(10); S.C. Stat. § 49-21-20; Mass. Gen. Laws ch. 21§§ 8C-8D.

out of state interests are discriminated against.* Ordinarily the regulation will stand if it is an evenhanded regulation to effectuate some legitimate public interest, and it only incidentally impacts interstate commerce. The Dormant Commerce Clause then involves a balancing of interests, taking into account alternatives with lesser impacts.† A regulation meeting the above criteria will be upheld unless the burden on interstate commerce is clearly excessive when contrasted with the local benefit.‡

5.10 RIGHTS APPURTENANT TO SPECIFIC LAND (SEE TERMS APPURTENANT)

In some jurisdictions, water used for a particular purpose such as irrigating a certain plot of land, may not be reallocated for another plot of land without filing of a new application.§ In these circumstances the water is said to be appurtenant to the land. Making water appurtenant to the land is believed to discourage speculation and inflated initial water rights aimed at encouraging later development. In most jurisdictions, a transfer of the land carries the water rights with it, unless the rights are specifically exempted in the contract of sale or some related document.¶

5.11 SEWAGE**

Domestic use includes sewage for used water in the home. In 2000, the United States consumption of water for domestic use was approximately 3590 million gallons per day.†† Once again California leads the nation in consumption with approximately 286 million gallons per day, nearly 8 percent of the nations total.‡‡ Nearly 84 percent of the population in 2000 was supplied by public water supplies while the remaining 16 percent relied on self-supplied ground water.§§

---

* The commerce authority is considered concurrent between the federal and state governments, and where Congress has not spoken, states may regulate within certain limits. See *Gibbons v. Ogden*, 22 U.S. (9 Wheat) 1, 6 L.Ed. 23 (1824). One of those limitations denies the states any right to discriminate against out of state interests with a few exceptions, such as market participation or public health and safety. The purpose of the Dormant Commerce Clause is to protect "interstate commerce." The clause is a combination of judicial interpretation of Congress' commerce power and the "Equal Protection Clause" of the 15th Amendment, though it is possible for a state law to violate only one of these clauses and not the other. See *Metropolitan Life Insurance Co. v. Ward*, 470 U.S. 869, 105 S.Ct. 1676, 84 L.Ed.2d 751 (1985). If state regulation conflicts with federal regulation the "Supremacy Clause" applies and the federal regulation trumps. U.S. Const. Art. V § 2.

† *Pike v. Bruce Church, Inc.*, 397 U.S. 137, 90 S.Ct. 844, 25 L.Ed.2d 174 (1970).

‡ *Id.*

§ See *Salt River Valley Users' Association v. Kovacovich*, 3 Ariz. App. 28, 411 P.2d 201 (1966).

¶ Utah Code Ann. § 73-1-11; *Russell v. Irish*, 118 P. 501 (Idaho 1911); *Frank v. Hicks*, 35 P. 475 (Wyo. 1894). In Colorado the question depends on the intent of the grantor (*Bessemer Irrigating Ditch Co. v. Woolley*, 76 P. 1053 (Colo. 1904)). In Utah the court ruled that a right does not become appurtenant on transfer until the state engineer issues a certificate. *Little v. Greene & Weed Inv.*, 839 P.2d 791 (Utah 1992).

** For a discussion of sewage and industrial waste treatment impacts on water quality, see 25 Am Jur Proof of Facts, 256–66, Water Pollution §§ 14–24.

†† http://ga.water.usgs.gov/edu/wudo.html.

‡‡ http://pubs.usgs.gov/circ/2004/circ1268/htdocs/table02.html.

§§ http://ga.water.usgs.gov/edu/wudo.html.

5.12   CONSUMPTION (SEE ALSO 8.25 "WATER CONSUMPTION IN THE UNITED STATES," APPENDIX A3, TABLES 5 AND 6)

Overall water consumption in the United States in 2000 was about 408,000 including both fresh and saline water supplies.* Fresh water accounts for the majority of this amount (approximately 345,000 million gallons per day). Saline water accounts for the remaining 62,300 million gallons used per day.†

A distinction can be made between consumptive and nonconsumptive uses of water. A consumptive use is one that uses up the water so that it is not available for other uses. A nonconsumptive use does not use up the water, so that it remains available for further uses. Such nonconsumptive uses may include aesthetics, fishing, swimming, boating, and other recreational uses, hydroelectric power generation, and so on.

5.13   RECREATION INCLUDING FISHING AND HUNTING‡ (SEE ALSO "PUBLIC RIGHTS," CHAPTER 2, 5.6)

Typically most recreational activities consume very little of the water that is withdrawn in the nation. They do have an impact, such as when pollutants are spilled into the water, vegetation is damaged and the ecosystem disrupted.

5.14   HYDROELECTRIC POWER§ (SEE ALSO APPENDIX 3, TABLE 12)

Most energy in the United States is produced by fossil fuels, but hydroelectric power remains a major source accounting for 9.8 percent of all power produced in the United States.¶ In 1992, the United States ranked second only to Canada in production of hydroelectric power in the world. Hydroelectric power is very big in the Pacific Northwest (Idaho, Washington, and Oregon are the three greatest producers).** Hydroelectric power accounts for 100 percent of the power in Idaho, 90 percent in Oregon, and 85 percent in Washington. Typically, water used for this purpose is not consumed. It merely moves huge turbines built into dams along large rivers, then the water continues on downstream. The dams built for this purpose may still have a major environmental impact as fish have greater difficulty crossing the dam to spawn. Furthermore, the flow of the water is usually altered by these structures, in order to cause a water buildup creating greater force as the water falls from higher

---

* http://pubs.usgs.gov/circ/2004/circ1268/htdocs/table02.html.
† *Id.*
‡ For a discussion of the effects of water pollution on the recreational usefulness of water bodies, see 25 Am Jur Proof of Facts, 235, Water Pollution.
§ *People v. Menagas*, 367 Ill. 330, 11 N.E.2d 403, 113 A.L.R. 1276 (1937) (considered a property right).
¶ http://ga.water.usgs.gov/edu/wuhy.html.
** http://ga.water.usgs.gov/edu/wuhy.html.

altitudes.* The advantages of hydroelectric power are that fuel is not consumed or burned and water is a renewable resource provided by nature. Hydroelectric power is one of the cleanest sources available.

It should be noted that other power generation generally requires water for cooling purposes, much of which is vented and lost in the form of steam.

5.15   PROTECTING PRIOR DAMS FOR HYDROELECTRIC POWER

Courts have generally protected the investments of companies building dams for the purpose of generating power from interference of newer dams for the same or similar purposes. Such dams often retain much of the river's flow for days at a time. Many statutes protect this interference with the river's natural flow.†

5.16   MUNICIPAL USE

Most jurisdictions provide special priority to municipal uses.‡ Many western states also require municipalities to apply for a permit.§ Some require estimates for growth. Municipalities in many jurisdictions can reserve rights for future uses.¶

5.17   DOMESTIC USES

Many states exempt at least some domestic uses of water from permit requirements.** Definitions of domestic uses may vary slightly from state to state. In some states,

---

* Federal permits are ordinarily required for the construction of hydroelectric power plants. Exceptions can be found in 16 U.S.C.A. 2705 (d) small power plants not exceeding 5 megawatts production or those on private land not exceeding 15 megawatts (40 megawatts in the case of certain state facilities). 16 U.S.C.A. 823a. A small hydroelectric power plant is described in 16 U.S.C.A. 2708 (small hydroelectric power plants not exceeding 30 megawatts, constructed at existing dams or using free flow with no dams). FERC rules for the exemptions can be found in 18 C.F.R. §§ 4.30(2), (29); 4.90–4.96; 4.101–4.108. FERC regularly controls and regulates rates these and other companies can charge for power. See 16 U.S.C.A. 824a-3. The issuance of an exemption does not automatically confer the necessary water rights, which must come from the states. Twin Falls Canal Co., Order Granting Exemption From Licensing, 35 FERC para. 35, 62, 104 (1986) (denying as unnecessary a state request for subordination to future upstream development).

† A Wisconsin statute allowed as little as 25 percent of the river's natural flow to pass.

‡ See Montana Code Ann. § 85-2-316.

§ Ariz. Rev. Stat. Ann. § 45-152(B)(4); Cal Water Code §§ 1460–1464 (municipal first in right regardless of first in time); Nev. Rev. Stat. Ann. § 533.340; Rev. Code Wash. Ann. § 90.03.260; Or. Rev. Stat. Ann. § 225.290 and § 537.190 (2)(commission may approve application for municipal use to exclusion of all subsequent appropriations); Colo. Rev. Stat. § 37-92-301(4)(conditional decree to protect water project during long construction phase).

¶ I.e., Arizona, California, Nevada, Oklahoma, Oregon, and Washington by statute, and Idaho, Wyoming, by court decisions. Colorado allows this by statute and court decision. Montana allows reservations of government for any future beneficial use. Most of these states need not put the water to beneficial use right away. See David H. Getches, Water Law, 3rd, at 116 (1997).

** Western states that exempt domestic uses in some circumstances include: Alaska (less than 500 gallons per day), CA (riparian uses with no priority), Mississippi (single household), Nevada (wells), Oklahoma (less than 5 acre feet per year and groundwater), OR (natural springs that would not leave land they flow on and rainwater collection), South Dakota (less than 25.920 gallons per day), Texas (not transferable, largely riparian, and given priority in shortage), and Washington (5,000 gallons per day or less for single homes, should be registered for sake of priority).

domestic uses, though exempt from the permit requirement, are considered vested rights,* with priority dates. In others, they are entirely outside the appropriation system.† Some states give preferences to domestic uses ahead of other uses.‡ Even in riparian states, domestic uses are allowed to the extent that they do not significantly alter the water.§ Under common law, a riparian user could use the water for any legitimate purpose as long as he or she did not interfere with the rights of others.¶

5.18   STOCK WATERING**

Another use that is often exempt from permit requirements is stock watering. This means allowing cattle to drink from a body of water. Some states, which require permits for domestic uses, still exempt stock watering.†† Where permits are required, the fees are generally less than other uses.‡‡

---

* I.e., Washington.
† I.e., Oklahoma.
‡ I.e., Texas.
§ *Gould v. Eaton*, 117 Cal 539, 49 P. 577 ( ); *McCarter v. Hudson County Water Co.* 70 N.J. 695, 65 A. 489 (1906); *Scranton Gas & Water Co. v. Delaware,L&W.R. Co.*, 240 Pa. 604, 88 A. 24 (1913).
¶ *Indian Refining Co. v. Ambraw River Drainage Dist.*, 1 F. Supp. 937 (D.C. Ill. 1913); *De Witt v. Bissell*, 77 Conn. 530, 60 A. 113 (); *Glenn v. Crescent Coal Co.*, 145 Ky 137, 140 S.W. 43 (); *A.L. Lakey Co. v. Kalamazoo*, 138 Mich. 644, 101 N.W. 841 (); *Scranton Gas & Water Co.v. Delaware, L. & W. R., Co.*, 240 Pa. 604, 88 A. 24 (1917); *Purcellville v. Potts*, 179 Va. 514, 19 S.E.2d 700 (1942); *Gaston v. Mace*, 33 W.Va. 14, 10 S.E. 60 (1889).
** For discussion of the quality needs of livestock, see 25 Am Jur Proof of Facts p. 246 Water Pollution § 8.
†† I.e., Utah and Idaho. Utah allows cattle to drink but vests no right. See also *Strobel v. Kerr Salt Co.* 164 N.Y. 303, 58 N.E. 142 (1900) (this case is from a riparian jurisdiction, though most appropriation states differ little in practice). In most states, in-stream uses are allowed, though some only allow them for government agencies for environmental purposes. See *State Dept. of Parks v. Idaho Dept. of Water Admin.*, 96 Idaho 40, 530 P.2d 924 (1974) (authorizing in-stream appropriations by state agencies). Diversion is seldom an issue regarding stock watering. See *Steptoe Live Stock Co. v. Gulley*, 53 Nev. 163, 295 P. 772 (1931) (stock watering allowed under appropriation system though no diversion).
‡‡ New Mexico requires a $5 fee for such rights.

6 Riparian Rights*

6.1 METHODS GENERALLY

Water rights are determined in the United States by two major methods. The first is termed riparian rights, which dominate primarily in the eastern United States. The other is called apportioned rights (also known as appropriation), which exist in various forms throughout the western United States. Water rights are riparian if they exist as a result of the land touching on the body of water. They are an appropriation if the rights exist independent of any riparian land ownership. There is some diversity among the jurisdictions as to the precise definition and extent of riparian lands. The move of most riparian jurisdictions toward permit systems has created many similarities between riparian and appropriation jurisdictions.

---

* See Appendix 3, Table 16 for list of riparian states.

6.2 RIPARIAN RIGHTS*

The riparian doctrine† holds that landowners whose land borders a body of water have certain rights‡ to use the water, including a right to have the water pass in its "natural flow."§ Twenty-nine states¶ have systems rooted in riparian rights.** Ten

---

* For a discussion of Louisiana's civil code, often considered a riparian system, see Chapter 13.

† *Kapp v. Hansen*, 79 S.D. 279, 111 N.W.2d 333 (1961); *Brainard v. State*, 12 S.W.3d 6 (Tex. 1999), reh'g overruled, (Jan. 6, 2000); *City of Barstow v. Mojave Water Agency*, 23 Cal. 4th 1224, 99 Cal. Rptr. 2d 294, 5 P.3d 853 (2000); *Florio v. State ex rel. Epperson*, 119 So. 2d 305, 80 A.L.R.2d 1117 (Fla. Dist. Ct. App. 2d Dist. 1960); *Farnes v. Lane*, 281 Minn. 222, 161 N.W.2d 297 (1968); Wehby v. Turpin, 710 So. 2d 1243 (Ala. 1998) (distinguishing between riparian and littoral rights); *Gregg Neck Yacht Club, Inc. v. County Com'rs of Kent County*, 137 Md. App. 732, 769 A.2d 982 (2001); *Stoesser v. Shore Drive Partnership*, 172 Wis. 2d 660, 494 N.W.2d 204 (1993); *Matter of Deadman Creek Drainage Basin in Spokane County*, 103 Wash. 2d 686, 694 P.2d 1071 (1985); *Ames Lake Community Club v. State*, 69 Wash. 2d 769, 420 P.2d 363 (1966); *Yates v. City of Milwaukee*, 77 U.S. 497, 19 L. Ed. 984 (1870); *Bino v. City of Hurley*, 273 Wis. 10, 76 N.W.2d 571, 56 A.L.R.2d 778 (1956) (riparian's right to use shoreline is a property right); *Port Clinton Associates v. Board of Selectmen of Town of Clinton*, 217 Conn. 588, 587 A.2d 126 (1991) (riparian property rights limited by public rights to access). Note a riparian's right to water ceases after is passes his land. *Kennebunk, Kennebunkport and Wells Water Dist. v. Maine Turnpike Authority*, 147 Me. 149, 84 A.2d 433 (1951); *Freed v. Miami Beach Pier Corporation*, 93 Fla. 888, 112 So. 841, 52 A.L.R. 1177 (1927). The following cases discuss division of property lines regarding the exercise of riparian rights. *Water Street Associates Ltd. Partnership v. Innopak Plastics Corp.*, 230 Conn. 764, 646 A.2d 790 (1994) (not always according to land boundaries); *Miller v. Baker*, 68 Wash. 19, 122 P. 604 (1912) (where land actually abuts stream); *Holyoke Co. v. Lyman*, 82 U.S. 500, 21 L. Ed. 133 (1872) (own side of river to center); *Warren v. Westbrook Mfg. Co.*, 86 Me. 32, 29 A. 927 (1893) (regarding channels on each side of an island); *Wisniewski v. Gemmill*, 123 N.H. 701, 465 A.2d 875 (1983); *R.W. Docks & Slips v. State*, 244 Wis. 2d 497, 2001 WI 73, 628 N.W.2d 781 (2001) (riparian rights listed and discussed); *Ritter v. Standal*, 98 Idaho 446, 566 P.2d 769 (1977) (right to unobstructed access to navigable waters); *Okaw Drainage Dist. of Champaign and Douglas County, Ill. v. National Distillers and Chemical Corp.*, 882 F.2d 1241 (7th Cir. 1989); *Sundell v. Town of New London*, 119 N.H. 839, 409 A.2d 1315 (1979) (recreational use, wharf pier, etc).

‡ Some states grant riparian owners a right to an unobstructed view. *Treuting v. Bridge and Park Commission of City of Biloxi*, 199 So. 2d 627 (Miss. 1967). *Hite v. Town of Luray*, 175 Va. 218, 8 S.E.2d 369 (1940) (riparian rights to use water passing land are property rights not easements); *Stratton v. Mt. Hermon Boys' School*, 216 Mass. 83, 103 N.E. 87 (1913); *Scranton Gas & Water Co. v. Delaware, L. & W. R. Co.*, 240 Pa. 604, 88 A. 24 (1913)(riparian rights to use water are qualified); *Bollinger v. Henry*, 375 S.W.2d 161 (Mo. 1964); *Harvey Realty Co. v. Borough of Wallingford*, 111 Conn. 352, 150 A. 60 (1930).

§ Most jurisdictions do not allow riparians to use water on nonriparian lands. *Harvey Realty Co. v. Borough of Wallingford*, 111 Conn. 352, 150 A. 60 (1930). Some riparian jurisdictions allow water to be used on nonriparian land so long as there is no harm to other riparians (or at the very least to the one complaining). *Stanton v. Trustees of St. Joseph's College*, 254 A.2d 597 (Me. 1969). It is important to recognize that a plaintiff must have standing to bring suit in court. Ordinarily, standing requires some harm to the plaintiff. *Stratton v. Mt. Hermon Boys' School*, 216 Mass. 83, 103 N.E. 87 (1913); *Shain v. Veneman*, 278 F. Supp. 2d 1006 (S.D. Iowa 2003) (did not show injury or redress, lacked standing). Nonriparians may acquire riparian rights but only to the extent the riparian has them to give. They are still liable for any harm to other riparians. *Wagner v. Purity Water Co.*, 241 Pa. 328, 88 A. 484 (1913).

¶ See Appendix 3, Table 14.

** It is left to the states to determine what system of water law to maintain and what if any state rights to grant landowners. *Federal Power Commission v. Niagara Mohawk Power Corp.*, 347 U.S. 239, 74 S. Ct. 487, 98 L. Ed. 666 (1954); *Thurston v. City of Portsmouth*, 205 Va. 909, 140 S.E.2d 678 (1965); *Harrell v. City of Conway*, 224 Ark. 100, 271 S.W.2d 924 (1954); *Priewe v. Wisconsin State Land & Improvement Co.*, 93 Wis. 534, 67 N.W. 918 (1896); *Matter of Chumstick Creek Drainage Basin in Chelan County*, 103 Wash. 2d 698, 694 P.2d 1065 (1985); *State v. Zawistowski*, 95 Wis. 2d 250, 290 N.W.2d 303 (1980) (using common law to determine riparian water rights); *Colorado v. New Mexico*, 459 U.S. 176, 103 S. Ct. 539, 74 L. Ed. 2d 348 (1982) (riparian doctrine primarily in eastern states).

others have systems that combine riparian and prior appropriation doctrines (see 7.25, "Hybrid Systems"). The riparian rights doctrine is tempered with a reasonable use standard to protect other riparian users farther downstream (see 4.5, "Reasonable Use Rule"). Today the standard is to allow all riparian users to use the water in a manner considered reasonable relative to all other rights holders. Most states now require riparian landowners to acquire permits before using water. No riparian state is governed solely by common law today (this means that there are always statutes and regulations involved with the rights). Note: In all states, regardless of the system, riparian landowners have rights to use the surface of the water.* A riparian owner whose rights have been injured may bring an action for damages.†

6.3   DISUSE OF RIPARIAN RIGHTS

Ordinarily, riparian rights are not lost by disuse. These rights exist on the basis of land ownership and not priority of prior use.‡

6.4   THE SOURCE OF TITLE TEST

In western states where prior appropriation also exists, the riparian lands of states still recognizing them are generally diminishing. The "source of title" test states that the riparian right extends only to the smallest tract in the chain of title.§ Many states that once recognized riparian rights no longer do. (See 4.2, "State Ownership Rights and Federal Reserved Rights," and 7.4, "The 'Colorado Doctrine'")

6.5   UNITY OF TITLE

A contrasting test to the "source of title" test, used in some states, is the "unity of title" test.¶ Later purchases of land bordering on earlier owned riparian land in no way restrict the rights of the landowner. Under this test, riparian rights exist, independent of the extent of the landowner's holdings. The only thing necessary to entitle a landowner to water rights is to show that some piece of land owned by the landowner borders a stream. As a point of interest, very few questions about the extent of riparian lands have arisen in the eastern states. The case law largely comes from western states that either once recognized riparian rights or still recognize them to some extent after adoption of the prior appropriations system (see 7.25, "Hybrid Systems").

---

* *Harris v. Brooks*, 225 Ark. 436, 283 S.W.2d 129, 54 A.L.R.2d 1440 (1955) (right to swim); *Freed v. Miami Beach Pier Corporation*, 93 Fla. 888, 112 So. 841, 52 A.L.R. 1177 (1927); *Thiesen v. Gulf, F. & A. Ry. Co.*, 75 Fla. 28, 78 So. 491 (1917) (right to bath in navigable waters).
† *Wisniewski v. Gemmill*, 123 N.H. 701, 465 A.2d 875 (1983); *Snyder v. Callaghan*, 168 W. Va. 265, 284 S.E.2d 241 (1981).
‡ *Anaheim Union Water Co. v. Fuller*, 150 Cal. 327, 88 P. 978 (1907) (plaintiffs may enjoin illegal uses in order to protect their riparian rights); see also *Yearsley v. Carter*, 270 P. 804 (Wash. 1928); *Watkins Land Co. v. Clements*, 86 S.W. 733 (Tex. 1905).
§ *Boehmer v. Big Rock Creek Irrigation Dist.*, 48 P. 908 (Cal. 1897).
¶ *Jones v. Conn*, 64 P. 855 (Or. 1901); *Clark v. Allaman*, 80 P. 571 (Kan. 1905); Restatement (Second) of Torts, § 843, comment c (1979).

6.6 ARTIFICIAL WATERCOURSES AND RIPARIAN RIGHTS*

An artificial watercourse dredged into the soil to provide access to water does not ordinarily carry with it any riparian rights.† Riparian rights would not normally be alienable or severable from the land, which borders the water.‡ This rule may not always apply where a right of way is granted or reserved for access to the water or where lands are held by tenants in common.§

6.7 CITY RIPARIAN RIGHTS

The majority rule is that a city on a waterfront is not considered a riparian proprietor with a right to take water for public water supply.¶ There is some authority for the opposite view, though.**

6.8 DRAINAGE AND UNWANTED RUNOFF (SEE ALSO 11.8 "LAW OF DRAINAGE")

Often landowners are not trying to trap excess water, but drain it. In such situations, jurisdictions vary considerably. There are two basic rules in the law of drainage. The first is the "common enemy rule."†† Under this rule the owner of land may protect himself as best he can by building dikes and drainage to keep water off his land, directing it to other lands. The other major rule is the "civil law" rule.‡‡ This rule subjects the land to a servitude for natural runoff across it. Under this rule a landowner may not prevent the water from crossing his land. Both of these rules have been applied with some flexibility by courts. For example, the "civil law" rule may be modified to allow landowners to divert water flow if damage is minimal or to prevent severe flooding. Note that these rules apply only to surface water runoff and not watercourses.

6.9 USES NOT INTIMATELY ASSOCIATED WITH THE WATER

In a Washington case involving a small lake on private property, the court denied defendants the right to build 180 feet out into the lake.§§ The court found this use

---

* See 78 Am Jur 2d, Waters §§ 195–99.

† *Thompson v. Enz*, 379 Mich. 667, 154 N.W.2d 473 (1967). Note: On appeal the court determined that other riparian owners and developers had also dredge canals to provide access in the past. The court found that the lake could not sustain the new burden and had reached the breaking point, such that the practice was still disallowed. See § 4.5.

‡ *Id.;* but see *Strong v. Baldwin*, 97 P. 178 (Cal. 1908) (holding that a portion of riparian land sold with the same rights to use water as the retained lands bordering the water retained riparian rights even though it did not border the water).

§ *Rancho Santa Margarita v. Vail*, 81 P.2d 533 (Cal. 1938).

¶ *Emporia v. Soden*, 25 Kan. 588 (1881); *Town of Purcellville v. Potts,* 19 S.E.2d 700 (Va. 1942);

** *Canton v. Shock*, 63 N.E. 600 (Ohio 1902) (expressing the minority view that a city may hold riparian rights for public water supply).

†† *Argyelan v. Haviland*, 435 N.E.2d 973 (Ind. 1982); *Johnson v. Whitten*, 384 A.2d 698 (Me. 1978) (see § 11.8).

‡‡ *Dekle v. Vann*, 182 So.2d 885 (Ala. 1966); *Robinson v. Belanger*, 52 N.W.2d 538 (Mich. 1952).

§§ *Bach v. Sarich*, 445 P.2d 648 (Wash. 1968). See § 4.4.

not intimately associated with the water and therefore not a riparian use at all. This ruling was based on the "common use rule" (discussed briefly in 4.6).

6.10 MINIMUM LAKE LEVELS

As with many appropriation states, many riparian states also require minimum water levels in certain lakes and reservoirs.*

6.11 RIPARIAN RIGHTS TO TRANSPORT WATER DOWNSTREAM

The Pennsylvania Supreme Court found that a power company seeking to send water downstream for use in a power plant had no right to do so. Riparian rights apply only to the natural flows of the stream.†

6.12 REASONABLE USE‡ OF LIMITED WATER (SEE ALSO 4.5 "REASONABLE USE RULE" AND 11.5 "RESTATEMENT OF TORTS [SECOND]")

Many courts out west have divided limited water resources according to the amount of land owned by the various rights holders and allocated a percentage of the water supply equal to the percentage of the total overall land interests.§ Even while riparian rights are less common in the west today, sharing of shortages pro rata is widely practiced in western irrigation districts among permit and rights holders. All leading cases requiring restrictions of riparian flows were due to drought conditions and low flows. When flows are above normal, each riparian may take as much water as he wants so long as he does not cause any harm.¶ There has been a shift from protecting better uses toward protecting prior uses.** Very few riparian cases actually mention priority, but the results almost always favor a prior user over a newer user.††

---

* Minn. Stat. Ann. § 103G.401; Mich. Comp. Laws Ann. § 281.63; Vt. Stat. Ann. Tit. 10 §§ 905, 1421–26; Ind. Code Ann. § 13-2-13-1. Sixteen eastern states gave consideration to minimum flow levels of some kind: Conn., Fla., Ill., Ind., ME, Mass., Mississippi, Tenn., NH, NJ, NY, OH, SC, VA, WV, and WI.

† *Alburger v. Philadelphia Elec. Co.*, 535 A.2d 729 (Pa. 1988).

‡ *State of Colo. v. State of Kan.*, 320 U.S. 383, 64 S.Ct. 176, 88 L. Ed. 116 (1943); *Stewart v. Bridges*, 249 Ga. 626, 292 S.E.2d 702 (1982); *Harris v. Brooks*, 225 Ark. 436, 283 S.W.2d 129, 54 A.L.R.2d 1440 (1955) (littoral rights in lakes); *Taylor v. Tampa Coal Co.*, 46 So. 2d 392 (Fla. 1950) (littoral rights in lakes); *Rice v. Naimish*, 8 Mich. App. 698, 155 N.W.2d 370 (1967); *In re Waters of Long Valley Creek Stream System*, 25 Cal. 3d 339, 158 Cal. Rptr. 350, 599 P.2d 656 (1979); *Lummis v. Lilly*, 385 Mass. 41, 429 N.E.2d 1146 (1982).

§ See *Hunter Land Co. v. Laugenour*, 250 P. 41 (Wash. 1926); *Wiggins v. Muscupiabe Land & Water Co.*, 45 P. 160 (Cal. 1896); but see *Harris v. Harrison*, 29 P. 325 (Cal. 1892) (allowing two irrigators use of the full flow of the stream for 3½ days per week); *Southern California Inv. Co. v. Wilshire*, 77 P. 767 (Cal 1904) (holding it was an error to divide water solely on the basis of the proportion of the frontage of the owners' lands on the stream).

¶ *Half Moon Bay Land Co. v. Cowell*, 160 P. 675 (Cal. 1916).

** Restatement (Second) of Torts § 850A (one of the factors to be considered in determining reasonable use is the protection of "existing values"). This equates to a rule of priority, though not to the extent of that used in prior appropriation jurisdictions. Other factors to be considered may trump the priority factor. See also §§ 4.5 and 11.5.

†† For an opposite result, see *Joslin v. Marin Mun. Water Dist.*, 429 P.2d 889 (Cal. 1967) (a rare case where a prior right was extinguished by a new use with no compensation).

6.13   ECONOMICS IN RESOLVING DISPUTES

In many jurisdictions, a newer user will simply buy out the prior user's water right because its value toward the new use is worth the price to compensate the prior user's loss. If the new use is not worth the compensation price then the new user will ordinarily fail in his appropriation.*

6.14   PRIORITY OF USES (SEE ALSO 7.14 "PREFERENCES")

Many jurisdictions have established priority systems† to protect certain uses ahead of others.‡ Often public water supply interests have a high priority.§

6.15   EASTERN PERMIT SYSTEMS (SEE ALSO 7.21 "PERMITS" (PRIOR APPROPRIATION))

As of 1995, at least fourteen eastern states¶ had adopted some type of permit system. The system ordinarily requires riparian users to register their uses and rights.** Unregistered uses are presumed to be abandoned.†† Although there are significant differences between riparian and prior appropriation states, in many ways the eastern permit systems resemble those commonly found in the west. It helps quantify

---

* The Pareto criteria state that an exchange can be made that benefits someone and injures no one. "When the exchange can no longer be made, the situation becomes one of Pareto optimality." *Black's* Law Dictionary, Seventh Edition, 1999. See *Strobel v. Kerr Salt Co.*, 58 N.E. 142 (N.Y. 1900); *United States v. Gerlach Live Stock Co.*, 339 U.S. 725, 70 S.Ct. 995, 94 L.Ed. 1231 (1950) (granting compensation to a small user for their loss due to a federal dam). See §§ 3.11 and 16.5.

† *Brown v. Ellingson*, 224 So. 2d 391 (Fla. Dist. Ct. App. 2d Dist. 1969) (priority of use set by statute).

‡ *Brummund v. Vogel*, 184 Neb. 415, 168 N.W.2d 24 (1969) (domestic use ahead of agriculture).

§ See *Adams v. Greenwich Water Co.*, 138 Conn. 205, 83 A.2d 177 (1951); *Taylor v. Tampa Coal Co.*, 46 So. 2d 392 (Fla. 1950); *Pierce v. Riley*, 35 Mich. App. 122, 192 N.W.2d 366 (1971), following *Thompson v. Enz*, 379 Mich. 667, 154 N.W.2d 473(1967).

¶ See Conn. Gen. Stat. Ann. §§ 22a-365 to –378; Del. Code Ann. tit. 7, §§ 6001-6031; Fla. Stat. Ann. §§ 373.012 to .619; Ga. Code Ann. §§ 12-5-20 to –31, 12-5-43-53; Iowa Code Ann. §§ 455B.261 to -.281; Ky. Rev. Stat. Ann. §§ 151.010 to -.600; Md. Nat. Res. Code Ann. §§ 8-101 to –204, 8-801 to –814; Mass. Gen. Laws Ann. Ch., 21G, §§ 1 to 19; Minn. Stat. Ann. §§ 103G.255 to -.315; Miss. Code Ann. §§ 51-1-5 to –55; N.J. Stat. Ann. §§ 58:1A-1 to –17; N.C. Gen. Stat. §§ 143-215.11 to –215.22; Va. Code Ann. §§ 62.1-242 to–253; Wis. Stat. Ann. §§ 30.18, 30.28, 30.292, 30.294, 30.298, 144.026. Iowa and Florida have extensive permit systems, while some sixteen other riparian states at least supplement their common law with some kind of permit system. Note that nineteen eastern states had enacted permitting or registration requirements that applied to both surface and groundwater. Unused rights are considered abandoned and extinguished. The states include Conn., Del., Fla., Ga., Ill., Ind., Ken., ME, Maryland, Mississippi, NJ, NY, NC, Ohio, SC, Tenn., VA, and WI. Mich., Ohio, Penn., and VT had pending legislation as of 1990. As of 1990, eighteen eastern states defined acceptable uses for surface and ground water: Conn., Del., Fla., Ga., Ill., Ind., Ken., ME, Maryland, Mississippi, NH, NY, NC, Ohio, SC, Tenn., VA, and WI. Fifteen eastern states had restricted the area of use for riparian rights, or diversions: Conn., Fla., Ga., Ill., Ind., Ken., ME, Mass., Mich., NJ, NY, NC, OH, SC, and WI. NH had legislation proposed as of 1990. Indiana, Vermont, and West Virginia require permits only in "critical areas." Arkansas, Illinois, Iowa, Kentucky, Maryland, and Minnesota set priorities during times of shortage, generally preferring domestic uses.

** Permits often amount to a form of priority system. They tend to blur the distinction between riparian and appropriation jurisdictions. *Omernick v. Department of Natural Resources*, 71 Wis. 2d 370, 238 N.W.2d 114 (1976) (establishing a form of priority in Wis.).

†† See *Village of Tequesta v. Jupiter Inlet Corporation*, 371 So.2d 663 (1979) (ruling that corporation must use the permit system to effectuate water rights and no taking could occur of unregistered rights).

existing rights and uses, and it controls and limits new uses.* Some states, such as Florida, appear to authorize nonriparian uses of water.† No state specifically limits uses to riparian properties, though most do not specifically authorize nonriparian uses either. It is not clear what effect statutes creating a permit system have on existing riparian rights. Some states require the use to be registered and a permit issued, though many do not say and uses are subject to judicial interpretation whether a permit is necessary for continued use. Some states exempt certain uses from the permit process or grant preferences for those uses.‡

6.16 THE REGULATED RIPARIAN MODEL WATER CODE§ (SEE ALSO 11.9)

In 1997, the Water Law Committee of the American Society of Civil Engineers completed work on the code, which reflects the transition toward regulatory codes in riparian states. Professor Joseph Dellapenna headed work on the code. The code attempts to allocate water rights in the most efficient and beneficial manner, using a permit system. It retains the reasonable use principle also reflected elsewhere in water law, and it preserves minimum flow levels.

6.17 WATER POLLUTION¶ (SEE ALSO 10.2 "CLEAN WATER ACT GENERALLY" FOR MODERN VIEW)

Uses, which pollute the waters, were traditionally only inhibited if they posed an unreasonable burden on other riparian uses.** Traditionally, waters were used to wash away, soaps, dye, sawdust, and other industrial products. Ordinarily, use is considered unreasonable if the cost is greater than the burden, in light of the uniform custom of the country. With shortages of suitable water, it is likely that pollution will continue to be regulated with greater frequency, in the future.††

---

* Courts ordinarily have power to quantify rights to water in excess of that needed for natural rights according to demands of equity. *Hidalgo County Water Imp. Dist. No. Two v. Cameron County Water Control & Imp. Dist. No. Five*, 250 S.W.2d 941 (Tex. Civ. App. San Antonio 1952) (judicial apportionment of riparian water).

† Fla. Stat. Ann. § 373.223(2): "The governing board or department may authorize the holder of a use permit to transport and use ground or surface water beyond overlying land, across county boundaries, or outside the watershed from which it is taken . . ." See also Ark. Code. Ann. § 15-22-304(b); Wis. Stat. Ann. § 30.18(6)(b).

‡ I.e., Kentucky and Maryland (agriculture) or Wisconsin (certain agricultural crops); Wisconsin and Michigan (mining listed as public interest); Georgia, Maine, and North Carolina (grant miners right of access); Minnesota (hearings to determine if mining is in public interest before permit).

§ A portion of the code can be found in Joseph L. Sax, Robert H. Abrams, Barton H. Thompson Jr., and John D. Leshy, *Legal Control of Water Resources: Cases and Materials,* 80–90 (3rd ed., 2000).

¶ For more on water pollution, see *25 POF, Water Pollution-Sewage and Industrial Wastes* §§ 1–51 at 233–332 (1970) with 1989 update.

** *Snow v. Parsons*, 28 Vt. 459 (1856) (discussing the legality of tanning company's by-products entering the stream).

†† See *25 POF, Water Pollution* § 1, stating that water pollution and the adequacy of water supply are a single problem.

6.18   FIRST IN TIME* (FOR APPROPRIATION, SEE 7.12 "RULE OF PRIORITY")

Water is ordinarily not to be unreasonably restricted in its use by an earlier user, to the detriment of all other users. This is a divergence from the law in England, which protects the riparian's right to use water in its free-flowing condition.† The English rule permitted uses, which were largely nonconsumptive and provided for the return of any water withdrawn. American jurisdictions generally permit some reasonable consumptive uses.

6.19   NATURAL FLOW (SEE ALSO 4.7 "THE NATURAL FLOW THEORY")

Many riparian states once protected the rights of riparians to enjoy the water in its natural flow.‡ This ordinarily means that material alterations to the flow of the water are forbidden. With the onset of the industrial revolution, reasonable use and permits, this has changed.§ Today, some courts continue to use "natural flow" language, but nearly all have moved to "reasonable use" doctrines in practice. Most riparian states now use permit-systems, making traditional "natural flow doctrines" obsolete.

6.20   RIPARIAN PRESCRIPTIVE RIGHTS (SEE ALSO 3.7 "PRESCRIPTION")

Prescriptive rights in riparian states are relatively common.¶ It is, ordinarily, only possible to acquire such rights against down stream riparians.** Use is not generally adverse against lower riparians unless it harms the rights of lower riparians.†† In addition to riparian states, most hybrid states used to recognize prescriptive rights, too.‡‡ It is worth mentioning that the trend seems to be moving away from prescriptive rights in hybrid jurisdictions.§§

---

* *Benton v. Johncox*, 17 Wash. 277, 49 P. 495 (1897) (riparian rights attach as of date when land settled).
† *Martin v. Bigelow*, 2 Aik. 184 (Vt. 1827). The English rule would often prevent later uses that interfered with the earlier use by limiting the water flow. See § 6.19, "Natural Flow."
‡ *Farrell v. Richards*, 30 N.J. Eq. 511 (N.J. Ch. 1879).
§ *Tyler v. Wilkinson*, 24 F.Cas. 472 (C.C.R.I. 1827).
¶ *Tyler v. Wilkinson*, 24 F.Cas. 472 (C.C.R.I. 1827).
** See *Dontanello v. Gust*, 86 Wash. 268, 150 P. 420 (Wash. 1915) (rights acquired against upper riparian because device built on upper owne'rs land). Adverse uses for less than the statutory period vest no rights. *Martin v. Bigelow*, 16 Am. Dec. 696 (Vt. 1827).
†† *Pabst v. Finmand*, 190 Cal. 124, 211 P. 11 (Cal. 1922). Some states do recognize illegal use as adverse regardless of harm.
‡‡ Getches, David H., *Water Rights,* 3rd ed. (1997), 71–72.
§§ See R.C.W.A. 90.14.220, "No rights to use of surface or ground waters of the state affecting either appropriated or unappropriated waters thereof may be acquired by prescription or adverse use." (passed in 1967).

6.21   ACCRETION AND AVULSION

Accretion* and avulsion† relate to the buildup and loss of soil within a body of water. This is the natural process by which a river-channel or shoreline changes over time.‡ Different rules often apply when the change is not natural.§ A related term, *reliction,* refers to land that becomes dry by the removal of waters instead of the buildup of soil.¶ The rule related to accretion is said to protect the riparian owner's access to water.**

---

* "The increase of riparian land by the gradual deposit, by water, of solid material, whether mud, sand, or sediment, so as to cause that to become dry land which was before covered by water." 78 Am Jur 2d, Waters § 406.

† "[a] sudden and perceptible loss or addition to land by the action of water, or a sudden change in the bed or course of a stream." 78 Am Jur 2d, Waters § 406. Avulsion is the result of very rapid change whereas accretion occurs over longer periods of time. Many courts use a different rule to determine ownership involving avulsion. See *Kinkead v. Turgeon,* 74 Neb. 580, 109 NW 744 (1906).

‡ See generally, 21 POF 2d, *Change in Shoreline* §§ 1–42, at 154–250 (1980) with 1989 update; 78 Am Jur 2d, Waters §§ 406–27 with 1999 update.

§ See 78 Am Jur 2d, Waters § 410. See also *Michailson v. Silver Beach Improv. Asso.,* 342 Mass 251, 173 N.E.2d 273(1961).

¶ 78 Am Jur 2d, Waters § 406.

** 78 Am Jur 2d, Waters § 410.

6.2.1 ACCRETION AND AVULSION

7 Prior Appropriation*

7.1 PRIOR APPROPRIATION (FOR A CLOSER LOOK AT INDIVIDUAL SYSTEMS, SEE CHAPTER 17)

The prior appropriation system† began in the American West, where much of the land was owned by the government. Because the government owned the land, few private owners had riparian rights. Appropriation began largely with early miners who sought water for their operations on federal lands. They simply followed the same rules they used for the minerals they competed for: "first in time, first in right." The earliest user could use to the exclusion of others. Early courts recognized these customs, and soon the system was applied to farmers and others where it seemed to work well. Whereas riparian law depends on land ownership, appropriation law relies on water usage. To perfect a water right, a person must 1) comply with all statutory requirements, and 2) put the water to a beneficial purpose. The right remains valid so long as the use lasts. The arid landscape of the West was another reason for the appropriation system. Water needs could not always be located on the lands bordering water, and local needs were best served by not allowing the fortunate few next to the water to monopolize it. Appropriation rights may be transferred to the extent that they do not prejudice the vested rights of others. Of the so-called "pure" appropriation states, only Colorado does not require a permit‡ to appropriate water.§ One of the drawbacks of appropriation is that it can feed into collective-action problems, preventing optimal use of water.¶

---

* See Table 17 for list of Prior Appropriation States. See Form 3 for water-right application.
† Ownership of land grants no rights in a pure appropriation jurisdiction. Water flowing in natural condition in-stream is unowned and held by the state for appropriation. Natural streams exclude unconnected underground water (traditionally all underground water, though this is rapidly changing), rain-water, melting snow, and other condensation (otherwise known as diffused surface waters) as it passes over surface of the earth and before entering any river or lake. A property right is acquired only when water is applied to a beneficial use (not wasted) with due diligence. Most jurisdictions also require an administrative permit. Appropriated water carries no limitations on the place of use. Appropriations are not to be held merely as investments. They must be used or forfeited. Priority is as of date water is first applied to a beneficial use. When there is a shortage of water, junior appropriators must cease their diversions first (there is no equitable sharing in a pure appropriation system). See Joseph L. Sax, Robert H. Abrams, Barton H. Thompson Jr., and John D. Leshy, *Legal Control of Water Resources: Cases and Materials,* 98-100 (3rd ed. 2000). There is considerable diversity in the laws of the several appropriation states, and pure appropriation, in the strictest sense is uncommon, if not unheard of.
‡ Many leading cases in water law come from Colorado because of its unique system involving court instead of administrative agencies in recording water rights, which has lead to an increase in the number of adjudications in state.
§ See Appendix 3, Table 15.
¶ See 12.1–12.7. Nearly all jurisdictions have some protections in their statutes to prevent certain types of waste. For an example, see 7.9.

7.2   PRIOR APPROPRIATION'S HISTORIC ROOTS

The origin of prior appropriation rights in the United States can be traced to early case law in California.* Gold miners used the water in their mining activities. Because they ordinarily did not own the land they mined, they had no riparian rights. The limited amount of water in parts of the West also required special consideration. The solution was to apply the same doctrine used to settle mining claims. Failure to use the water constituted an abandonment of the right just as failure to work the mine did. From these early traditions, modern water law in the American west was born. Seeds of prior appropriation were present in 1847 in Utah. The Mormon church approved the custom of diverting water by group and putting it to beneficial uses. The church supervised these efforts. An 1880 statute recognized these rights acquired by appropriation. According to Gould and Grant's textbook on water law, the principle of priority in time appears to have been recognized by custom before a statute on the matter existed.† The primary idea rising out of this system is one of state regulation, such that permits, licenses, and concessions are not to be issued to the detriment of prior users. The Desert Land Act of 1877 allows the reclamation of lands within several western states and territories.‡ It permits reclamation§ subject to prior appropriation of waters.

7.3   THE "CALIFORNIA DOCTRINE"

Despite the number of California cases applying appropriation rights, the California court found that the United States patent to riparian land carried with it riparian rights.¶ In California the riparian rights of landowners are subject to prior appropriation rights of others before the patent for the land was issued. Today only California, Nebraska, and Oklahoma recognize the possibility of new riparian uses in the West.

7.4   THE "COLORADO DOCTRINE"

Nine western states follow pure appropriation law, known as the "Colorado Doctrine."** In six states said to follow the "Californian Doctrine," above, all new rights are appropriation, though some current rights are traced to historic riparian uses.††

---

* *Irwin v. Phillips*, 5. Cal. 140 (1855). See also *Eddy v. Simpson*, 3 Cal. 249 (1853) (sometimes called the first case to state the doctrine of prior appropriation).

† George A Gould. and Douglas L. Grant, *Cases and Materials on Water Law*, 5th ed., 18 (1995) (quoting 1 Wells A. Hutchins, *Water Rights Laws in the Nineteen Western States* 163 (1971).

‡ California, Oregon, Nevada, Washington, Idaho, Montana, Utah, Wyoming, Arizona, New Mexico, and North and South Dakota. Colorado has since been added.

§ See Chapter 16 on Reclamation.

¶ *Lux v. Haggin,* 69 Cal. 255, 10 P. 674 (1886) (known as the "California Doctrine," now used in only California, Nebraska, and Oklahoma).

** E.g., Alaska, Arizona, Colorado, Idaho, Montana, Nevada, New Mexico, Utah, and Wyoming.

†† E.g., Kansas, North Dakota, Oregon, South Dakota, Texas, Washington.

7.5 DIVERSION* (SEE ALSO 5.4, "IN-STREAM VALUES," AND 5.7, "WATER AS A NATURAL RESOURCE")

In some western states, the right to appropriate water (or divert it) is limited by the demands of public interests, such as use by for recreation or by fish and wildlife.† The state's definition of a diversion has often made a difference in determining whether an appropriation was made.‡ In Utah, Nevada, and New Mexico, reaping the benefits of natural overflow and percolation did not constitute an appropriation of the water.§ In Oregon, a right to the water is acquired by natural overflow.¶ This rule was codified in Colorado.** All western states have some form of "no injury" rule. This means that changes in diversion that would harm junior appropriators are generally not allowed.††

Where animals do their own digging, some states have upheld stock water appropriations.‡‡ In Utah the animals are allowed to drink, but no right is vested.§§ As a practical matter, today in-stream water rights in most states are initiated by permit

---

* Ordinarily, diversion is an outdated requirement for perfecting a water right. Most states allow rights to in-stream uses. Only California specifically holds that in the absence of a diversion no right may be vested. See *California Trout, Inc. v. State Water Resources Control Board*, 90 Cal. App. 3d 816, 153 Cal. Rptr. 672 (1979); but see Cal. Water Code § 1707, allowing conventional right to be converted into an in-stream right. Alaska and Arizona specifically allow in-stream appropriations. Most states are not so explicit. For examples of cases eliminating the diversion requirement at least in part, see *State v. Morros*, 766 P.2d 263 (Nev. 1988); *Steptoe Live Stock Co. v. Gulley*, 53 Nev. 163, 295 P. 772 (1931). In New Mexico, such appropriations have been determined to be legal, but as of 1999 no such appropriation was yet made. Many states explicitly allow state agencies to appropriate flows. See Arizona, Alaska, Idaho, Montana, Nebraska, Nevada, Oregon, Utah, Colorado, and Wyoming. See Table 20. Montana held that diversion was necessary for state in-stream rights for recreation and fishing uses if the rights predated 1973 when these uses were recognized as beneficial uses. See *In re Adjudication of Dearborn Drainage Area*, 234 Mont. 331, 766 P.2d 228 (1988). See Form 8, Change point of diversion.

† *Nebraska Game and Parks Commission v. The 25 Corporation, Inc.*, 236 Neb. 671, 463 N.W.2d 591 (1990).

‡ See *Nebraska Game and Parks Commission v. The 25 Corporation, Inc.*, 236 Neb. 671, 463 N.W.2d 591 (1990) (holding that Nebraska appropriation must be governed by public interest and finding in-stream flows subject to that interest are allowed, though not specified). Colorado found that a natural dam intended to divert water back to its natural channel was sufficient to meet the diversion requirement. *City of Thornton v. City of Fort Collins*, 830 P.2d 915 (Colo. 1992).

§ *Hardy v. Beaver County Irrigation Co.*, 234 P. 524 (Utah 1924); *Walsh v. Wallace*, 67 P. 914 (Nev. 1902); *State ex rel. Reynolds v. Miranda*, 493 P.2d 409 (N.M. 1972).

¶ *In re Silvies River*, 237 P. 322 (Ore. 1925) (the court did suggest that in time economic necessity might require a controlled diversion). This case allowed natural irrigation of farmland. Montana also authorizes this. California, Colorado, Idaho, and Nevada consider stock watering to be an appropriation.

** See *Broad Run Inv. Co. v. Deuel & Snyder Improvement Co.*, 108 P. 755 (Colo. 1910).

†† See *Southeastern Colorado Water Conservancy Dist. v. Fort Lyon Canal Co.*, 720 P.2d 133 (Colo. 1986); *W.S. Ranch Co. v. Kaiser Stell Corp*, 79 N.M. 65, 439 P.2d 714 (1968).

‡‡ *Live Sock Co. v. Gulley*, 295 P. 772 (Nev. 1931); *Hunter v. United States*, 388 F.2d 148 (9th Cir. 1967); *Stevenson v. Steele*, 453 P.2d 819 (Id. 1969); and *England v. Ally Ong Hing*, 459 P.2d 498 (Ariz. 1969).

§§ *Adams v. Portage Irrigation Reservoir & Power Co.*, 72 P.2d 648 (Utah 1937). The general rule today in most jurisdictions is that no formal diversion is necessary for an appropriation. See *In re Adjudication of Dearborn Drainage Area*, 234 Mont. 331, 766 P.2d 228 (1988) (use for fish and wildlife by state requires diversion prior to 1973 because not previously seen as beneficial use); *State v. Morros*, 766 P.2d 263 (Nev. 1988) (state may appropriate for wildlife watering); *State Dept. of Parks v. Idaho Dept. of Water Admin.*, 96 Idaho 40, 530 P.2d 924 (1974) (in-stream appropriations by state agency allowed).

application.* The granting of rights still ordinarily requires some intent and purpose.† Even in Colorado, which has no permit law, intent and notice are important.‡ Many states use methods other than appropriations to protect in stream flows. Such methods include minimum flow requirements.§ The greatest concerns over in stream appropriations include fear of harm to current rights, fear of harm to future growth, and the belief that water should be used on the land.¶ Many states restrict the time in which structures can be built.** Most states require an application to change the point of diversion.††

7.6   BENEFICIAL USE AND LOSS OF RIGHTS‡‡ (SEE ALSO 6.13, "ECONOMICS IN RESOLVING DISPUTES," AND 6.14, "PRIORITY OF USES;" SEE 5.3, "ABANDONMENT OF RIGHTS," AND 9.4, "EFFECT OF NONUSE")

Beneficial use is the basis for allocation of water in most western states. It is premised on the idea that water cannot be absolutely owned, only used.§§ It is a theme repeated over and over throughout the statutes of these states.¶¶ The Washington

---

* Several states authorize some type of instream uses by statute, usually by state agencies, I.e., Alaska, Arizona, California, Colorado, Hawaii, Idaho, Kansas, Montana, Nebraska, Nevada, Oklahoma, Oregon, Utah, Colorado, Washington, and Wyoming. These tend to have junior priorities, because the statutes authorizing them are recent. Montana also is among several states that allow leasing of water rights for in-stream use. See Mont. Code Ann. §§ 85-2-102, 402, 404. Colorado requires these appropriations to be dedicated to the state. See Colo. Rev. Stat. § 37-92-102(3). California allows conversion of traditional right to in-stream uses. See Cal Water Code § 1707. See also Or. Rev. Stat. § 537-350 (in-stream right has same legal status as other rights; also providing for abandonment of in-stream uses); R.C.W.A. 90.03.345, reservation of water for certain purposes including maintenance of minimum flows constitutes appropriation within meaning of chapter.

† *Power v. Switzer*, 55 P. 32 (Mont. 1898) (turning of creek from its banks for no apparent purpose was not an appropriation). See also *In the Matter of Dearborn Drainage Area*, 766 P.2d 228 (Mont. 1988) (idea of appropriation rejected because there was no intent to appropriate at the time).

‡ *City of Aspen v. Colorado River Water Conservation Dist.*, 696 P.2d 758 (Colo. 1985) (acts on the land were not required to perfect a water right); *In the Matter of the Applications for Water Rights of the Upper Gunnison River Water Conservancy Dist.*, 838 P.2d 840 (Colo. 1992) (acts of negotiating and executing a contract are sufficient to show intent and provide notice).

§ A number of states allow in stream appropriations but subordinate them to other future uses, e.g., Idaho. Some minimum flow requirements may flow from federal laws such as the Endangered Species Act. See § 10.10. See also R.C.W.A. 90.03.247, minimum flows and levels.

¶ Joseph L. Sax, Robert H. Abrams, Barton H. Thompson Jr., John D. Leshy, *Legal Control of Water Resources: Cases and Materials,* 117 (3rd ed. 2000).

** E.g. Arizona, Idaho, Nebraska, Nevada, New Mexico, Oklahoma. Oregon, Texas, and Wyoming. This list is not exclusive.

†† See Form 8

‡‡ See Table 21 on Beneficial Uses. See Form 4 Beneficial Use, Form 5 Forfeiture.

§§ *Melville v. Salt Lake County*, 570 P.2d 687 (Utah 1977).

¶¶ Ariz. Rev. Stat. Ann. § 45-131("Beneficial use shall be the basis, measure and limit to the use of water."); Kans. Stat. Ann. § 82a-713; Neb Rev. Stat. § 46-238; Nev. Rev. Stat. § 533.035; N.M. Const. Art. 16, § 3; N.M. Stat. Ann. § 72-1-2; N.D. Cent. Code § 61-04-01.2; S.D. Codified Laws Ann. § 46-1-8; Utah Code Ann. § 73-3-12; Wyo. Stat. § 41-2-101. See also Colo Const. Art. XVI. § 6; Colo. Rev. Stat. § 37-92-103 (4). Many other states also use beneficial use, including virtually all western, appropriation states. These uses may be defined by statute but are not always so specifically defined. Idaho, for example, uses beneficial use but has no comprehensive statutory definition. Other state definitions are broad and nonspecific, e.g., that of South Dakota. Some examples of uses specified by statutes are domestic uses, municipal, irrigation, industrial, stock watering, power, mining, recreation, and fish and wildlife. Not all states define the same uses as beneficial. For a breakdown of beneficial uses by state, see David H. Getches, Water Law in a Nut Shell, 98 (3rd Ed. 1997)

Supreme Court ruled on the definition of "reasonable use" as an element of "beneficial use,"* saying, "beneficial use is a term of art." The court listed two principal elements of a water right: 1) purposes or types of activities for which water might be used, and 2) the amount or measure of the water right as determined by the amount reasonably necessary (water duty) for the purpose toward which the water use is put.† In determining the second element, courts apply "reasonable use."‡ While duty in Washington was not a hard-and-fast measure, in many states it is.§ As a result of the beneficial use rule, water must be put to use and not stored indefinitely for speculative purposes.¶ Ordinarily, water not used is considered abandoned** and reverts back to the state.†† Different rules exist for making this determination in various jurisdictions. No junior will lose rights by abandonment if no water is available.‡‡

7.7 RESERVOIRS AND WATER STORAGE§§ (SEE ALSO 7.22)

Beneficial use of water would be limited to short periods of time, where runoff was present, without water-storage facilities. Storage helps maximize the benefits of water use. Two types of storage exist: on channel and off channel. On channel storage means that the storage facility is part of the stream or channel. Off channel means that water is diverted away from the stream or channel to another location. No legal distinction is made between the two types of storage. Most dams are considered on channel facilities and constitute a diversion of water for legal purposes of perfecting a water right. A permit to store water is obtained from the same agency or court that oversees other water rights. Some states require separate permits for storage and for beneficial use.¶¶ Such an approach recognizes that often the group or individual storing water and the one making the beneficial use of it are not the same. The process often involves reservoir companies and individual irrigators, for example. Often the facilities (dams, reservoirs, and so on) must also be approved by the state engineer

---

* RCW 90.03; RCW 90.54
† *State of Wash. Dept. of Ecology v. Grimes*, 121 Wash. 2d 459, 852 P.2d 1044 (1993).
‡ *Id.*
§ See Neb. Rev. Stat. § 46-231(1 c.f.s. [cubic feet per second] for 70 acres); S.D. Codified Laws § 46-5-6 (limiting appropriations for irrigation to 1 c.f.s. for 70 acres and 3 acre-feet per acre per year); Idaho Code § 42-202 (1 c.f.s. for 50 acres); Wyo. Stat. § 41-4-317 (1 c.f.s. for 70 acres); Cal Water Code § 1004 (2½ acre feet per acre per year for irrigation of uncultivated areas not devoted to cultivated crops).
¶ See § 5.3 Abandonment of Rights.
** See § 5.3.
†† Several states require separate permits for storage and application for beneficial use, e.g., Arizona, Nevada, and Wyoming.
‡‡ Note that water is not forfeit for failure to use it, if a junior cannot use his appropriation due to shortages. Most states also make reasonable allowances for other short-term, unforeseeable excuses for failing to use the water, so that the right will not be lost in these circumstances. Many states require a period of time, such as five consecutive years of nonuse before a right is forfeit. Some states require an action of some kind before a right may be forfeit. See § 5.3.
§§ See Form 10 Water Storage Application. See 78 Am Jur 2d, Waters §§ 200–01, 205.
¶¶ E.g., Arizona, Nevada, Wyoming (Wyo. Stat. Ann. § 41-3-302), and Nebraska. Cal. Water Code § 1242 classifies underground storage as a beneficial use if after the water is applied to some beneficial purpose. Wyoming requires excess water stored in a reservoir be made available at reasonable rates to irrigators. Wyo. Stat. Ann. § 41-2-325.

or water resources director (state agent in charge of water rights). Most states exempt small facilities from permit requirements.

7.8 THE ONE-FILL RULE

Storage is generally limited by what is known as the one-fill rule. This means that the holder of a reservoir may not continually refill it throughout the year to hold more than its single use capacity. Put another way, the reservoir may only be filled once per year.* This rule is primarily applied to irrigation uses and not often to other uses because of the inefficiency of its strict application.† Application tò a hydroelectric dam, for example, would severely limit the dam's utility. Since a dam does not usually constitute a consumptive use, this would make little sense.

7.9 EFFICIENCY AND WASTING WATER (SEE ALSO 4.5, 4.6, 6.12, AND 8.15 ON REASONABLE USE RULE)

Some states such as California allow only a vested right to reasonably use water but not to waste it.‡ Other States view retroactive application of a "duty" statute as unconstitutional interference with a vested right.§ Such interference would ordinarily require compensation.¶ In some situations, the water lost from evaporation and seepage during travel to the storage facility may be great enough that the waste is considered unreasonable. The rights holder may be required to build pipes, tunnels, and so on to slow or stop the loss. This can be expensive, but where water is scarce and demand is high a court is likely to require such measures. Improvements in technology may be an economic advantage to large companies that use them. In a statement by the Utah-Idaho Sugar Company, they indicated that they had achieved a substantial reduction in water use and increased the acres irrigated by about one sixth when they switched to a sprinkler system instead of gravity-powered irrigation.** The costs of upgrading may be such that it is reasonable not to upgrade in

---

* *Windsor Reservoir & Canal Co. v. Lake Supply Ditch Co.*, 44 Colo. 214, 98 P. 729 (1908).
† For an example of the inefficiency of strict application, see *Denver v. Northern Colorado Water Conservancy Dist.*, 276 P.2d 992 (Colo. 1954) (forbidding use of a dam by the city for storage beyond the one-use capacity and requiring larger tunnels to be built at taxpayer expense to do what might other wise have been done much cheaper with smaller tunnels). See § 7.11.
‡ *Imperial Irrigation District v. State Water Resources Control Board*, 225 Cal. App.3d 548, 275 Cal. Rptr. 250 (1990) (refusing to recognize companies right to full volume of water, where some of that water was wasted through seepage and various other inefficient means of storage). The California court remarked that simply because a use is beneficial, it is not necessarily reasonable. See Cal. Water Code § 10902(b)(defining efficient water management practices); see also Colo Rev. Stat. § 37-92-103(4) (defining "beneficial use" as including amount "reasonable and appropriate." See Neb. Rev. St. § 46-231; S.D. Codified Laws § 46-5-6; 82 Okl. St. § 105.12 (all limiting the amount of water that may be applied to land for irrigation). Many states have statutes require efficient facilities, I.e., Alaska, Colorado, Idaho, Oregon, and South Dakota. Most court decisions follow this line of thought, requiring reasonable efficiency. See *Glenn Dale Ranches, Inc. v. Shaub*, 94 Idaho 585, 494 P.2d 1029 (Idaho 1972).
§ *Enterprise Irrigation Dist. v. Willis*, 284 N.W. 326 (Neb. 1938).
¶ See § 2.1 on eminent domain
** George A. Gould, and Douglas L. Grant, *Cases and Materials on Water Law* Fifth Ed., 41 (1995).

some situations.* With water supplies increasingly below demand, there is increased incentive for states to conserve.†

7.10 WATER RIGHTS INITIATED BY TRESPASS (ALSO KNOWN AS PRESCRIPTIVE RIGHTS; SEE 3.7, 6.20, AND 13.7)

In some states, water rights initiated by trespass may be valid. As a general rule, an individual must own or have a right to use the land on which the right would be exercised in order to obtain a water right.‡ Where a private party has the right of condemnation the lack of access to water is not a violation of this rule.§ Prescriptive rights to water are almost unheard of where permits are required to appropriate water and abandonment statutes exist. They have also been difficult to prove in jurisdictions that recognize them, given the less-than-obvious nature of the uses and imprecision of measurements accompanying these uses.

7.11 REASONABLE DILIGENCE

Many states date the priority of a water right back in time, to the date of the first step toward execution of the right, conditional upon reasonable diligence in construction.¶ This has occasionally been a problem when changes in the plans become necessary.** A Colorado court stated that a finding of diligence requires "an intention to use the water, coupled with concrete action amounting to diligent efforts to finalize the intended appropriation."†† Another Colorado court stated that the purpose of diligence is to prevent hoarding of undeveloped water.‡‡ In states where a permit is required, the priority date is the date of the application for a permit. Ordinarily, the permit process involves surveys before the application is submitted.§§ In New Mexico a notice of intention may be filed, before the application, fixing the priority date, if the application is filed.¶¶ Most states also limit the time available for construction

---

* See *Colorado v. New Mexico*, 459 U.S. 176, 103 S.Ct. 539, 74 L.Ed.2d 348 (1982) (Supreme Court states undoubtedly there is evidence in the record of large losses due to seepage and evaporation during transport, but it is a leap after observing that losses occur to conclude that they are wasteful). See § 2.2.

† A Washington State provision provides that "it is the policy of the state to promote the use of the public waters in a fashion which provides for obtaining maximum net benefits arising from both diversionary uses of the state's public waters and the retention of waters within streams and lakes in sufficient quality to protect in-stream and natural values and rights." R.C.W.A. 90.03.005.

‡ *Lemmon v. Hardy*, 95 Idaho 778, 519 P.2d 1168 (1974) (ruling that a water right be post dated to the date of an amended application because speculation is not a valid ground for obtaining a right).

§ See *Kaiser Steel Corp. v. W.S. Ranch Co.*, 81 N.M. 414, 467 P2d 986 (1970). See § 7.20.

¶ *City and County of Denver v. Northern Colorado Water Conservancy District*, 130 Colo. 375, 276 P.2d 992 (1954) (note: the state legislature in Colorado has defined "reasonable diligence," as of 1990, in line with the dissent of this case. See Colo. Rev. Stat. § 37-92-301(b)). See § 7.8.

** *Id.* (Note: Colorado is one of a few states that still do not require a permit.)

†† *Colorado River Water Conservation Dist. v. Denver*, 640 P.2d 1139 (Colo. 1982).

‡‡ *Public Serv. Co. v. Blue River Irrigation Co.*, 753 P.2d 737 (Colo. 1988).

§§ E.g. Neb. Rev. Stat. § 46-238; Ariz. Rev. Stat. Ann. § 45-150; N.M. Stat. Ann. § 72-5-1; Wyo. Stat. § 41-4-506; Idaho Code § 42-204; S.D. Codified Laws 46-5-25; Nev. Rev. Stat. § 533.390.

¶¶ N.M. Stat. Ann. § 72-5-1.

and development. They ordinarily make provisions for extensions of time but this often involves a fee and such extensions are usually limited.

7.12   RULE OF PRIORITY

The normal rule of priority is one of first in time.* The first to perfect a water right has priority over later perfected rights.† When water is not sufficient to supply all appropriators, senior appropriators make a call and junior users must shut down their diversions to enable senior rights holders to fill their needs. This is termed "calling the river." Some states treat ground- and surface water as connected. In such states, wells may be shut down to preserve a senior's surface rights. If shutting down the wells will not immediately help the senior, by restoring water levels, the uses may continue. This is because shutting down the junior uses would only hurt those users and benefit no one. Where needs are seasonal and water levels cannot be restored in time, the call is likely to have a greater impact during the next season.‡ This is most likely to occur due to the time lapse, as water seeps through the ground.

7.13   SEASONAL PRIORITIES

These exist in some states and entitle the holder of such rights to take water only during certain times or seasons of the year.§ The modern rule is that unless the claim incorporated the time limit, no time limit exists for the use of water.¶

7.14   PREFERENCE

In times of shortage, use of a preference system means that water priority is given according to the statutory preference of a particular use over another use of lower

---

* See R.C.W.A. 90.03.010, "the first in time shall be first in right."
† See *State Ex Rel. Cary v. Cochran*, 138 Neb. 163, 292 N.W. 239 (1940) (requiring the state to enforce priority of rights during a time of water shortage); but see Colo. Rev. Stat. § 37-92-102 (2) (d), stating that "no reduction of any lawful diversion because of the operation of the priority system shall be permitted unless such reduction would increase the amount of water available to and required by water rights having senior priorities." This law would not affect later priorities downstream that would have no effect on senior priorities, whereas upstream juniors might. One court even found it unnecessary to determine if there was water for appropriation, citing priority. See *Benz v. Water Resources Com'n*, 94 Or. App. 73, 764 P.2d 594 (Or. App. 1988) (it is not necessary to determine if a river is fully appropriated since senior rights must be satisfied before juniors in any event). (See also § 7.24.) Seniority usually also protects the quality of the water necessary for the prior reasonable use, though this matter is not often litigated, probably because there are numerous state and federal water-quality statutes already dealing with these issues.
‡ Note that as of May 2007, 760 groundwater users in eastern Idaho received warnings that they may need to shut down due to insufficient supply. These notices were sent out so that two trout farms with senior rights could continue operating. These actions were taken in response to calls from senior users in 2005. This example shows how long it can take to remedy a problem with water delivery to senior holders. Low snowpacks and drought forecasts are blamed for the shortage. It is often difficult to quickly remedy a problem dealing with shortages of surface water due to ground withdrawals, since it takes time for the water table to rise again.
§ See *Oliver v. Skinner*, 226 P.2d 507 (Ore. 1951) (an appropriation for irrigation of hay lands, never used after a certain date cannot be used for crops requiring a later date).
¶ See *Harkey v. Smith*, 247 P. 550 (N.M. 1926).

priority.* Consideration is for how the water is used as opposed to the date of priority.† Another form of preference comes where individual applicants compete for the same right in unappropriated water. Junior appropriators are often said to have a vested right to their appropriations as against seniors who change their use to the detriment of the junior.‡ There are numerous exceptions to this doctrine.§

7.15 ROTATION

This is a variation of the administration of water rights. Where there is sufficient water for all of several small rights holders, each may take a turn, called a rotation, accessing the total water flow for a short period of time until the next holder's turn.¶

7.16 WATERCOURSE**

A watercourse is generally considered to consist of a bed and banks with well-defined boundaries and a water flow.†† This matters because diffused surface waters, or those not in a recognized watercourse are not ordinarily regulated by the state.‡‡ Diffused waters consist of rainfall and snowmelt runoff up until the point in time when it joins with a recognized body of water or watercourse. There is no requirement for a watercourse to flow continuously. Where water cuts out an unmistakable channel that regularly flows with water, even though the channel may be dry for parts of the year, it may well be considered a watercourse and governed by the same rules.§§ There is considerable variation among the states in applying a definition to watercourses.¶¶

---

* This may work in theory but is seldom strictly applied because it would interfere with the prior appropriation system of first in time. Some jurisdictions require condemnation of water by more preferred user, I.e. Idaho, Kansas, Nebraska, and Wyoming. See David H. Getches, *Water Law,* 3rd Ed. at 105 (1997). Some states only give preference among uses with the same date, I.e., Alaska, Arizona, California, Nebraska, North Dakota, and Texas.

† See S.D. Codified Laws Ann. § 46-1-5; Mont. Code Ann. § 85-2-316; Utah Code Ann. § 73-3-21 (giving preference to domestic use but maintaining priority of time as between the same or similar uses); see also *Phillips v. Gardiner,* 2 Or. App. 423, 469 P.2d 42 (1970).

‡ *Farmers Highline Canal & Reservoir Co. v. Golden,* 129 Colo. 575, 272 P.2d 629 (1954).

§ See *Metropolitan Denver Sewage Disposal Dist. No. 1 v. Farmers Reservoir & Irrigation Co.,* 499 P.2d 1190 (Colo. 1972).

¶ See Wyo. Stat. § 41-3-612; *McCoy v. Huntley,* 119 P. 481 (Or. 1911); *Big Cottonwood Tanner Ditch Co. v. Shurtliff,* 164 P. 856 (Utah 1917).

** For Restatement (Second) of Torts definition see § 1.7 fn.

†† *State v. Hiber,* 48 Wyo. 172, 44 P.2d 1005 (1935); *Johnson v. Board of County Com'rs of Pratt County,* 259 Kan. 305, 913 P.2d 119 (1996); *Maddocks v. Giles,* 1999 ME 63, 728 A.2d 150 (Me. 1999) (defining a watercourse where water appears to flow in a particular direction and by regular channel with bed, banks, and sides).

‡‡ For an exception see OR. Rev. Stat. § 537.800 (applying permit system to spring water and seepage). See also *Norden v. State,* 329 Or. 641, 996 P.2d 958 (2000) (use of permit system regarding seepage and spring water).

§§ *Id.*

¶¶ See *Hoefs v. Short,* 273 S.W. 785 (Tex. 1925) (stating that all that is necessary is that there must be sufficient water at sufficient intervals as may make the stream practicable for irrigation); *Oklahoma Water Resources Bd. v. Central Oklahoma Master conservancy Dist.,* 464 P.2d 748 (Okla. 1969) (finding that while diffused surface waters may be capture by a landowner, when they enter a well-defined channel they lose their original character and become a constituent part of the stream or body).

Many western state constitutions and statutes define watercourse.* Some states consider all water as belonging to the state.† In Alaska, the attorney general ruled that appropriation statutes even apply to glacial ice (in solid ice form). Some states define streams that exist entirely on private land as private property, while streams arising on public land are subject to appropriation.‡ One state even applies the appropriation doctrine to seepage and springs.§

7.17   PHREATOPHYTES AND DEVELOPED WATER (SEE ALSO 8.8, "SURFACING STREAMS")

Some plants and mosses absorb a high amount of water¶ along the riverbanks and then discharge much of it back into the atmosphere. It is theorized that the removal of such vegetation will increase water flow in streams and rivers. It is estimated that such vegetation now cover nearly 16 million acres and discharge between 20 and 25 million acre-feet of water annually. The environmental impact of removing this foliage is not fully known, and conservation groups often oppose removal of these plants. Courts have found that actions, of this kind, which increase the flow in streams and rivers, do not give rise to water rights to the increased flow, outside of the established priority system.** This is often referred to as developed water, or water that did not previously exist in the priority system.

7.18   ARTIFICIALLY INDUCED PRECIPITATION (SEE ALSO 14.2)

This is a largely unproved field of science, and while some mild success does seem likely, ordinarily artificial precipitation is subject to the same rules as natural precipitation.††

7.19   DIVERSION EFFICIENCY (SEE ALSO 7.9)

Ordinarily, absolute efficiency is not necessary to exercise a diversion. Reasonable efficiency is all that is required.‡‡ If a diversion requires an unreasonable amount

---

* Colo. Const. Art. 16 § 5 (natural streams); N.M. Const. Art. 16 § 2 (perennial or torrential); N.D. Cent. Code § 61-01-01 (excludes diffused surface waters); Ariz. Rev. Stat. Ann. § 45-131A (excludes diffused surface waters); also see Nev. Rev. Stat. § 533.025 (includes all waters in state); Or. Rev. Stat. § 537.110 (all waters in state); Utah Code Ann. § 73-1-1 (all waters in state belongs to the state); see also Cal. Water Code S 1252.1.

† See *Melville v. Salt Lake County*, 570 P.2d 687 (Utah 1977) (No one owns or can own water in the state. One can only acquire a right to use water).

‡ See Wells A. Hutchins, *The California Law of Water Rights*, 407–13 (1956).

§ OR. Rev. Stat. § 537.800.

¶ See http://ga.water.usgs.gov/edu/watercycletranspiration.html.

** *R.J.A., Inc. v. Water Users Association of District No. 6*, 690 P.2d 823 (1984) (denying a water right for developed water and stating that such a scheme is for the legislature to establish not the courts); but see *People v. Shirokow*, 605 P.2d 859 (Cal. 1980) (requiring the eradication of phreatophytes as a condition for receiving a permit to appropriate water). See § 7.21.

†† See Cal. Water Code § 401

‡‡ *State Ex Rel. Crowley v. District Court*, 108 Mont. 89, 88 P.2d 23 (1939). Some states limit the amount of water available for certain purposes to prevent waste.

of undiverted water, it may be considered inefficient and unprotected to the extent other's rights are unreasonably affected.

7.20 ACCESS TO WATER BY CROSSING THE LAND OF ANOTHER

When it becomes necessary for one individual to cross the land of another to make use of a water right, most jurisdictions allow for a right-of-way.* If this right of way is considered a taking, then just compensation is usually required to the injured party.† Statutes allowing for a right of way have been held constitutional under the 14th Amendment of the United States Constitution.‡ In many eastern states, where riparian rights are exclusive, condemnation is not readily allowed.

7.21 PERMITS (SEE ALSO 3.7, "PRESCRIPTION," 6.15, "EASTERN PERMIT SYSTEMS," AND 10.5, "CWA PERMIT SYSTEM")

In the western states, permits are often required before any person may divert water for use. Of the western states recognizing the appropriation doctrine, only Colorado does not use the permit system.§ Ordinarily, the permit system requires anyone wishing to have a claim recognized to file an application in accordance with state law.¶ Those who do not file are usually not given priority under the law and may find that they have no recognized rights to water in the state. In a way, the permit system is much like the recording statutes for real-estate title.

Where an individual purchases real estate and fails to record, such that the original owner resells the property to another buyer, who does record, the second buyer may end up with legal title if he purchased in good faith and without notice. Recording the right places others on notice of the use and allows the state to reasonably rely on the record in making further allocations. Because many states have statutes or constitutional provisions providing that all water belongs to the state initially, prior uses not recorded are generally not recognized. In states where unappropriated water belongs to the state, prescriptive rights generally fail because one may not adversely take public property.** Some states provide express exceptions to the permit requirements for some small appropriations.†† State procedures for obtaining a permit vary.‡‡

---

* *Bower v. Big Horn Canal Association*, 77 Wyo. 80, 307 P.2d 593 (1957) (plaintiff has right to condemn right-of-way to collect seepage from defendant's operations but not to insist on defendant's continued use of the ditches).
† *Kaiser Steel Corporation v. W.S. Ranch Company*, 81 N.M. 414, 467 P.2d 986 (1970) (see § 7.10); see N.M. Stat. Ann. § 75-1-3.
‡ *Clark v. Nash*, 198 U.S. 361, 25 S.Ct. 676, 49 L.Ed. 1085 (1905).
§ Colorado Const. Art. XVI, §§ 5, 6. The Colorado Constitution requires 1) unappropriated water, 2) of a natural stream, 3) a diversion from that stream, and 4) beneficial use.
¶ *Wyoming Hereford Ranch v. Hammond Packing Co.*, 33 Wyo. 14, 236 P. 764 (1925).
** *People v. Shirokow*, 605 P.2d 859 (Cal. 1980); see also *Hoadly v. San Francisco*, 50 Cal. 265, 274–76 (1875). See § 7.17.
†† See Cal. Water Code §§ 1228–1229.2; Tex. Water Code Ann. §§ 11.142, 11.143; S.D. Codified Laws Ann. § 46-5-8.
‡‡ See section in the back of this book titled "Western States' Water Departments and Permitting Agencies" for links to many of the forms and procedures in the various jurisdictions.

The typical process includes filing an application containing basic information about the proposed use, its location, and the quantity of water desired as well as other relevant information needed to approve the application. Most applications are available from the regulating agencies within a state.* Once an application is received, it is usually required that notice be given to the public. Following this period of notice, the application may be either approved or challenged. A challenge typically means that there will be a hearing where parties are allowed to present evidence and witnesses. The hearing officer will weigh the evidence and make a decision on the record. If the application is approved, a permit is issued. All jurisdictions† have some form of statutory criteria for approving applications. These criteria typically include public interest, beneficial use, availability of water, and no harm to prior users.

7.22   DUAL PERMITS (SEE 7.7)

Many states have what are known as dual permit requirements. This means that a person wishing to store water must apply for a permit to build the storage facility and the person wishing to use the water on a particular property must apply for a secondary permit to use the water.‡

7.23   COLORADO APPROPRIATION (SEE 17.5)

The Colorado Constitution is believed to forbid the permit system.§ As a result, most contested water rights are decided in the courts based on evidence presented at trial. The Water Right and Adjudication Act of 1969 carved the state into seven water divisions, covering different major water basins. Each division has one district judge assigned as a water judge. The court in effect becomes an administrative agency, maintaining complete records and supervisory authority over water rights¶. Anyone making a claim of a water right must file with the water clerk assigned to assist the judge over the basin in which his or her claim resides. The application must give a detailed description of the ruling sought. Rights are given priority based on the date of their adjudication.

7.24   DOUBLE APPROPRIATIONS AND THE AVAILABILITY OF WATER

Many states have laws that forbid appropriation of water downstream in amounts that would exceed the maximum allocated rights upstream.** In other words, the law assumes the maximum allotted use, even though this maximum use may not truly happen.

---

* See list of western state agencies in "Water Codes and Water Departments" section.
† See section on Colorado System for details about that states judicial hearings, § 7.23.
‡ See Ariz. Rev. Stat. Ann. § 45–151; Or. Rev. Stat. § 537.400; Wash. Rev. Code Ann. § 90.03.370; Wyo. Stat. § 41-3-301.
§ "The right to divert the unappropriated waters of any natural stream to beneficial uses shall never be denied." Colorado Const. Art. XVI, Sec. 6.
¶ Colo. Rev. Stat. § 37-92-210, 203, and 204.
** Tex. Water Code Ann. § 11.134 (b)(2) & (3); *Lower Colorado River Authority v. Texas Department of Water Resources*, 28 Tex. Sup. Ct. J. 420, 689 S.W.2d 873 (1984); see also Mont. Code Ann. § 85-2-311; Neb. Rev. Stat. § 46-235; Wyo. Stat. § 41-4-503. For an opposite approach, see *Benz v. Water Resources Comm'n*, 764 P.2d 594, 599 (Or. 1988).

7.25 HYBRID SYSTEMS* (SEE 6.2, "RIPARIAN RIGHTS," 7.1, "PRIOR APPROPRIATION," 7.3, "THE 'CALIFORNIA DOCTRINE'")

Riparian rights were initially recognized in several states out West. These states eventually converted to the appropriations system but preserved the riparian rights that were already in place.† This is now known as a hybrid system. Today, ten states use a combination of riparian and appropriation doctrines. It is worth noting that much of the leading case law for riparian practice comes from these hybrid jurisdictions in the West, where prior appropriation law is also a significant presence. For many of these states, neither the riparian system nor the prior appropriation system is entirely fitting. These states tend to lie on the coast, or border the arid lands of the West.

7.26 PUEBLO RIGHTS

Some settlements in the southwest United States derived certain rights from their Spanish origins.‡ These rights predate annexation into the United States (Treaty of Guadalupe Hidalgo, 1848).§ The rights were based on the needs of the settlement and grow with the settlement. Most important, they exist outside the priority system.¶ Local towns, called pueblos, were at the center of such rights. The original states affected included Arizona, California, Colorado, New Mexico, and Texas.** Water was considered an important public asset to be governed for the benefit of the whole community. The problem with pueblo rights is their potential to take away vested rights from other appropriators as the city's demands grow. Such rights are frowned upon today in almost all jurisdictions but do exist, often restricted by reasonable use to meet the needs of the city.†† Only California continues to recognize such rights.

---

* See Table 18 for list of hybrid states.
† See Appendix 3, Table 16. Where federal patent contain no reference to riparian rights the matter depends on state law. *Brewer-Elliott Oil & Gas Co. v. U.S.*, 260 U.S. 77, 43 S. Ct. 60, 67 L. Ed. 140 (1922). See R.C.W.A. 90.03.010, "subject to existing rights all waters within the state belong to the public, and any right thereto, or to the use thereof, shall be hereafter acquired only by appropriation for a beneficial use and in the manner provided and not otherwise."
‡ See *State Ex Rel. Martinez v. City of Las Vegas*, 118 N.M. 257, 880 P.2d 868 (1994) (refusing to recognize pueblo rights for the city of Las Vegas). The decision in *Martinez* was ultimately upheld by the New Mexico Supreme Court in *New Mexico v. Martinez*, 2004 WL 1039867, Docket No. 22,283 (holding that pueblo rights were inconsistent with the state's system of prior appropriation).
§ The Treaty of Guadalupe Hildago required the United States to protect property rights, which had been created by its predecessor. This included pueblo rights, which existed at the time of the treaty.
¶ See *New Mexico v. Aamodt*, 537 F.2d 1102 (10th Cir. 1976) cert. Denied 429 U.S. 1121 (1977) (holding that *Winters* rights do not apply to Pueblo Indians because of their fee ownership of their land, but because their water rights predate any white settlers, essentially holding the same effect of *Winters*). See also 618 F.Supp. 993 (D.N.M. 1985)(holding that pueblo Indians have aboriginal title to their traditional lands).
** Today they are of primary importance in California, New Mexico, and Texas. See *Los Angeles v. San Fernando*, 14 Cal.3d 199, 123 Cal.Rptr. 1, 537 P.2d 1250 (1975) (for discussion of Pueblo Rights of Los Angeles).
†† See *Los Angeles v. San Fernando*, 14 Cal.3d 199, 123 Cal.Rptr. 1, 537 P.2d 1250 (1975); *Cartwright v. Public Serv. Co. of New Mexico*, 343 P.2d 654 (N.M. 1958).

7.27   PUBLIC INTEREST REVIEW (SEE 7.30)

Sixteen western states have some form of public interest review of the permit process.* Many of these statutes are vague. This often makes it difficult for administrators, who must create standards for review, where lacking in legislation. California courts have stated that public interest is the primary standard for issuing new permits. The problem was that there were few initial standards in California law to define and determine the public interest. The Model Water Code gives some guidelines.†
It protects interests such as aesthetics; fish, wildlife; other environmental interests; navigation; recreation; and public water supplies. An Alaska statute requires that alternate potential uses be considered.‡

7.28   ARTIFICIAL WATER TRANSPORTED IN STREAM

When water is artificially introduced to a stream for transport further downstream, the one introducing the water may have a right to withdraw an equal amount farther downstream.§ These rules may consider evaporation and other transport losses when determining the amount of water that can be withdrawn downstream. This is not a big deal if the total water lost in the stream changes little, but if the stream loses more water when it carries more then it may become an issue.

---

* Alaska Stat. §§ 46.15.040, -.080(a); Ariz. Rev. Stat. Ann. §§ 45–142, –143; Cal Water Code §§ 1225, 2155; Idaho Code §§ 42–201, 203A, 203C; Kan. Stat §§ 82a–705, –711; Mont. Code Ann. §§ 85-2-302, 311(2); Neb. Rev. Stat §§ 46–233, –234, –2,116; Nev. Rev. Stat. §§ 533.325, .370 (3), 534.040(1); N.M. Stat. Ann. §§ 72-5-1, -6, -7, 72-12-3,3E; N.D. Cent. Code §§ 61-04-02, 06; Or. Rev. Stat. §§ 537.130–70(4); S.D. Comp. Laws Ann. §§ 46-1-15, -2A-9, -5-10, -6-3; Tex. Water Code Ann. §§ 11.121, -134(3); Utah Code Ann. §§ 73-3-1, –8(1); Wash. Rev. Code Ann. §§ 90.03.250, .290, -44.050, -44.060; Wyo. Stat. §§ 41-4-503, -3-930 to –932. Utah specifically allows permits to be issued on a temporary basis. A North Dakota case, *United Plainsmen Ass'n v. North Dakota State Water Conservation Comm'n,* 247 N.W.2d 457 (N.D. 1976), held that public interest review must occur before an energy permit could be issued.
† A Model Water Code (1958) at 179 as cited in Joseph L. Sax, Robert H. Abrams, Barton H. Thompson Jr., John D. Leshy, *Legal Control of Water Resources: Cases and Materials,* 188 (3rd ed. 2000).
‡ Alaska Stat. § 46.15.080 (5). See also Idaho Code § 42-1501; Alaska Stat. § 46.15.080; Tex. Water code Ann. § 11.1501 (stating that permits must be consistent with regional water plan); Mont. Code Ann. § 85-2-311. See also *Young & Norton v. Hinderlider,* 15 N.M. 666, 110 P. 1045 (1910) (stating that public interest should be read broadly to secure the greatest possible benefit); *Skokal v. Dunn,* 109 Idaho 330, 707 P.2d 441 (1985) (defining public interest); *Stempel v. Department of Water Resources,* 82 Wash. 2d 109, 508 P.2d 166 (1973) (holding that environmental statutes elevated importance of environmental issues).
§ *Rock Creek Ditch & Flume Co. v. Miller,* 93 Mont. 248, 17 P.2d 1074, 89 A.L.R. 200 (1933). The burden of proof is on the one claiming to have introduced the water. *Smith v. Duff,* 39 Mont. 382, 102 P. 984 (1909); *Herriman Irr. Co. v. Butterfield Min. Co.,* 19 Utah 453, 57 P. 537 (1899).

7.29 REUSE OF WATER BY ORIGINAL APPROPRIATOR

Under ordinary circumstances, most jurisdictions allow recapture of unused water*
by the original appropriator so long as it is still on the appropriator's land. This ordinarily changes once the water re-enters the public domain.†

7.30 PROTECTING WATER BASINS OF ORIGIN (SEE 5.9 AND 7.27)

Several states require the impact of inter-basin transfers on the basin of origin to be
assessed under public interest review.‡

7.31 RIGHTS TO WATER QUALITY (SEE ALSO 10.2)

Seniority protects both the right to use a specific amount of water and the right to a
quality of water necessary for the designated use.§

---

* *Bower v. Big Horn Canal Association*, 77 Wyo. 80, 307 P.2d 593 (1957); *Cleaver v. Judd*, 238 Or. 266,
393 P.2d 193 (1964); *State Dept. of Ecology v. U.S. Bureau of Reclamation*, 827 P.2d 275 (Wash. 1992)
(second use of water by original owner is within rights); *Stevens v. Oakdale Irrigation Dist.*, 13 Cal.
2d 343, 90 P.2d 58 (1939) (defendant allowed to recapture water from its operations on its own land);
Estate of Paul Steed v. New Escalante Irrigation Co., 846 P.2d 1223 (Utah, 1992) (holding that water
belongs to original owner so long as it is in his control on his land); *Arizona Public Service, Co. v.
Long*, 160 Ariz. 429, 773 P.2d 988 (1989) (holding that cities may put sewage effluent to any reasonable
use they see fit). See also *Reynolds v. City of Roswell*, 99 N.M. 84, 654 P.2d 537 (1982) (regardless of
origin, water left over after treating sewage is wastewater not surface water); *Lambeye v. Garcia*, 18
Ariz. 178, 157 P. 977 (1916); Wedgworth v. Wedgworth, 20 Ariz. 518, 181 P. 952 (1919) (holding that
wastewater before it returns to a natural channel is not subject to appropriation); *Thayer v. Rawlins*,
594 P.2d 951 (Wyo. 1979) (junior appropriator using wastewater takes chance on continued flow);
Southeastern Colorado Conservancy Dist. v. Shelton Farms, Inc., 187 Colo. 181, 529 P.2d 1321 (1974)
(developed water, that added to natural supply only because of efforts of developers, is not subject to
appropriation); *Elgin v. Weatherstone*, 123 Wash. 429, 212 P. 562 (1923) (holding that foreign water,
once abandoned by developer, does not become part of the natural flow of the drainage and may be
used by the first to capture it); *Dodge v. Ellensburg Water Co.*, 46 Wash. App. 77, 729 P.2d 631 (1986)
(allowing a later user in time to take water at an earlier point than one who had been taking the discharge). It is estimated that 2 percent of California's water use is recycled municipal water. Joseph
L. Sax, Robert H. Abrams, Barton H. Thompson Jr., and John D. Leshy, *Legal Control of Water
Resources: Cases and Materials,* 164 (3rd ed. 2000). More than 100, 000 acre-feet of sewage effluent
is used in Arizona. Sax at 175. For an exception to the usual rule see Colo. Rev. Stat. § 37-82-102, distinguishing between tributary and nontributary seepage and forbidding recapture of tributary seepage,
even on original users own land. See *Ranson v. Boulder*, 161 Colo. 478, 424 P.2d 122 (1967) (forbidding
recapture on user's own land before others could rely on seepage).

† *Wyoming Hereford Ranch v. Hammond Packing Co.*, 236 P. 764 (Wyo. 1925) (that part of release to
stream became public water, subject to appropriation again); *Jones v. Warmsprings Irr. Dist.*, 162 Or.
186, 91 P.2d 542 (1939); *Northport Irr. Dist. v. Jess*, 215 Neb. 152, 337 N.W.2d 733 (1983); *United
States v. Haga*, 276 Fed. 41 (D. Idaho 1921); Fuss v. Franks, 610 P.2d 17 (Wyo. 1980).

‡ See Or. Rev. Stat. §§ 537.801 et seq.; Neb. Rev. Stat. § 46-289; Wy. Stat. § 41-2-121(a)(ii)(E)(VIII).
Several eastern states also have similar legislation. Minn. Stat. § 103G.265(2); Conn. Gen. Stat.
§ 22a-369(10); S.C. Stat. § 49-21-20; Mass. Gen. Laws ch. 21§§ 8C-8D.

§ Issues regarding the quality of water are not often litigated, probably because of the numerous state
and federal laws already regulating water quality. See generally § 10.2 on the Clean Water Act.

7.32 INCREASES IN USE AND BACKDATING
(SEE ALSO 7.11 "REASONABLE DILIGENCE")

When the intention is to expand or change a use from the onset, such increases in water may be backdated to the time of the original use, so long as due diligence is applied.*

7.33 WATER MASTERS

Water districts, subject to frequent conflict and competition among users for limited resources, may also be subject to the appointment of a water master, by the state, to handle complex issues of supply and demand, by monitoring withdrawals and uses from a source, and ensuring that prior appropriators acquire their rightful appropriations.†

7.34 DIFFUSED SURFACE WATERS (SEE ALSO 18.6)

Diffused surface waters are those that collect on the ground but do not flow in defined banks or paths called channels.‡ These waters include runoff from rainfall. Most states do not apply the appropriations doctrine to waters outside a defined watercourse.§ The Restatement of Torts (second) addresses these waters, defining them as "water from rain, melting snow, springs or seepage, or detached from subsiding floods, which lies or flows on the surface of the earth but does not form a part of a watercourse or lake."¶ Water diverted from lake or stream for irrigation or other purposes is not diffuse surface water and cannot be diverted across the land of another without permission.**

7.35 RATE OF DIVERSION

Many states†† limit the amount of water that can be applied over a given time to a given measurement of land. This limit is sometimes called "duty of water."‡‡

---

* *McPhee v. Kelsey*, 44 Or. 193, 74 P. 401 (1903); *Foster v. Foster*, 107 Or. 355, 213 P. 895 (1923); *Oliver v. Skinner*, 190 Or. 423, 226 P.2d 507 (1951); *Farmers Highline Canal & Reservoir Co. v. City of Golden*, 129 Colo. 575, 272 P.2d 629 (1954) (increased use only backdated to the extent of use contemplated at the time of appropriation). These issues are not often litigated. Joseph L. Sax, *Legal Control of Water Resources,* 122 (3rd Ed. 2000).
† See R.C.W.A. 90.03.060.
‡ Water bodies must ordinarily be permanent at least in the sense that they are regularly or intermittently filled in whole or in part with water in an easily defined space.
§ Greater authority is claimed by Alaska, Montana, Nevada, Oregon, Texas, and Utah over waters in their boundaries. Only Utah and Colorado interpret state authority to exist beyond natural watercourses. See David H. Getches, *Water Law,* 3rd Ed. at 107 (1997).
¶ Restatement (Second) of Torts, § 846.
** This type of unused water is often called tailwater. See *Loosli v. Heseman*, 66 Idaho 469, 162 P.2s 393 (Idaho 1945). As a point of interest, this right may be acquired by prescription. David H. Getches, *Water Law,* 3rd Ed, at 294 (1997).
†† South Dakota, Wyoming, and Nebraska allow application at a rate of one cubic foot per second (cfs) for every seventy acres. Idaho allows 1cfs per eighty acres. There are also limits on annual use for land in many of these states.
‡‡ Duty of Water is defined as "the amount of water that through careful management and use, without wastage, is reasonably required to be applied to a tract of land for a length of time that is adequate to produce the maximum amount of the crops that are ordinarily grown there."

8 Groundwater*

8.1 GROUNDWATER†

Most groundwater is stored in the pores and interstices of rock formations. Its movement is largely a function of geological conditions. The amount of open space in a rock determines how porous the rock is. There is no single definition‡ for groundwater, but generally it is water that exists in the interstices of rocks and is capable of being withdrawn through wells. Ordinarily, this water exists only below the saturation point of the rock and soil. Where the rock or soil are not fully saturated, removal is difficult.

8.2 PERMEABILITY

The ability of rock to pass water through its pores and open spaces is called its permeability. Rock is considered saturated when it can no longer store more water in its pores.

8.3 MAJOR DOCTRINES REGARDING PERCOLATING WATER (SEE 8.13 TO 8.16)

The English Common-law Rule§: "…the person who owns the surface may dig therein, and apply all that is there found to his own purposes at his free will and pleasure; and if, in the exercise of such right, he intercepts or drains off the water collected from underground springs in his neighbor's well, this inconvenience to his neighbor falls within the description of damnum absque injuria, which cannot become the ground of an action."¶

---

* See Form 7 Well permit application; Figs 1 Routes of Groundwater Contamination, 2 Ground Saturation, 3 Ground Water Storage, 4 Ground Water Flow Paths, 7 Well Types, 8 Aquifers and Wells, 9 Artesian Wells, 20 Subsurface Diversion Wells, 21 Subsurface Diversion, 22 Impacts of Groundwater Depletion, 23 Salt Water Intrusion, 24 Land Subsidence, 26 Groundwater Withdrawals in the United States, 29 East Coast per capita Groundwater Withdrawals in United States, 32 Age of Groundwater. For discussion of law regarding springs and wells see 78 Am Jur 2d, Waters §§ 146–150; See §§ 151–154 regarding underground streams; and §§ 155–163 regarding Artesian Basins.
† The Restatement (Second) of Torts § 845 defines ground water as, water that is naturally flowing or resting beneath the earth's surface.
‡ See Rev. Code Wash. (ARCW) § 90.44.035 (3) (for one such definition).
§ This rule survives only in a few eastern states and Texas. See § 8.14.
¶ *State v. Michels Pipeline Construction, Inc.*, 63 Wis. 278, 217 N.W.2d 339 (1974).

Reasonable Use*: "In some states, the rule of the common law followed in early decisions has given way to the doctrine of reasonable use limiting the right of a landowner to percolating water in his land to such an amount of water as may be necessary for some useful or beneficial purpose in connection with the land from which it is taken, not restricting his right to use the water for any useful purpose on his own land, and not restricting his right to use it elsewhere in the absence of proof of injury to adjoining landowners."†

Correlative Rights‡: "Under the rule of correlative rights, the rights of all landowners over a common basin, saturated strata, or underground reservoir are coequal or correlative, and one cannot extract more than his share of the water, even for use on his own land, where others' rights are injured thereby."§

Restatement of the Law of Torts¶: "A possessor of land or his grantee who withdraws ground water from the land and uses it for a beneficial purpose is not subject to liability for interference with the use of water by another, unless (a) the withdrawal of water causes unreasonable harm through lowering the water table or reducing artesian pressure, (b) the withdrawal of ground water exceeds the proprietor's reasonable share of the annual supply or total store of ground water, or (c) the withdrawal of the ground water has a direct and substantial effect upon a watercourse or lake and unreasonably causes harm to a person entitled to the use of its water."**

8.4   AQUIFERS

An aquifer is an underground reservoir of water, which occurs when significant amounts of water can be extracted due to saturation of the rock and soil. Aquifers may be either confined or unconfined. Unconfined aquifers must be pumped out because they are under pressure equal to the atmosphere. Confined aquifers are often called artesian aquifers and are under great pressure, such that water is forced to the surface once a well is drilled. These occur when impermeable rock above and below squeeze the water, causing it to rise when wells are drilled. They often occur when part of the water is stored at an elevation above the well.

8.5   ARTESIAN WELLS

An artesian well is one that is under pressure, such that water rises to the surface on its own without the need for pumping. Most states require that one tapping into such

---

* This rule has been widely adopted in the eastern United States but is largely unnecessary in western states where permits are required for withdrawal. Arizona adheres to this rule; Oklahoma adopted it then abandoned it thirteen years later in 1949.

† *Michels*, 63 Wis. 278 (quoting Corpus Juris Secundum).

‡ California is the leading state for this doctrine. See *Katz v. Walkinshaw*, 74 P. 766, 772 (Cal. 1903). See § 8.15.

§ *Michels*, 63 Wis. 278 (quoting Corpus Juris Secundum).

¶ The restatement has been followed in Michigan, Ohio, and Wisconsin. It was rejected in Indiana, which stayed with the old English Rule.

** Restatement of the Law of Torts § 858 A.

a well cap it off to prevent waste. They may also require licensed well diggers and may state well specifications.*

8.6 LIABILITY FOR GROUND SUBSIDENCE AND LOWER WELLS (SEE 11.1 THROUGH 11.10)

Some courts have held that landowners, who withdraw water from wells, on their own land, sufficient to cause subsidence on other lands in the general area, are liable for damages.† Damage from land subsidence was estimated by the United States Water Resources Council to be $32 million in the Houston-Galveston area alone.‡ The Restatement (Second) of Torts imposes liability when a well "unreasonably" causes harm to another well by lowering the water level or pressure.§

8.7 UNDERGROUND STREAMS

These are defined as waters that flow within relatively certain boundaries, "as a constant stream in a known and well defined natural channel."¶ Such streams are often subject to the law of surface water rather than groundwater.**

8.8 SURFACING STREAMS

In one case, an Arizona court found that a spring which surfaced had prior to surfacing been percolating groundwater eventually flowing into the same creek. An individual wishing to acquire rights to the newly surfaced stream applied because groundwater in the state was not subject to appropriation rules making the surfaced spring water a new source. The court denied the application because another law forbids appropriations that would negatively impact prior appropriations.†† Where groundwater becomes surface water and then returns to groundwater again the

---

* See *Current Creek Irrigation Co. v. Andrews*, 9 Utah 2d 324, 344 P.2d 528 (1959) (addressing whether prior users of an artesian well have a vested right to the pressure. The court found for the prior rights holder against the junior pumpers). Note: this ruling resulted from Utah law (Utah Code Ann. § 73-3-23), expressly granting a right of replacement to any senior whose quantity or quality of appropriated water is diminished by a junior. See also *Wayman v. Murray City Corp.*, 458 P.2d 861 (Utah 1969) (finding that a city changing its point of diversion in such a way as to interfere with private water rights to an artesian well must pay the expenses of the private users). See § 8.20. For a different result, see *Erickson v. Crookston Waterworks Power & Light Co.*, 117 N.W. 435 (Minn. 1908) (finding that the private owner can be required to deepen his well for the common good). The Restatement (Second) of Torts § 858 (1) (a), imposes liability if a well "unreasonably" interferes with another by lowering the water pressure.

† *Friendship Development Co. v. Smith-Southwest Industries, Inc.*, 576 S.W.2d 21 (Texas 1978); see *State v. Michels Pipeline Construction, Inc.*, 63 Wis. 278, 217 N.W.2d 339 (1974) (holding that a sewer company could not drain water to the detriment of others in order to lay sewer lines and adopting the position of Restatement of the Law of Torts).

‡ James W. Johnson, The 1980 Arizona Groundwater Administration and Management: A Minerals Industry Perspective, 26 Rocky Mtn. Min. L. Inst. 1031, 1032 n. 3 (1980).

§ Restatement (Second) of Torts, § 858 (1) (a).

¶ *Hayes v. Adams*, 109 Or. 51, 218 P. 933 (Or. 1923).

** See *Herriman Irrigation Co. v. Keel*, 25 Utah 96, 69 P. 719 (Utah 1902).

†† *Collier v. Arizona Dept. of Water Resources*, 722 P.2d 363 (Ariz. App. 1986).

ownership question is complicated. Some jurisdictions treat groundwater the same as surface water and avoid this problem.* Others use different standards to determine ownership of each.

8.9 STREAM FLOW

Surface streams may be classified as gaining (effluent) from groundwater or losing (influent) to it. Discharge from groundwater may be natural, such as when it seeps into lakes and rivers, or artificial such as when a well is dug. The interplay between surface and groundwater can create confusion in state water law when surface water and groundwater are treated separately.†

8.10 CONE OF INFLUENCE‡

When a well is drilled and water siphoned from the underlying body, it may impact other wells by causing a depression in the water table. This cone shaped depression is called the cone of influence. Additionally, quick depletion of an aquifer may impact the future usefulness of the aquifer as a source of water. Salt water may displace the depleted freshwater, or subsidence may occur. Subsidence is the sinking of the surface land above the aquifer. This restricts the amount of space below and limits the water that can be stored within. Some jurisdictions provide legal remedies against the pumper for damages caused to the lands and property of others. Most pumpers do not feel the economic damages from their pumping personally.§ Actions may be brought under negligence, nuisance, groundwater law principals, and the obligation for subjacent support.¶

8.11 RACE TO THE BOTTOM

When multiple users have access to the same aquifer, they must compete for the water and there is an incentive to waste the resource on marginal uses rather than run the risk that another user will take it before the first has a chance to use it. This race to the bottom is often called the *tragedy of the commons*, a collective action problem that may only be avoidable if use of the resource is regulated and controlled.

---

* See Utah Code Ann. § 73-1-1; Wash Rev. Code Ann. (ARCW) § 90.44.020.
† Colorado and New Mexico have discovered problems with their separate systems for surface and groundwater, where junior pumpers have depleted sources to surface waters used by senior permit holders. Both states have changed to an integrated system. See *Pecos County Water Control & Improvement Dist. v. Williams*, 271 S.W.2d 503 (Tex. Civ. App. 1954). AZ uses a test to determine whether to treat ground water under the appropriation doctrine or reasonable use. If a well reduces a stream by 50 percent or more of the amount pumped over 90 days of continuous pumping, then it is treated under the appropriation doctrine. See *Maricopa County Mun. Water Conservation Dist. No. One v. Southwest Cotton Co.*, 4 P.2d 369 (Ariz. 1931). The Restatement (Second) of Torts § 858 (1) (c), treats ground and surface waters as a single source when withdrawals directly effect the stream in a substantial way.
‡ See Table 22, Land Subsidence.
§ See *Friendswood Dev. Co. v. Smith-South-west Indus., Inc.*, 576 S.W.2d 21 (Tex. 1978).
¶ *Henderson v. Wade Sand & Gravel Co., Inc.*, 388 So.2d 900 (Ala. 1980) (applying nuisance law).

8.12 COLLECTIVE-ACTION THEORY (SEE CHAPTER 12, "SOCIAL THEORY")

This theory postulates that rational self-interested individuals will not act effectively in pursuit of public goods (those that are indivisible or non-excludable).* Mancur Olson argues that in most instances a group will not act effectively in pursuit of common interests. Olson shows that each member will have a rational incentive to want to take a free ride and hope that other members of the group will make the opposite choice. The problem exists when each individual weighs a series of potential costs and benefits from different courses of action and weighs the risks that others will act selfishly or selflessly. The real problem of collective action is in determining under what circumstances a rational individual will act in concert with others for the collective good. Olson postulated that "in a large group in which no single individual's contribution makes a perceptible difference to the group as a whole, or the burden or benefit of any single member of the group, it is certain that a collective good will not be provided unless there is coercion or some outside inducement that will lead the members of the group to act in their common interest." This is where statutory law comes into play.

8.13 GROUNDWATER LAW

Some states recognize the rights of the overlying landowner to undergroundwater. Many will restrict the ability of the landowner to pump to reasonable uses, however, especially if the pool is beneath the land of more than one riparian owner. Generally speaking, those with rights to groundwater have a right to protect it from pollution.

8.14 ABSOLUTE OWNERSHIP DOCTRINE

Under this doctrine, a landowner may withdraw any water found beneath his property.† This is based on the ancient tradition that a landowner is entitled to the air above and land beneath his property infinitely. Water was considered part of the land. This doctrine is still said to be the law in Connecticut, Georgia, Illinois, Indiana, Maryland, Massachusetts, Mississippi, Rhode Island, and the District of Columbia.‡ Nearly all of these jurisdictions temper the effect of the doctrine with legislation or common law rules today. Under the strict rule not even willful injury by withdrawal was actionable. Today all jurisdictions offer some protection against willful injury. Many states following the rule today also allow damages for land subsidence.

---

* Daniel Little, *Varieties of Social Explanation.* (1991). For further explanation, see the work of Mancur Olson (1965).
† *Acton v. Blundell*, 12 Mees. & W. 324, 152 Eng. Rep 1223 (Eng. 1843).
‡ Note that South Dakota previously used the absolute ownership rule and later changed to public ownership of unallocated groundwater. The court in *Knight v. Grimes*, 127 N.W.2d 708 (S.D. 1964), found it acceptable for the state to act in the public interest by abolishing unexercised absolute ownership rights. A similar statute was upheld in Kansas. See *Baumann v. Smrha*, 145 F. Supp. 617 (D. Kan.) aff'd mem., 352 U.S. 863. 77 S.Ct. 96, 1 L.Ed.2d 73 (1956).

8.15   CORRELATIVE RIGHTS

Rights to groundwater under this theory are still determined by land ownership. The difference is that overlying landowners are limited to a reasonable share of the aquifer supply.* In California, surplus water left over after landowners' uses, were allocated according to the appropriation rule.

8.16   PRIOR APPROPRIATION RIGHTS TO GROUNDWATER†

This doctrine recognizes legal rights, giving priority to the first person to make beneficial use of the water. This was partially enacted in order to protect investments into purchases of land, irrigation equipment, and drilling wells. This rule allows for damages to prior users by new pumpers. The rule is not strictly applied, in order that new users may have access to necessary water. The extent to which a new user is allowed to withdraw is determined by the state in which the user lives. Most states regulate use in order to protect the supply and optimize new economic uses. States using this method include Alaska, Colorado, Idaho, Kansas, Montana, Nevada, New Mexico, North Dakota, Oregon, South Dakota, Utah, Washington, and Wyoming. For an example of liability under this rule, see *Current Creek Irrigations Co. v. Andrews*, 9 Utah 2d 324, 344 P.2d 528 (Utah 1959).‡

8.17   CONDEMNATION OF GROUNDWATER

Whether this constitutes a taking or not may vary from state to state. Many states view the right to groundwater as inseparable from the land above. When the state restricts usage, it ordinarily exercises only a police power for the common good. When the state seizes the land to access the water itself for the public good, there is an issue of compensation. In Nebraska, a court found that Nebraska's doctrine of reasonable use required compensation for the seizure of land and water. The court reasoned that, " the right to use groundwater does not float in a vacuum of abstraction but exists only in reference to and results from ownership of the overlying land."§ In order to seize the water, the state had to seize some part of the land resulting in an economic loss to the landowner.¶

---

* *Katz v. Walkinshaw*, 141 Cal. 116, 74 P. 766 (Cal. 1903). See 8.3.
† Twelve western states use this doctrine for groundwater. For those with separate codes for each see Idaho Code §§ 42-226 to –240; Nev. Rev. Stat. §§ 534.010-350; N.M. Stat. Ann. §§ 72-12-1 to –28; Or. Rev. Stat. §§ 537.505-795; S.D. Codified Laws Ann. §§ 46-6-1 to –31; Wash. Rev. Code Ann. §§ 90.44.010-450; Wyo. Stat. §§ 41-3-901 to –938. For those which use a single code for both see Alaska Stat. §§ 46.15.010-270; Kan. Stat. Ann. §§ 82a-701 to –731; Mont. Code Ann. §§ 85-2-101 to –520; N.D. Cent. Code §§ 61-01-01, 61-04-01 to –32; Utah Code Ann. §§ 73-3-1 to –29. Typically these codes require permits to appropriate. Some minor exceptions exist.
‡ See also *Parker v. Wallentine*, 103 Idaho 506, 650 P.2d 648 (Idaho 1982) (application in reasonable use state showing prior appropriation law still applies to some wells in those states).
§ *Sorenson v. Lower Niorara Natural Resources Dist.*, 376 N.W.2d 539 (Neb. 1985).
¶ Nebraska used the reasonable rights doctrine associated with many riparian states. This result might be different to some degree in a prior appropriation jurisdiction for unappropriated rights.

8.18 PUBLIC RESOURCE

Most states (including most of those listed above) recognize no private ownership in groundwater and consider it public property. These states subject groundwater to management in the public interest. Rights to use the water are generally created by permit. Because property owners are not considered owners of the water beneath their land in these states, no taking occurs when laws are changed and prior rights lost.*

8.19 REASONABLE USE DOCTRINE IN GROUNDWATER (SEE 4.5 AND 4.6)

Better known as the "American Rule,"† the reasonable use rule is largely the same as applied to surface waters in riparian states.‡

8.20 "ECONOMIC REACH" RULE (SEE 6.13)

This is an attempt to strike a balance between junior and senior rights. The court requires that use by a senior well owner must be reasonably adequate in light of historic and economic use.§ This rule invites inquiry into the economic value of the competing uses. This rule has been applied in Colorado, Idaho,¶ and Utah.** A primary concern of this rule is the extent to which seniors should be required to improve their pumping measures when juniors lower the water level. The senior cannot be required to improve his facilities beyond the economic value, after consideration of all factors involved.

8.21 PUBLIC INTEREST REVIEW

Only Colorado and Oklahoma, among appropriation states, do not adhere to some form of public interest review in issuing water rights. The criteria and specificity of the statutes varies considerably from giving almost no guidance to the administrator†† to great specificity.‡‡ These statutes attempt to balance use in a manner consistent with the public interest and usually weigh the costs and benefits to the public of such allocations.

---

* *Town of Chino Valley v. City of Prescott*, 131 Ariz. 78, 638 P.2d 1324 (Ariz. 1981).

† *Adams v. Lang*, 553 So.2d 89, (Ala. 1989).

‡ *Forbell v. City of New York*, 164 N.Y. 522, 58 N.E. 644 (N.Y. 1900); Ball v. U.S., 1 Cl. Ct. 180 (1982) (reasonable use allows landowner to make lawful use of percolating waters on own land even to detriment of neighbor).

§ *City of Colorado Springs v. Bender*, 148 Colo. 458, 366 P.2d 552 (Colo.1961).

¶ *Baker v. Ore-Ida Foods, Inc.*, 95 Idaho 575, 513 P.2d 627 (Idaho 1973). See § 8.23.

** *Wayman v. Murray City Corp.*, 23 Utah 2d 97, 458 P.2d 861 (Utah 1969). See § 8.5.

†† See *Young & Norton v. Hinderlider*, 15 N.M. 666, 110 P. 1045 (1910).

‡‡ Alaska Stat. § 46.15.080(b); for express cost benefit analysis see Wash. Rev. Code Ann. § 90.54.020 (2) "Allocation of waters among potential uses and users shall be based generally on the securing of the maximum net benefits for the people of the State." For discussion of opportunity costs see *Tanner v. Bacon*, 136 P.2d 957 (Utah 1943) (Wolfe concurring).

8.22   RETURN FLOW AND ARTIFICIAL RECHARGE (SEE 8.24)

Water that is used or transported on surface lands often returns a portion to the ground through seepage.* Water such as treated or partially treated sewage may be reinserted into the ground (known as recharge). Soil mechanisms help further treat the water. In some cases water is inserted into the ground for storage purposes.†

8.23   "SAFE YIELD"

The term "safe yield" has numerous different definitions in different jurisdictions. Generally it refers to sustainable withdrawal. Some jurisdictions refer to recharge and others refer to acceptable consequences.‡

8.24   RECHARGE (SEE 8.22)

Recharge occurs when water from another source (also known as foreign or imported water) is introduced into the ground to replenish the groundwater supply. Most western states have statutes authorizing recharge of some form.§

8.25   WATER CONSUMPTION IN THE UNITED STATES (SEE ALSO 5.12)

By far the vast majority of water consumed is for hydroelectric power, which takes up to 48 percent percent. Irrigation accounts for 34 percent and public use only 11 percent. Most

---

* A nonriparian landowner may sue to enjoin an illegal use of surface waters, when those surface waters supply the property owner's land through percolation. *Miller v. Bay Cities Water Co.*, 157 Cal. 256, 107 P. 115 (1910).

† In New Mexico, ground water supplying a river was found not to be a new source when a user with a prior surface right sought to acquire part of that right by well do to lowering of the river from other pumping, but an attempted change in the point of diversion. See *Templeton v. Pecos Valley Artesian Conservancy District*, 65 N.M. 59, 332 P.2d 465 (1958). In 1966, another New Mexico court found that water diverted into an aquifer ceased to retain its identity as private and became a public resource. See *Kelley v. Carlsbad Irrigation Dist.*, 415 P.2d 849 (N.M. 1966). The court in New Mexico has interpreted the Templeton Doctrine as allowing, in some instances, the rights of surface users to groundwater that feeds the surface water, to relate back in priority to the date of the surface right. In some instances wells are shut down to protect prior surface rights. Shutting down the well does not automatically lead to increases in surface flow though. There is usually a lag time between closure of the well and feeling the effects of the closure. For a discussion of these issues, see George A. Gould and Douglas L. Grant, *Cases and Materials on Water Law Fifth Edition*, 374–78 (1995).

‡ See Idaho Code § 42-237a(g); S.D. Codified Laws § 46-6-3.1; Tex. Water Code §§ 52.051 to -.064; Neb. Rev. Stat. §§ 46-656 to –674.20; Ariz. Rev. Stat. Ann. §§ 45-437, -544, -453; Okla. Stat. tit. 82, §§ 1020.1-1020.22; see also *Baker v. Ore-Ida Foods, Inc.* 513 P.2d 627 (Idaho 1973) (discussed in § 8.20); *Doherty v. Oregon Water Resources Director*, 308 Or. 543, 783 P.2d 519 (1989). Most appropriation states reject absolute protection of senior appropriators in their historic methods of diversion. The general idea is to require reason in the appropriation. This is generally an attempt to balance economic interests with protection of senior appropriators. Nearly all of these states integrate surface and ground water priorities to some extent.

§ See Neb. Rev. Stat. §§ 46-233, -240 to –242, -295, -296, -299; Utah Code Ann. §§ 73-3-2, 73-3b-101 to –402; Wash Rev. Code Ann. §§ 90.44.035, -.040, -.130; see also *Jensen v. Dept. of Ecology*, 685 P.2d 1068 (Wash. 1984) (percolating water introduced by Bureau of Reclamation for irrigation did not become public water when commingled because Bureau intended to recapture it); Ariz. Rev. Stat. §§ 45-801.01 to -.898.01 (possibly the most extensive recharge statute); N.M. Stat. Ann. §§ 72-12A-3, -8 (one of the least extensive recharge statutes).

of the remaining water is used by industry, mining, and for livestock.* In 2000, the United States population reached almost 300 million and used nearly 350 billion gallons of water per day, of which nearly 250 billion gallons originated as surface water.†

8.26 COMMON METHODS OF CONTROLLING SALTWATER INTRUSION (SEE FIGURE 20)

Over pumping near saline water bodies can result in intrusion of salt water into the basin and aquifer. This may be controlled in any number of ways including the following list: 1) reduction of pumping or rearrangement of pumping patterns; 2) recharge of the basin (ordinarily with imported water) to raise the groundwater level above sea level; 3) creation of a coastal fresh water ridge through injection wells and spreading basins; 4) construction of an artificial subsurface physical barrier; or 5) creation of a pumping trough along the coast. ‡

8.27 RELATION OF SURFACE AND GROUNDWATER RESOURCES§

Most states today recognize at least some connection between groundwater and surface streams.¶ An increasing number of cases are recognizing the connection between ground and surface water and allowing surface rights holders a remedy for harm caused by later groundwater withdrawals.** A nonriparian owner may also have rights against illegal surface users, when that water supplies the nonriparian tract through percolation.††

8.28 "CRITICAL AREA" DESIGNATION

Many states,‡‡ particularly in the west have designated certain areas as "critical areas," in order to manage the areas impacted by overdrafts. These areas are usually defined by the area of an aquifer.

---

* http://pubs.usgs.gov/circ/2004/circ1268/htdocs/table02.html (see Table 2 in Appendix 3).
† http://pubs.usgs.gov/circ/2004/circ1268/htdocs/table01.html (see Table 1 in Appendix 3).
‡ Dan Tarlock, James N. Corbridge Jr., David H. Getches, *Water Resource Management: A Casebook in Law and Public Policy,* 564-565 (University Casebook Series 4th ed. 1993)
§ See Fig 2, Ground Saturation.
¶ Many states integrate management of ground and surface waters, including New Mexico, Oregon, Utah, Washington, and Wyoming. In Nevada, ground and surface waters are treated as separate on paper but are usually coordinated in practice. California also treats them separately under state law but they are usually integrated at the local level (these two states are correlative use states). Water that is not part of a stream flow in California is treated under the common law. Groundwater in Colorado is presumed to be tributary to surface water unless proven otherwise. Idaho uses a form of integrated management though domestic wells are exempted from the permit system. Montana uses optimal beneficial use as the basis for regulating groundwater and requires permits for any use. Arizona uses reasonable use as the basis of regulating groundwater.
** *Gallegos v. Colorado Ground Water Commission,* No. 05SA253 (Colo. 2006); *Montana Trout Unlimited v. Montana Department of Natural Resources and Conservation,* No. 05-069 2006 MT 72 (Mont. 2006).
†† *Miller v. Bay Cities Water Co.,* 157 Cal. 256, 107 P. 115 (1910).
‡‡ I.e., Arizona, California, Colorado, Hawaii, Idaho, Kansas, Nebraska, Montana, Nevada, New Mexico, Oklahoma, Oregon, Texas, Washington, and Wyoming. Programs vary among these several states, though the areas are usually overseen by the state engineer or equivalent state officers.

9 Native American Rights

9.1 THE *WINTERS* DOCTRINE

Native American water rights do not fall entirely in either the riparian or appropriations school of thought. In a 1908 case, the United States Supreme Court found after non-Indian settlers diverted water away from the reservation, that water rights were reserved for the Indians by necessary implication when the 1888 treaty was signed.* The Court found it unreasonable to believe that Indians would reserve land for farming without also reserving water to make such use possible. In the years following the decision in *Winters,* the United States represented the Indian tribes on many occasions that seemingly compromised the *Winters* rights. In many instances, these decrees still control Indian rights to water.† The Supreme Court held in 1963 that neither Congress nor the president could have intended to establish reservations without reserving for those reservations water necessary to make them habitable.‡ The *general rule* is that the creation of reservations necessarily carried with it a reservation of water necessary to sustain the reservation for human habitation.

9.2 QUANTITY OF WATER RESERVED

In *Arizona v. California* it was argued that quantity should be limited to amounts likely to be needed by the sparse Indian population.§ The Supreme Court rejected this argument and decided that the Indians were entitled to the amount of water sufficient to irrigate all the practicably irrigable acreage of the reservation. As a practical note, it was claimed that the government omitted some of the reservation lands, and the Court refused to reopen the matter after the initial irrigable acreage was determined.¶

9.3 COMPETING USERS WITH PRIOR USES

Competing users whose use of water predated the creation of the reservation have priority against the rights of Indians on the reservation. Water rights exist as of the creation of the applicable portion of the reservation.

---

* *Winters v. United States,* 207 U.S. 564, 28 S.Ct. 207, 52 L.Ed. 340 (1908); *U.S. v. City of Challis,* 133 Idaho 525, 988 P.2d 1199 (1999); *Totemoff v. State,* 905 P.2d 954 (Alaska 1995).

† *United States v. Gila Valley Irr. Dist.,* 920 F.Supp. 1444 (D.Ariz. 1996), affirmed 117 F.3d 425 (9th Cir. 1997).

‡ *Arizona v. California,* 373 U.S. 546 (1963).

§ *Navajo Development Co., Inc. v. Sanderson,* 655 P.2d 1374 (Colo. 1982) (if more than reservation desired, govt. must acquire as anyone else would).

¶ *Arizona v. California,* 460 U.S. 605, 103 S.Ct. 1382, 75 L.Ed.2d 318 (1983). For recent continuation of this conflict see §§ 2.1, 2.2.

9.4   EFFECT OF NONUSE

Winters rights to water are not lost by nonuse as other water rights often are.

9.5   COMPARISON OF INDIAN RIGHTS TO BOTH RIPARIAN AND APPROPRIATION RIGHTS

Indian rights apply mostly to land neighboring the waterway, similar to riparian rights, but also have a date of appropriation and must give way to the rights of prior users like appropriation rights.

9.6   INDIAN AGRICULTURAL WATER USAGE FOR NONAGRICULTURAL PURPOSES

While the purpose of the reserved water was originally for agriculture, the special master appointed in *Arizona v. California* stated that the water did not have to be used for agricultural purposes. A tribe may use its water for industrial or other purposes.*

9.7   NONAGRICULTURAL RESERVED WATER

Where a nonagricultural use is considered integral to the tribe, such as fishing, enough water is reserved to sustain that purpose.†

9.8   THE PURPOSE OF INDIAN RESERVATIONS LOOSELY CONSTRUED

The purpose of Indian reservations is broadly interpreted to encourage Indian self-sufficiency.‡ This differs from methods of construing the purpose of other federal lands.§

9.9   INDIAN RIGHTS IN GROUNDWATER

Though most of the decisions have dealt with surface water, Indian rights in groundwater also exist.¶

9.10   INDIAN RIGHTS TO OFF-RESERVATION WATER

In dire circumstances a broad interpretation of *Winters* might include rights to waters that do not touch the reservation land. This has not been favorably met thus far and on at least one occasion the Department of Justice has refused to fight for such rights to secure Indian hunting rights granted by a treaty.**

---

* *Colville Confederated Tribes v. Walton*, 647 F.2d 42 (9th Cir.), cert. denied, 454 U.S. 1092, 102 S.Ct. 657, 70 L.Ed.2d 630 (1981).
† *United States v. Adair*, 723 F.2d 1394 (9th Cir. 1983), cert. denied, *Oregon v. United States*, 467 U.S. 1252, 104 S.Ct. 3536, 82 L.Ed.2d 841 (1984). See § 9.14, 9.15.
‡ *United States v. Finch*, 548 F.2d 822 (9th Cir. 1976), vacated on other grounds *Finch v. United States*, 433 U.S. 676, 97 S.Ct.2909, 53 L.Ed.2d 1048 (1977).
§ *United States v. New Mexico*, 438 U.S. 696, 98 S.Ct. 3012, 57 L.Ed.2d 1052 (1978).
¶ *Cappaert v. United States*, 426 U.S. 128, 96 S.Ct. 2062, 48 L.Ed.2d 523 (1976).
** *Shoshone-Bannock Tribes v. Reno*, 56 F.3d 1476 (D.C.Cir.1995).

9.11 ENFORCING INDIAN WATER RIGHTS

While the rule was established in *Winters*, enforcing it has proven difficult. The government has proved resistive to the efforts of many tribes to ascertain and assert their *Winters* rights to unused water and has refused to represent the tribes in some situations. Litigation is costly, as are surveys to establish irrigable lands and determine the amount of reserved water. Opponents argue that granting the Indians water rights that have been long unused disrupts the appropriations system and prevents certainty.

9.12 ALLOTTED LANDS AND THE *WALTON* RULE

When land on a reservation is allotted, that portion of the irrigable water rights is said to follow the allotted land.* The *Powers* case indicates that a non-Indian obtaining allotted land receives the same rights, but the nature and extent of those rights has yet to be fully considered. It appears that unlike an Indian holder of allotted land, the non-Indian loses his rights if he fails to use them fully.† This is known as the *Walton Rule* and only applies when land is bought by a non-Indian from an Indian and not to homesteaders who purchase from the government.

9.13 LEASE OR SALE OF *WINTERS* RIGHTS

Indians may lease land and rights to non-Indians.‡ It is less certain whether they may lease the water to non-Indians for use on other lands. It seems likely that water may be leased for such purposes under *Winters*. It is likely that an outright sale of water rights separate from the land may violate the trust principles of *Winters* because it would possibly threaten the continued existence of the tribal land base.

9.14 *WINTERS* RIGHTS AND THE FIFTH AMENDMENT

When government takes water rights, the payment of compensation for the taking may depend on whether the reservation was created by executive order or treaty. Those created by executive order do not have recognized property rights and therefore require no compensation. Rights established by treaty or statute do carry Fifth Amendment recognition, and therefore must be compensated when taken.§

---

* *United States v. Powers*, 305 U.S. 527, 59 S.Ct. 344, 83 L.Ed. 330 (1939).

† *Colville Confederated Tribes v. Walton*, 647 F.2d 42 (9th Cir. 1981), appeal after remand 752 F.2d 397 (9th Cir. 1985), cert denied, 475 U.S. 1010 (1986).

‡ *Skeem v. United States*, 273 Fed. 93 (9th Cir. 1921).

§ *United States v. Adair*, 723 F.2d 1394 (9th Cir. 1983), cert. denied, 467 U.S. 1252 (1984). See § 9.7. It is worth noting that when the United States acts pursuant to its commerce authority over navigation, no compensation is necessary even when the rights interfered with were not related to navigation. See *United States v. Cherokee Nation of Oklahoma*, 480 U.S. 700, 107 S.Ct. 1487, 94 L.Ed.2d 704 (1987) (holding that the tribe was not entitled to compensation for injuries suffered by the loss of mineral rights related to ownership of the bed, even though they owned the bed and the rights were unrelated to navigation, because Congress' authority over navigable waters is absolute). See navigable servitude, § 3.11; see also § 1.8.

9.15 ABORIGINAL LANDS

When Native Americans claim as part of the reservation aboriginal lands, the date of the right to water for such lands is time immemorial. The rights are still only those necessary to provide the Indians with a moderate living from fishing and hunting, however, and not those enjoyed in aboriginal times.*

9.16 FEDERAL COURT AS FORUM

The federal government holds legal title to *Winters* rights in trust for the tribes and seldom consents to being sued in state court. Since the federal government is a necessary party and must consent to being sued, most Indian water-rights disputes occur in federal court.† It is still possible under the *McCarran Amendment* passed in 1952 for a state court to hear these cases.‡ The nature and extent of reserved Indian water rights remain matters of federal law.§

---

* *United States v. Adair*, 723 F.2d 1394 (9th Cir. 1983), cert. denied, 467 U.S. 1252 (1984). See § 9.7.

† See 28 U.S.C.A. § 1345 for jurisdiction.

‡ 43 U.S.C.A. § 666. See also Appendix 3; U.S. v. State of Or., 44 F.3d 758 (9th Cir. 1994) (state decisions are reviewable by the U.S. Supreme Court); *Arizona v. San Carlos Apache Tribe of Arizona*, 463 U.S. 545, 103 S. Ct. 3201, 77 L. Ed. 2d 837 (1983); *State ex rel. Greely v. Confederated Salish and Kootenai Tribes of Flathead Reservation*, 219 Mont. 76, 712 P.2d 754 (1985) (rights in state courts not diminished).

§ *Colorado River Water Conservation Dist. v. United States*, 424 U.S. 800, 96 S.Ct. 1236, 47 L.Ed.2d 483 (1976).

10 Environmental Regulation*

10.1 NEPA†

The primary value of NEPA is to require an Environmental Assessment (EA)‡ or an Environmental Impact Statement (EIS).§ NEPA also sets national policy regarding the environment to promote the health and welfare of man.¶ Many states now have similar legislation, with similar goals and policy statements to NEPA.** The Supreme Court has stated that NEPA is largely procedural and not primarily substantive in its nature.†† It is possible that federal action involving the issuance of a permit may be enough to activate NEPA's requirements.‡‡ NEPA functions to require a cost-benefit analysis for major federal projects.§§ This analysis is subject to

---

* See Appendix 2 Generally for Statutory References; Table 13 Priority Pollutants; Form 1 Application for Army Corps Permit; Figs 1 Typical Routes of Groundwater Contamination, 5, Hydrologic Cycle, 6 Water Cycle, 11 pH of Precipitation, 14 Reverse Osmosis, 15 Distillation, 16 Distillation, 17 Distillation, 18 Energy Costs of Distillation, 19 Desalinization Costs, 23 Salt Water Intrusion

† For a detailed analysis of federal procedure under NEPA see 11 Fed Proc, L Ed § 32:16-32:40; for forms see also 9 Fed Proc Forms 29.

‡ An environmental assessment is a concise public report intended to provide sufficient evidence to show whether an environmental impact statement is needed. It is often prepared to aid in compliance with NEPA or to assist in the preparation of an EIS. An EA is much abbreviated when compared to an EIS, which can be quite extensive. If an agency determines that no EIS is needed, they must then prepare a finding of no significant impact (FONSI). The Supreme Court and federal agencies have largely accepted the Council on Environmental Quality (CEQ) as the appropriate body for issuing rules regarding NEPA. See Executive Order 11514 (issued 1970). CEQ regulations are published in the CFR. See 40 CFR §§ 1501.1-1508.28.

§ 42 U.S.C.A. 4332 (NEPA § 102 (2) (c)). Agencies are required to address major environmental impacts and alternatives. The EIS should be used to inform federal agency decisions. For the definition of "major federal actions" see CEQ regulations, listing considerations used in making the determination. The judicial standard for both which alternatives should be considered and what is major is one of reasonableness.

¶ 42 U.S.C.A. 4321 (NEPA § 2), 42 U.S.C.A. 4331 (NEPA § 101).

** See Cal. Pub. Res. Code §§ 21000-21177; Wash. Rev. Code Ann., ch. 43.21C. The issuance of a license or permit may activate these state code requirements. See *Stempel v. Department of Water Resources*, 508 P.2d 166 (Wash. 1973). See § 10.11. Many state codes are more substantive than NEPA. See Cal. Pub. Res. Code § 21081; N.Y. Envtl. Conserv. Law § 8-0109; Minn. Stat. Ann. § 116D.04, subd. 6. The Washington act listed above does not contain a substantive portion but the court has stated that where there is no important benefit to offset environmental damage, they are warranted in holding the government action violates the substantive portion of SEPA. See *Ullock v. City of Bremerton*, 565 P.2d 1179 (Wash. Ct. App. 1977); *Polygon Corp. v. City of Seattle*, 578 P.2d 1179 (Wash. 1978).

†† *Vermont Yankee Nuclear Power Corp. v. NRDC*, 435 U.S. 519, 98 S.Ct. 1197, 55 L.Ed.2d 460 (1978). See also *Strycker's* Bay Neighborhood Council v. Karlen, 444 U.S. 223, 100 S.Ct. 497, 67 L.Ed.2d 433 (1980); *Robertson v. Methow Valley Citizens Council*, 490 U.S. 332, 109 S.Ct. 1835, 104 L.Ed.2d 351 (1989) (while gov't must consider mitigation measures, there is no requirement that such measures be implemented when issuing a permit).

‡‡ See *Zabel v. Tabb*, 430 F.2d 199 (5th Cir. 1970) (holding NEPA applicable to Refuse Act permits). See 10.15. It should be noted that both EPA and the Corps of Engineers have their own regulations, which require some form of public interest review as well. 33 C.F.R. § 320.4(a)(1).

§§ 25 Fed Proc L Ed § 57:402, Navigable Waters: Judicial Review.

judicial review* but limited to the question of whether or not the agency considered all environmental factors before reaching its decision.†

10.2   CLEAN WATER ACT GENERALLY‡ (FOR HISTORY, SEE 10.18; SEE 6.17 "WATER POLLUTION" FOR COMMON LAW RULE)

Most water pollution§ is industrial, agricultural, or municipal. Between 1980 and 1995 the number of rivers and streams with levels of fecal coliform bacteria more than 200 cells per 100 million ranged from a low of fifteen in 1991 to 35 in 1995.¶ There were a total of 8,315 oil spills in U.S. waters in 1998, consisting of more than 800 million gallons. This was less than the more than 9,000 spills and 3.1 billion gallons in 1996.** Of these spills, the greatest amount of pollution was in the rivers and canals of the United States (1944 spills and 280 million gallons in 1998).†† The United States also had more than 16,000 wastewater treatment facilities servicing almost 200 million people (1996).‡‡ This discharge is treated to reduce the level of contaminants, but the reduction and degree of treatment vary from facility to facility. To help control these and other pollutants, in the 1970s Congress enacted the Clean Water Act. The CWA is primarily concerned with surface water restoration and does not cover groundwater directly.§§

---

* *Defense Fund v. Corps of Engineers of United States Army*, 492 F.2d 1123 (C.A.5 Miss 1974); *Environmental Defense Fund, Inc. v. Froehlke*, 473 F.2d 346 (C.A.8 Ark 1972); *Liby Rod & Gun Club v. Poteat*, 457 F.Supp 1177 (D.C. Mont. 1978), affd in part and revd in part on other grounds 594 F.2d 742 (C.A.9 Mont).

† *Environmental Defense Fund, Inc. v. Froehlke*, 473 F.2d 346 (C.A.8 Ark 1972).

‡ For detailed analysis of federal procedure under the CWA, see 11 Fed Proc, L Ed § 32:258-32:424; for forms see also 9 Fed Proc Forms 29. For look at the typical routes of groundwater pollution, see Figure 1. For discussion of the water-quality needs of plants, see 25 Am Jur Proof of Facts p. 237, Water Pollution § 3. For discussion of the needs of humans and animals, see 25 Am Jur Proof of Facts p. 239, Water Pollution § 6. For a discussion of the self-cleansing capacity of lakes and streams, see 25 Am Jur Proof of Facts p. 252-254, Water Pollution § 12.

§ For more on Water Pollution, see 25 POF, Water Pollution—Sewage and Industrial Wastes §§ 1–51 at 233–332 (1970) with 1989 update.

¶ Statistical Abstract of the United States, 234 No. 389 (Dep't Commerce ed. Hoover Business Press 120th ed. 2000).

** Statistical Abstract of the United States, 234 No. 390 (Dep't Commerce ed. Hoover Business Press 120th ed. 2000).

†† Id.

‡‡ Statistical Abstract of the United States, 234 No. 391 (Dep't Commerce ed. Hoover Business Press 120th ed. 2000).

§§ *FD & P Enterprises, Inc. v. U.S. Army Corps of Engineers*, 239 F. Supp. 2d 509 (D.N.J. 2003) (The CWA does not grant the Corps of Engineers jurisdiction over wetlands adjacent to non-navigable tributaries to navigable waters); *Solid Waste Agency of Northern Cook County v. U.S. Army Corps of Engineers*, 531 U.S. 159, 121 S. Ct. 675, 148 L. Ed. 2d 576 (2001) (Corps of Engineers rule extending definition of navigable waters under the CWA to include those traversed by migratory birds found to exceed Corps authority under the CWA); *U.S. v. Phillips*, 356 F.3d 1086 (9th Cir. 2004), opinion amended and superseded on denial of reh'g, 367 F.3d 846 (9th Cir. 2004), cert. denied, 125 S. Ct. 479 (U.S. 2004) (though not itself navigable, creek fits within Corps jurisdiction because it is tributary to navigable waters); *Rapanos v. United States Army Corp. of Engineers*, 547 U.S. 715, 126 S.Ct. 2208, 2006 U.S. Lexis 4887 (2006) (CWA applies only to permanent waters). See http://www.epa.gov/owow/ wetlands/pdf/RapanosGuidance6507.pdf, for the current status of the Corps of Engineers jurisdiction over waters under the CWA. Document located in the appendices. See Appendix 1.

10.3   GOALS AND POLICY

Section 101 of the act (CWA) lays out the goals and policy. The major goals include restoring water quality in the nation's rivers, lakes, and streams, monitoring, regulating, and restricting pollutants. The goals include provisions for protecting the rights of states, encouraging foreign nations to act responsibly, and administration by the EPA. The plan initially consisted of state-developed ambient water-quality standards, applicable to navigable waters. The primary focus was on human health, and enforcement took place only to prevent an imminent health hazard or when a discharge reduced the quality of the water below a certain ambient level. Because many states lacked the motivation to act, Congress changed course and in 1972 established a permit system aimed at "fishable and swimmable waters." The amendments applied a best practical control technology available (BPT) standard on all point sources.* A point source is defined as "any discernible, confined and discrete conveyance, including but not limited to any pipe, ditch, channel, tunnel, conduit, well, discrete fissure, container, rolling stock, concentrated animal feeding operation, or vessel or other floating craft from which pollutants are or may be discharged." The term excludes agricultural stormwater discharges and irrigation return flows.† By 1983 the standard rose to the best available technology economically achievable (BAT).‡

10.4   MAJOR POLLUTANTS§

Section 307 of the Act requires the EPA to maintain a list of toxic substances¶ and establish limits on them.** (For the complete list of priority pollutants, see Table 13 in Appendix 3.)††

10.5   CWA PERMIT SYSTEM (SEE 6.15 AND 7.21 FOR STATE SYSTEMS UNDER RIPARIAN AND APPROPRIATION LAW)

Section 402 of the act creates a permit system for pollution discharge. It is known as the National Pollutant Discharge Elimination System (NPDES). Permits are required for all point sources (see definition above, 10.3) discharging into navigable waters. Navigable waters refer to waters of the United States. Administration of the

---

* 33 U.S.C.A. § 1311(b)(2)(E) (FWPCA § 301). See Appendices 2 for text.
† 33 U.S.C.A. § 1362 (14) (FWPCA § 502). See Appendices 2 for text.
‡ 33 U.S.C.A. § 1311 (b)(2)(A) (FWPCA § 301). Best available technology economically achievable for pollutants other than publicly owned treatment works.
§ See Figure 1 for Typical Routes of Groundwater Contamination.
¶ Termed priority pollutants.
** 33 U.S.C.A. § 1317 (FWPCA § 307). Toxic and pretreatment effluent standards. See Table 13.
†† http://www.epa.gov/waterscience/criteria/wqcriteria.html.

water quality standards is left to the states, which may impose stricter standards if they wish.*

10.6   WASTEWATER MANAGEMENT

Section 208† of the act deals with wastewater management and treatment. The act calls for development of area wastewater management plans. Problem areas are designated and agencies created to oversee implementation of the treatment plan.

10.7   DREDGE AND FILL PERMITS

Sections 404 of the act requires permits from the Army Corps of Engineers for any discharge of dredge or fill materials into United States waters. The definitions are broadly interpreted to include nonobvious activities such as construction projects and flood-control activities.‡ The Corps must consider all "relevant factors" before a project under Section 404§ may be allowed. The Corps discretion is far-reaching but recent Supreme Court decisions appear to be cutting it back somewhat.¶ Few other

---

* 33 U.S.C.A. § 1341 (CWA § 401), *See PUD No. 1 of Jefferson County v. Washington Department of Ecology,* 511 U.S. 700, 114 S.Ct. 1900, 128 L.Ed.2d 716 (1994) (declaring that § 303 is incorporated into § 301(mentioned with § 302 in § 401) by reference and therefore those standards are among "other limitations" by which states may ensure compliance under § 401) (also holding that minimum flow requirements under the above sections are valid). This decision greatly increased state power to restrict federal permits under the CWA by allowing both the designated standards under the act, and state standards. F.E.R.C. reacted by stating that Section 401 authorizes states only to make conditions related to water quality. Tunbridge Mill Corp., Order Issuing License, 68 F.E.R.C. para 61, 078 (1994). This opinion does not appear to modify the decision in PUD No. 1. At least one court has restricted the pollutants regulated to point sources. See *Oregon Natural Desert Ass*'n v. Dombeck, 172 F.3d 1092 (9th Cir. 1998).

† 33 U.S.C.A. § 1288. This plan requires the governors of each state to designate a local planning agency to develop the plans for wastewater treatment. These plans must identify nonpoint source pollution (see 10.3 for point source definition). In 1987 the act was amended again, directing the states to submit an assessment report to EPA, describing problems from nonpoint sources. See Section 319, 33 U.S.C.A. § 1329. Nonpoint sources account for most of the pollution in 65 percent of rivers and streams within the United States and exceed all other forms of pollution in most states. They account for 76 percent of pollutants in lakes. See George A. Gould and Douglas L. Grant, *Cases and Materials on Water Law,* 5th ed., 534 (1995). No effective method is built into the CWA for dealing with nonpoint sources. They are left largely to the states, which have relied mostly on voluntary programs. See Chapter 12 for a discussion of the problems of game theory and collective action.

‡ *United States v. Akers,* 785 F.2d 814 (9th Cir. 1986).

§ States may assume administration of section 404 permits if they have an approved program. See 43 C.F.R. §§ 233.70, 233.71.

¶ See *Rapanos v. United States Army Corp. of Engineers,* 547 U.S. 715 (2006). See § 1.8; Appendix 1. This case seems to be scaling back the earlier decision in *United States v. Holland,* 373 F.Supp. 665 (M.D.Fla. 1974), where a federal court found § 404's broad definition to be within the Commerce Clause authority. In *Natural Resources Defense Council v. Callaway,* 392 F.Supp. 685 (D.D.C. 1975), the federal court found an Army Corps of Engineers self-limitation on its statutory authority to be an unlawful act in derogation of their duties under section 404. The factors considered by the Corps include significant adverse effects of the discharge of pollutants on 1) human health; 2) aquatic life and other wildlife dependent on the aquatic ecosystem; 3) aquatic ecosystem diversity, productivity, and stability; and 4) recreation, aesthetic, and economic values. The Corps must give great weight to state and local agencies with stricter standards. No permit may be issued that violates state water-quality or environmental standards. State and local permits must also be acquired in order to construct.

environmental laws place so much discretion in a federal agency. Permits are reviewable by the EPA* and may be vetoed by that agency once issued.† It is through the Section 404 permit requirement that many other environmental laws are extended in reach to private individuals.‡ The granting of a permit is considered state action and may turn an otherwise private project into state action. Perhaps the furthest reaching of these statutes is the Endangered Species Act covered below.

10.8   THE WALLOP AMENDMENT

Water quality can be reduced by upstream uses that reduce the amount of water available downstream. This can cause high concentrations of pollutants downstream. There is a dichotomy between quantity and quality laws, which tends to create friction between the two interests. Allocation is largely a function of state law, while quality is ordinarily a function of federal law. In this way the friction is often between the state and federal governments.§ The Wallop Amendment was added to ensure that established water rights would not be defeated by the Clean Water Act.¶ In short, the act should not be construed to interfere with state rights to allocate water. The Wallop Amendment's purpose is to protect only those rights that are incidentally affected, and it has been of limited value when it comes to protecting state rights.**

10.8.1   WATER QUALITY TRADING

This is a market-based approach to cleaning up water pollution. Section 303 (d) of the CWA allows one point source, which decreases pollution more than required, to sell the extra to another point source, so that the second might increase the number of pollutants released at its facility.††

---

* See *James City County v. Environmental Protection Agency*, 12 F.3d 1330 (4th Cir. 1993) (upholding the veto of a 404 permit for a dam even though no threatened or endangered species were involved and EPA did not consider the city's need for water). There is a possible exception to the otherwise tough scrutiny of EPA review. A national permit for general use enables building unless E.S.A. or The National Wild and Scenic River System are implemented, when dealing with streams of less than five cubic feet per second average flows. See 33 C.F.R. pt. 330. Another exception to section 404 scrutiny is for construction of farm or stock ponds or irrigation ditches, or the maintenance of drainage ditches. 33 U.S.C.A. § 1344(f)(2). Note: Unlike section 10, section 404 permits must comply with EPA guidelines.

† See 33 U.S.C.A. § 1344; for definition of "navigable waters" under the act, see 33 U.S.C.A. § 1362 (7) (defining navigable waters as "the waters of the United States including the territorial seas."

‡ See § 10.1 NEPA.

§ See 33 U.S.C.A. § 1251 (g), The Wallop Amendment. "It is the policy of Congress that the authority of each State to allocate quantities of water within its jurisdiction shall not be superseded, abrogated, or otherwise impaired by this chapter," passed in response to western state fears that quality control laws might be used to usurp state allocation authority. But *see Riverside Irrigation District v. Andrews*, 758 F.2d 508 (1985) (restricting state authority to authorize construction of a dam on creek, for environmental concerns, by refusing to grant a federal permit for dredge material, by Corps of Engineers). See § 10.10.

¶ 33 U.S.C.A. § 1251(g) (FWPCA § 101). § 1251 (a) (2) sets goals for water quality standards for the "protection and propagation of fish, shellfish and wildlife and provides for recreation in and on the water."

** *Riverside Irrigation Dist. v. Andrews*, 758 F.2d 508 (10th Cir. 1985).

†† http://yosemite.epa.gov/R10/OI.NSF/d9fbcd8fc7ce1c5d882564640065adff/e061bb2efbef6d54882566 950062b816?OpenDocument.

10.9   STATE AND LOCAL WETLANDS REGULATION

Generally, all regulation of wetlands by states and localities is restrictive. Most is more so than Section 404. Most still require permits be obtained from the appropriate state and local agencies. Most state programs begin with an inventory of natural resources. The permit process usually involves public notice, and contains standards and procedures for the application and appeal.*

10.10   ENDANGERED SPECIES ACT†

The Endangered Species Act was an effort by Congress to protect endangered and threatened plants, animals, and their habitats.‡ ESA § 2 (c)(2) states "it is . . . declared to be the policy . . . that Federal agencies shall cooperate with State and local agencies to resolve water resource issues in concert with conservation of endangered species."§ ESA § 7 generally prohibits federal actions that jeopardize the existence of any endangered species. The agency involved is required to consult U.S. Fish and Wildlife Service to determine the effects of any proposed action.¶

10.11   STATE WATER QUALITY STATUTES

All states have statutes regulating water quality to some degree. In some cases, these statutes may also coincide with public welfare requirements (particularly in the west) where water use is balanced with the needs of the public (see § 8.21 Public Interest Review). In some of these states the pollution of water is considered in light of public interests in clean water.**

---

* See Julian Conrad Juergensmeyer, *Urban Planning and Land Development Control Law,* 2nd ed. § 13.14, 417 (1986).
† For a detailed analysis of federal procedure under the ESA, see 25 Fed Proc, L Ed § 56:2017-56:2057; 13 Fed Proc Forms 50.
‡ 16 U.S.C.A. § 1531 (ESA § 2).
§ 16 U.S.C.A. § 1531 (c)(2).
¶ See *Tennessee Valley Authority. v. Hill,* 437 U.S. 153, 98 S.Ct. 2279, 57 L.Ed.2d 117 (1978) (§ 404 permit denied because of potential impact on snail darter). In the snail darter case, ultimately the dam was finished based on legislation after a failed attempt to gain an exemption from the panel created in 16 U.S.C.A. § 1536(e). The dam did not fully accomplish what it was intended to do, and the snail darter in that area did die out. Other populations were ultimately discovered elsewhere, though, and the species was removed from the list. Note that permits may only be valid on the condition that individuals acting on them do not threaten to destroy or adversely modify the critical habitat of an endangered or threatened species. 33 C.F.R. § 330.4(b) (2). See also *Riverside Irrigation District v. Andrews,* 758 F.2d 508 (tenth Cir. 1985). See § 10.8.
** See *Stempel v. Department of Water Resources,* 508 P.2d 166 (Wash. 1973) (Water officials initially concluded "public welfare" related only to withdrawal of water and not pollution. See § 10.1. The Washington Supreme Court concluded differently in light of two new environmental statutes); RCW 43.21C (SEPA); RCW 90.54 (WRA). The department is now obligated to consider the full environmental and ecological impact of use. See also Cal. Water Code § 1257 (officials must consider the "relative benefit to be derived from . . . all beneficial uses of the water concerned . . .); Alaska Stat. § 46.15.080 (considering the impact on fish and game); Or. Rev. Stat. § 537.170(5) (a); Utah Code Ann. § 73-3-8; Wash. Rev. Code § 90.54.020. The Washington State Supreme Court addressed the issue of whether or not quality permits for existing rights could be conditioned on minimum stream levels in Public Utility District No. 1 of *Pend Oreille v. County V. State of Washington, Dept of Ecology,* 146 Wash.2d 778, 51 P.3d 744 (2002). There the court refused to find an abandonment or forfeiture of rights but did determine that the state could condition rights on minimum levels under certain conditions. See also R.C.W.A. 90.03.380.

10.12   STATE REGULATION

Much of the regulation of water use is based on environmental statutes.* States, like the federal government, have many different statutes on their books, intended to protect the environment. Although state laws regarding navigable waters may not present a standard lower than the federal law, they can sometimes use standards greater than the federal standard. The key concern is whether the state law interferes with a national policy.†

10.13   WETLANDS DEFINITION (SEE ALSO 18.10)

wetlands *n. pl.* swamps and other damp areas of land.‡ The Army Corp. of Engineers settled on the following definition: "those areas that are inundated or saturated by surface or ground water at a frequency and duration sufficient to support, and that under normal circumstances do support, a prevalence of vegetation typically adopted for life in saturated soil conditions. Wetlands generally include swamps, marshes, bogs, and similar areas."§

10.14   IMPORTANCE OF WETLANDS (SEE ALSO 10.23 AND 10.24 ON FLOODPLAINS)

Wetlands appear to be among our most important biospheres. They are home to numerous plants and animals and provide protections from natural disasters. They are also a significant source of other natural resources such as oil and gas. The rapid loss of wetlands can contribute to major disasters such as the one that occurred when Hurricane Katrina struck the coast of Louisiana and Mississippi.¶

---

* See *Muench v. Public Service Commission*, 261 Wis. 492, 53 N.W.2d 514 (1952).

† The notion that states may not discriminate against out-of-state interests and may not act to defeat a federal program is often referred to as a part of what is known as the Dormant Commerce Clause. The clause allows concurrent authority of states in matters of commerce, subject to limitations placed on the states by exercise of the federal authority over commerce. The Supremacy Clause requires that in matters where specific constitutional authority belongs to Congress, federal laws under the Constitution trump state laws, which may conflict. Congress is limited to the powers granted it in the Constitution when it comes to legislation, and the Commerce Clause is no exception. See *United States v. Morrison*, 529 U.S. 598, 120 S.Ct. 1740, 146 L.Ed.2d 658 (2000); *United States v. Lopez*, 514 U.S. 549, 115 S.Ct. 1624, 131 L.Ed.2d 626 (1995). See Appendix 1. This restriction may pose some potential problems for the federal government when attempting to implement its policies nationally, especially where non-navigable waters are concerned.

‡ Frank R. Abate, *The Oxford Pocket Dictionary and Thesaurus*, American Edition, 916 (1997).

§ 33 C.F.R. § 323.2 (c). Section 404 applies only to dredge and fill materials. For definition of dredge and fill materials, see 33 C.F.R. § 323.2 (l). The terms are limited generally to materials taken from or discharged into waters of the United States. The section exempts certain fill activities such as normal farming or ranching activity. These exemptions are construed narrowly, though. See *Avoyelles Sportsmen's League, Inc. v. Alexander*, 473 F.Supp. 525 (W.D.La. 1979).

¶ See also 10.23 regarding floodplains and FEMA.

10.15 WETLANDS DEVELOPMENT (SEE ALSO 18.10)

Wetlands are heavily regulated by the federal, state, and local governments. Development almost always requires a permit.* Permits at all levels must be obtained before work begins. Several federal acts affect interplay between the different levels of government in the permitting process. The first is The Rivers and Harbors Act of 1899,† which was designed to protect navigable waters, not wetlands. Jurisdiction is over only navigable waters and requires recommendation by the Chief of Engineers and authorization by the Secretary of the Army.‡ Other such acts include the Clean Water Act (see 10.2), and the Endangered Species Act (see 10.10).

10.16 SECTION 10 JURISDICTION (SEE 10.18 "THE RIVERS AND HARBORS ACT OF 1899")

Federal jurisdiction over tidal water extends to the mean high-water line, which is determined by using tidal cycle data.§

10.17 WILD AND SCENIC RIVERS ACT

In 1968, Congress declared a policy that certain rivers and their environments would be protected in their natural flow for scenic, recreational, or other historic

---

* Eastern coastal states generally limit regulation to dredging and filling of coastal wetlands; see Con Gen. Ann. §§ 22a-28 to 22a-15; Ga. Code Ann. §§ 43-2401 to 43-2413; 38 Me. Rev. Stat. Ann. §§ 471–76, 478; Md. Nat. Res. Code Ann. §§ 9-201, 9-202, 9-301 to 9-310; Mass. Gen. Laws Ann. Ch. 131 § 40A (Wetlands Protection Act); Miss. Code Ann. §§ 49-27-1 to 49-27-69; N.H. Rev. Stat. Ann. 483-A: 1 to6; N.J. Stat. Ann. 13:9-A-1 to 10; N.Y. Envir. Conserv. Law §§ 25-0101 to 25-0602; N.C. Gen. Stat. § 113–229; R.I. Gen. Laws 1978, § 11-46.1-1, §§ 2-1-13 to 2-1-25, 49-27-1 to 49-27-69; Va. Code Ann. 1978, §§ 62.1-13.1 to 62.1-13.20. See also Cal. Gov. Code §§ 66601 to 66661; Mich. Comp. Laws Ann. §§ 281–631 to 281–644; Minn. Stat. Ann. § 105-485; 10 Vt. Stat. Ann. SS 1421 to 1426; Wash. Rev. Code Ann. Ch. 90.58; Wis. Stat. Ann. SS 59.971, 144.26

† 33 U.S.C.A. §§ 401–18.

‡ *Id.* § 403. For definition of navigable waters used by the Army Corp. of Engineers see 33 C.F.R. § 329.4; To see if the water body comes within the definition see 33 C.F.R. 329.14. The army has traditionally used the definition, "those waters that are subject to the ebb and flow of the tide and/ or are presently used, or have been used in the past, or may be susceptible for use to transport interstate or foreign commerce." The Army jurisdiction was expanded in the late 1960s to include environmental impacts (see *Zabel v. Tabb,* 430 F.2d 199 (5th Cir. 1970) (see § 10.1); see also C.F.R. § 209.120 (allowing the corp. to consider public interest and environmental issues (superseded by 42 Fed. Reg. 37,133 (1977)) but has been retracted somewhat by recent court decisions (see § 1.8). For a complete discussion of permit procedures, see 25 Fed. Proc., L. Ed. § 57:229-57:247; for related forms see 13 Fed Proc Forms 51.

§ 33 C.F.R. § 329.12; for nontidal waters, see § 329.11 (subject to fluctuation based on rainfall, topography, etc. and more difficult to calculate because of the more numerous variables); see also *United States v. Cameron,* 466 F.Supp. 1099, 1111 (M.D.Fla. 1978) (Court found that the ordinary high-water line of nontidal water bodies is not readily susceptible to uniform and precise definition in recognizing government survey data). The Court in *United States v. Rio Grande Dam & Irrigation Co.,* 174 U.S. 690, 19 S.Ct 770, 43 L.Ed. 1136 (1899), stated that "it is not a prohibition of any obstruction to the navigation, but any obstruction to the navigable capacity, and anything, wherever done or however done, within the limits of the jurisdiction of the United States which tends to destroy navigable capacity . . ." See also *Sanitary Dist. v. United States,* 266 U.S. 405 (1925); *Wisconsin v. Illinois,* 278 U.S. 367, 49 S.Ct. 163, 73 L.Ed. 426 (1929) (reaffirming the use of section 10 to protect navigation on the Great Lakes from water diversion).

and environmental value.* The goal is to preserve the rivers in their free-flowing condition. It prohibits construction of dams or other structures that might in any way interfere with the rivers.† This is largely a land-management law.

10.18 THE RIVERS AND HARBORS ACT OF 1899 (FOR CLEAN WATER ACT SEE 10.2; SEE ALSO 10.15 AND 10.16 FOR MORE ON THIS ACT)

The Rivers and Harbors Act‡ was originally passed to prevent interference with navigation. It gained new life in 1970 when it was applied to all industrial discharges. The act makes it unlawful to discharge any refuse of any kind into navigable waters, other than that from streets or sewers, without a permit from the Army Corps of Engineers. The Environmental Protection Agency hands down guidelines for the Corps to follow. The act was so successful that it led to a series of amendments to the Water Pollution Control Act, in 1972.§ The Federal Pollution Control Act Amendments completely rewrote the original Water Pollution Control Act of 1948 and were dubbed the Clean Water Act.¶

Under the Rivers and Harbors Act, there is no private right of action.** A general intent to violate the act should be demonstrated in order to pursue criminal sanctions under the act.†† The federal government must prove the elements: 1) defendant caused fill to occur, 2) in navigable waters, and 3) without prior authorization from the Corps of Engineers.‡‡ These elements must be shown beyond a reasonable doubt.§§

---

* 16 U.S.C.A. § 1271

† Section 3 lists three types of eligible rivers or streams: 1) wild rivers, with no access other than trails; 2) scenic rivers, largely undeveloped; and 3) recreational rivers, with some development already on them. 16 U.S.C.A. 1273. More than 150 rivers were included by 1995, up from eight designated by the original act. Still others are being studied.

‡ 33 U.S.C.A. § 407

§ See 13 Fed Proc Forms § 51:30, *Navigable Waters: Citizen Suits.*

¶ 33 U.S.C.A. § 1251.

** *Sierra Club v. Andrus*, 610 F.2d 581 (C.A.9 Cal 1978) (private persons may not sue to enforce criminal provisions); *California v. Sierra Club*, 68 L.Ed.2d 101, 101 S.Ct. 1775 (U.S. 1981) (injured persons have no private right under the act to sue); *Citizens' Committee for Environmental Protection v. United States Coast Guard*, 456 F.Supp. 101 (D.C. NJ 1978); *Loveladies Property Owners Asso. v. Raab*, 430 F.Supp. 276 (D.C. NJ) (no reasonably implied right to civil action). But see *Vieux Carre Property Owners, Residents & Associations, Inc. v. Brown*, 875 F.2d 453 (1989) (finding standing for private individuals under the APA because historic preservation was within the zone of interests Congress sought to protect). Private parties suffering special injury because of unauthorized activity may sue under the act. *Sierra Club v. Andrus*, 610 F.2d 581 (1979). 78 Am Jur 2d, Waters § 109 with 1999 update. 25 Fed Proc L Ed § 258, *Prohibition on Private Enforcement.*

†† *United States v. Commodore Club, Inc.*, 418 F.Supp. 311 (D.C. Mich.).

‡‡ *United States v. Commodore Club, Inc.*, 418 F.Supp. 311 (D.C. Mich.); United States v. Osage Co., 414 F. Supp. 1097 (W.D. Pa. 1976). For more on the permit procedures under the Rivers and Harbors Act, see 25 Fed Proc L Ed §§ 57:229–57:247; see §§ 57:248–57:259 for enforcement proceedings.

§§ *United States v. Commodore Club, Inc.*, 418 F. Supp. 311 (E.D. Mich. 1976).

10.19   THE SAFE DRINKING WATER ACT*

In 1974 Congress passed the Safe Drinking Water Act.† The purpose of the act was to protect sources of drinking water. The act includes requirements that states control underground injections, which could pollute public water sources.‡ The act empowers the EPA to set levels for the maximum amount of contaminants allowed in drinking water when it is delivered to the consumer and the EPA can prevent the use of contaminated supplies.

10.20   THE TOXIC SUBSTANCES CONTROL ACT

This act§ regulates the substances that pose the greatest threat to human health when introduced into the water supply. Specifically, it regulates the manufacture, testing, and distribution of such substances.

10.21   THE RESOURCES CONSERVATION AND RECOVERY ACT

This act¶ regulates the present and future handling, treatment, storage, transport, and disposal of hazardous waste.

10.22   CERCLA

The Comprehensive Environmental Response, Compensation, and Liability Act** deals with cleanup of existing dumpsites, where hazardous chemicals have been stored and pose a threat. The act requires that owners of the sites notify EPA of the site's existence, and it gives EPA the right to clean up the site if the owner does not or no owner can be found. Additionally, the act creates a "Superfund" for cleanup costs and creates liability in owners for the costs of such cleanup. Since the cost of cleanup is likely to be considerable, restitution costs can shut down many poorly run storage facilities.

10.23   FLOODPLAINS

A floodplain is the flat area touching on the banks of the river, where floodwaters periodically flow and deposit soil. When these areas are developed, they create several problems. Annual floodwaters often cause severe damage to homes in floodplains. It is expensive to protect and drain the floodplain, and most measures to do so tend to alter the river's flow elsewhere, causing new environmental difficulties. Houses and buildings on active floodplains often must be raised to avoid the floodwaters and are

---

* For a discussion of public health and drinking water standards, see 25 Am Jur Proof of Facts, 240–46, Water Pollution § 7.
† 42 U.S.C.A. 300f–300j.
‡ 42 U.S.C.A. 300h.
§ 15 U.S.C.A. § 2601.
¶ 42 U.S.C.A. § 6921.
** 42 U.S.C.A. § 9601–9615, 9631–9657.

subject to high flood-insurance premiums. Also, as Hurricane Katrina showed us, alteration of the natural drainage system can prove disastrous for large communities living in these areas.

10.24   FEMA'S NEW FLOODPLAIN MAPS

In the wake of Hurricane Katrina, the Federal Emergency Management Agency (FEMA)* is revising its floodplain maps to include more area.† If approved, this will raise insurance rates and restrict development in these added areas. Some individuals will also be required to purchase flood insurance. The new maps are likely to make building in the floodplains cost-prohibitive by requiring new structures to be raised up.

---

* See http://www.floodsmart.gov/floodsmart/pages/index.jsp.
† See article by Ashley Bach and Karen Johnson, *Bigger Floodplain, Bigger Worries, Seattle Times,* available at http://seattletimes.nwsource.com/html/eastsidenews/2004041546_floodmaps29e.html November 28, 2007.

11 Nuisance and Tort Law

11.1 NUISANCE AND POLLUTION OF WATERS

At common law, a person is guilty of a nuisance when he or she pollutes a water source used by the public for drinking or cooking. It is no excuse that a municipal corporation provided inadequate drainage.*

11.2 NUISANCE AND OBSTRUCTION OF PUBLIC WATERWAYS†

Also at common law, it is a nuisance to obstruct a public waterway in such a way as to interfere with navigability. ‡ This could include construction of a dam, wharf, or bridge§ or by deposit of waste or rubbish.¶ The owner of the soil between the high- and low-water marks may use it for his own purposes, so long as those purposes do not interfere with the navigability of the waterway.** The interference must be significant.†† In the event that a boat accidentally sinks in the waterway, the owner is not guilty of a nuisance, even if it obstructs navigation. Therefore, public authorities may move the ship.‡‡ The obstruction of passage for fish is not normally considered a nuisance.§§ If the obstruction overflows the waterway and endangers public health, it is a nuisance.¶¶ When a lower landowner fills in a waterway and the

---

* Charles E. Torcia, *Wharton's Criminal Law 15th Edition* vol. 4, § 530 (1996); see *Jacksonville v. Doan*, 48 Ill. App. 247 (1892); *Commonwealth v. Kington Coal Co.*, 175 Ky. 780, 194 S.W. 1038 (1917); *Messersmidt v. People*, 46 Mich. 437, 9 N.W. 485 (1881); *Mergentheim v. State*, 107 Ind. 567, 8 N.E. 568 (1886); *State v. Smith*, 82 Iowa 423, 48 N.W. 727 (1891).

† See 78 Am Jur 2d, Waters, §§ 96–107 (interference with navigation), 203 (Dam as nuisance).

‡ Charles E. Torcia, *Wharton's Criminal Law 15th Edition* vol. 4, § 533 (1996); see *New Orleans, M. & T.R. Co. v. Mississippi*, 112 U.S. 12, 28 L.Ed. 619, 5 S. Ct. 19 (1884); *West Chicago S.R. Co. v. Illinois*, 201 US 506, 50 L. Ed. 845, 26 S. Ct. 518 (1906); *Olive v. State*, 86 Ala. 88, 5 So. 653 (1888); *Attorney Gen. Ex rel. Adams v. Tarr*, 148 Mass. 309, 19 N.E. 358 (1889); *State v. Hutchins*, 79 N.H. 132, 105 A. 519 (1919); *State v. Narrows Island Club*, 100 N.C. 477, 5 S.E. 411 (1888); *State ex rel. Lyon v. Columbia Water Power Co.*, 82 S.C. 181, 63 S.E. 884 (1909); *State v. Carpenter*, 68 Wis. 165, 31 N.W. 730 (1887); *Whorton v. Malone*, 209 W. Va. 384, 549 S.E.2d 57 (2001); see, e.g., Ariz. Rev. Stats. Ann. § 13-2917 (A); Minn. Stats Ann. S 609.74 (2). Diversion by an upper owner interfering with lower owner's rights is treated under nuisance law. This includes use of groundwater. *Henderson v. Wade Sand and Gravel Co., Inc.*, 388 So. 2d 900 (Ala. 1980). Injunction is an allowed remedy for wrongful interference with water rights, to prevent further interference.

§ *State v. Godfrey*, 12 Me. 361 (1835) (see § 11.4); *People v. Vanderbilt*, 26 N.Y. 287 (1863).

¶ *People v. Gold Run Ditch & Mining Co.*, 66 Cal. 138, 4 P 1152 (1884).

** Zug v. Commonwealth, 70 Pa. 138 (1871).

†† *State v. Wilson*, 42 Me. 9 (1856). It is enough that navigation be less convenient, secure, and expeditious. The waterway need not be fully blocked. See *State v. Narrows Island Club*, 100 N.C. 477 (1888).

‡‡ *McLean v. Mathews*, 7 Ill. App. 599 (1880).

§§ *Commonwealth v. Chapin*, 22 Mass 199 (1827).

¶¶ *State v. Close*, 35 Iowa 570 (1872); *Commonwealth v. Webb*, 27 Va. 726 (1828).

upper owner diverts flow into the bed that would have been sufficient for the diversion, the lower owner may not sue for damages against the upper owner.* Other diversions to the detriment of lower riparians, such as interfering with others water rights, are actionable.†

11.3   TEST OF NAVIGABILITY RELATING TO NUISANCE LAW

The test is whether the waterway is in fact navigable‡ or used commercially for transportation, even if for only part of the course or part of the year. §

11.4   STATE AUTHORITY TO OBSTRUCT (SEE FEDERALISM, CHAPTER 1)

Subject to federal authority of navigable waters of the United States, a state may authorize obstructions of a public waterway if in its judgment the public interest requires it to do so.¶ The federal government also has such powers.** A duly authorized obstruction cannot constitute a nuisance.††

11.5   THE RESTATEMENT OF TORTS (SECOND) (SEE 8.3 "GROUNDWATER DOCTRINES," 8.6, "LIABILITY FOR SUBSIDENCE")

Rules relating to the allocation of water have historically been assigned to the Restatement of Torts. Most cases involving riparian rights are after-the-fact cases to redress a harm already committed. These cases are often resolved using nuisance principles, which fall into tort law instead of property law.‡‡ The Restatement of

---

* *Smith v. King Creek Grazing Ass'n*, 671 P.2d 1107 (Idaho App. 1983); See also § 11.8, "Law of Drainage."

† Wong Nin v. City and County of Honolulu, 33 Haw. 379, 1935 WL 3373 (1935); *Woodland v. Lyon*, 78 Idaho 79, 298 P.2d 380 (1956).

‡ *United States v. Holt State Bank*, 270 U.S. 49, 70 L.Ed. 465, 46 S.Ct. 197 (1926) (see § 3.1); *United States v. Utah*, 283 U.S. 64, 75 L.Ed. 844, 51 S.Ct. 438 (1931); see also *Neterer v. State*, 98 Wash. 635, 168 P. 170 (1917). See also § 1.8.

§ *Economy Light & Power Co. v. United States*, 256 U.S. 113, 65 L.Ed. 847, 41 S.Ct. 409 (1921).

¶ *Gibson v. United States*, 166 U.S. 269, 41 L.Ed. 996, 17 S.Ct. 578 (1897).

** *United States v. Appalachian Electric Power Co.*, 311 U.S. 377, 85 L.Ed. 243, 61 S.Ct. 291 (1940) (The Court here defines the phrase "used or are susceptible of being used" as being met if a river is (a) presently used for commerce; (b) in the past was so used, though no longer; or (c) in the future could be so used. Simply stated, a waterway remains navigable once found to be so. See also § 1.7.

†† *State v. Godfrey*, 24 Me. 232 (1844). See § 11.2. Ordinarily, the grant must be specifically stated. An obstruction that exceeds the authority conferred is subject to abatement as a nuisance. See Charles E. Torcia, *Wharton's Criminal Law 15th Edition* vol. 4, § 534 (1996). See *Commonwealth v. Church*, 1 Pa. 105 (1845).

‡‡ See Restatement of Torts (Second) § 850 (establishing liability for unreasonable use that causes harm to another riparian's reasonable use of water). § 850A lists factors that affect the determination of reasonableness. They include purpose, suitability to the watercourse, economic value of the use, extent of the harm caused, ease of avoiding the harm by adjusting the use, practicality of adjustments to quantity of water, protection of existing uses and investments, and the demands of justice. See also §§ 4.5, 6.12.

Torts is a document written by the American Law Institute to help guide the development of tort law in America. It is frequently cited but is not binding on courts unless adopted by the legislature of a given state. To date, no state has adopted the Restatement in its entirety, with regard to water issues, but the document remains a valuable reference that can aid the reader in understanding the law and policy, since it's influence has tended to be quite significant.*

11.6 LIABILITY OF WATER PROVIDER IF SERVICE FAILS RESULTING IN LOSS BY FIRE

A large number of cases hold that the provider is not liable.† The reasoning seems to be that the defendants were not undertaking the extinguishing of fires by providing water.‡

11.7 LIABILITY FOR CONTAMINATED WATER

This seems to be an easy question, and courts routinely find liability for misfeasance in cases where contaminated water is supplied.§

---

* The relevant sections of the Restatement include: §§ 821B (public nuisance), 826 (trespass), 850 (riparian uses), 850A (reasonable use determination), 855 (reasonableness not affected by classification of rights as riparian or not), and 856 (riparian not liable for harming use by nonriparian with exceptions: grant from another riparian right created by government, and public rights), 858 (reasonable use for groundwater. See § 8.3). The Restatement uses a balancing test in § 826 to determine reasonableness of conduct (i.e., whether gravity of harm is greater than actor's utility). § 850 discusses harm by one riparian against another and uses reasonable use as the determiner. § 850A discusses the factors of reasonable use, including 1) purpose of use; 2) suitability of use to watercourse; 3) economic value of use; 4) social value of use; 5) extent and amount of harm; 6) practicality of avoiding the harm; 7) practicality of adjusting the amount of water used by all users; 8) protection of existing values and investments; and 9) justice of requiring one causing harm to bear the loss. See Getches, *Water Law,* 3rd, at 49. § 855 allows some use of water on nonriparian lands as reasonable. States following the Restatement lead here include Georgia, Kansas, Massachusetts, New Hampshire, New York, North Carolina, Oklahoma, Texas, and Vermont. See *Pyle v. Gilbert,* 245 Ga. 403, 265 S.E.2d 584 (Ga 1980). Getches, at 53.

† See *H.R. Moch Co. v. Rensselaer Water Co.,* 247 N.Y. 160, 159 N.E. 896 (1928); *Reimann v. Monmouth Consol Water Co.,* 9 N.J. 134, 87 A.2d 325 (1952); *Earl E. Roher Transfer & Storage Co. v. Hutchinson Water Co.,* 182 Kan. 546, 322 P.2d 810 (1958); *Cole v. Arizona Edison Co.,* 53 Ariz. 141, 86 P.2d 946 (1939); *Consolidated Biscuit Co. v. Illinois Power Co.,* 303 Ill. App. 80, 24 N.E.2d 582 (1939). Note that as of the 1970s only Florida, Kentucky, North Carolina, and Pennsylvania had recognized a tort duty. See William L. Prosser, *Law of Torts* 4th ed., 626 (1971). For liability, see *Mugge v. Tampa Water Works Co.,* 52 Fla. 371, 42 So. 81 (1906); *Harlan Water Co. v. Carter,* 220 Ky. 493, 295 S.W. 426 (1927); *Tobin v. Frankfort Water Co.,* 158 Ky. 348, 164 S.W.956 (1914); *Fisher v. Greensboro Water Supply Co.,* 128 N.C. 375, 38 S.E. 912 (1901); *Potter v. Carolina Water Co.,* 253 N.C.112, 116 S.E.2d 374 (1960); *Doyle v. South Pittsburgh Water Co.,* 414 Pa. 199, 199 A.2d 875 (1964).

‡ See William L. Prosser, *Law of Torts* 4th ed., 625 (1971).

§ See *Hayes v. Torrington Water Co.,* 88 Conn. 609, 92 A. 406 (1914). For liability for contaminated groundwater by sewage, see 6 POF2d p. 595–645 (1975) with 1989 update. For more on failure to properly maintain sewers, see also 8 POF2d *Municipality's* Failure to Maintain Sewers Properly, 101–44 (1976) with 1989 update.

11.8   LAW OF DRAINAGE* (SEE ALSO 6.8, "DRAINAGE AND UNWANTED RUNOFF")

Traditionally, two competing rules exist in this area of the law. The first is the civil law rule† makes it a potential tort to divert water from its natural flow. Few states are willing to strictly implement this rule and it is usually viewed flexibly as allowing some diversion to protect property when the damage caused is only slight and no viable alternative existed. See reasonable use below.

The second rule is often termed the common enemy doctrine. ‡ It treats surface water as a common enemy of all landowners and allows any necessary diversion to protect one's own property, even if the water ultimately damages another person's property. Most states that follow this doctrine have rules requiring landowners to avoid acting in negligence when diverting water from their lands.

As a general rule of thumb, one owner cannot divert the water from a watercourse so as to cause injury to another.§

Today nearly half the states have moved to a "reasonable use" rule,¶ which looks at the reasonableness of the defendant's conduct in contrast to the circumstances. Equity is a major consideration. Landowners must take reasonable precautions to minimize damage to others' property. With the onset of reasonable use, both the common law and civil law rules seem to meet in the middle. The only real difference is the variance in definitions of "reasonable use," with the common enemy states

---

* See § 11.2 for Rule Regarding Obstruction of the Bed by down stream owner. For more on the law of drainage, see 6 POF 2d 301; 78 Am Jur 2d, Waters §§ 119–139, 167, 169 with 1999 update. For more on surface drainage, see 6 POF2d p 301–44, *Unreasonable Alteration of Surface Drainage* (1975) with 1989 update.

† See *Benton City v. Adrian*, 50 Wash. App. 330, 748 P.2d 679 (1988) (nuisance or civil law rule); *Gross v. Connecticut Mutual Life Ins.*, 361 N.W.2d 259 (S.D. 1985); *Baily v. Floyd*, 416 So.2d 404 (Ala. 1982); *Equitable Life Assur. Soc. V. Tinsley Mill Village*, 249 Ga. 769, 294 S.E.2d 495 (1982). See also 78 Am Jur 2d, Waters § 121 with 1999 update.

‡ See *Argyelan v. Haviland*, 435 N.E.2d 973 (Ind. 1982) (using the common enemy doctrine). While ordinarily these doctrines apply to riparian lands, appropriation states may apply to them as well. See *White v. Pima County*, 161 Ariz. 90, 775 P.2d 1154 (App. 1989) (using common enemy doctrine); see also *Benton City* and *Weaver*. A case exists under the "common enemy doctrine" if 1) surface water is collected in an artificial channel or volume is increased to destructive quantities onto lower lying lands; 2) drainage is done in such a manner as to exceed natural capacity; or 3) landowner discharges surface waters onto lands where they would not naturally drain. See *Hoffman v. Koehler*, 757 S.W.2d 289 (Mo App. 1988). 78 Am Jur 2d, Waters § 120 with 1999 update.

§ See *Johnson v. Whitten*, 384 A.2d 698 (Me. 1978); see also § 6.8. A downstream diverter has been found liable to upstream owners where he negligently obstructed the waterway, causing the water to pool and flood upstream lands during heavy rain. *Bristol v. Rasmussen*, 249 Neb. 854, 547 N.W.2d 120 (1996). See also *Street v. Tackett*, 494 Sp.2d 13 (Ala. 1986)(upper owner may alter flow but not to cause undue burden on lower land). 78 Am Jur 2d, Waters §§ 122 with 1999 update.

¶ *McGlashan v. Spade Rockledge Terrace Condo Dev. Corp.*, 402 N.E.2d 1196 (Ohio 1980) (using reasonable use); *Hall v. Wood*, 443 So. 2d 834 (Miss. 1983) (using "reasonable use" rule); *Tulare Irrigation Dist. v. Lindsay-Strathmore Irrigation Dist.*, 3 Cal. 2d 489, 45 P.2d 972 (Cal. 1935) (reasonable use); *Stratton v. Mt. Hermon Boy's School*, 216 Mass. 83, 103 N.E. 87 (Mass. 1913) (reasonable use allowed on nonriparian land absent injury to other riparians, though in this case plaintiff showed injury); *Gutierrez v. Rio Rancho Estates, Inc.*, 605 P.2d 1154 (N.M. 1980). See also *Weaver v. Bishop*, 206 Cal. App. 3d 1351, 254 Cal. Rptr. 425 (1988) (reasonable use). See also § 11.5 Restatement of Torts (Second). The reasonable use rule is consistent with the views of the restatement. See also 78 Am Jur 2d, Waters §§ 122–23.

naturally tending to favor greater construction than the civil law states. The same rules apply to drainage of unwanted groundwater that apply to surface water.*

11.9 UNREASONABLE INJURY UNDER THE REGULATED RIPARIAN MODEL WATER CODE (SEE 6.16)

Section 2R-1-03 of the code states, "No person using the waters of the State shall cause unreasonable injury to other water uses made pursuant to valid water rights, regardless of whether the injury relates to the quality or the quantity impacts of the activity causing the injury." Unreasonable injury is defined as "an adverse material change in the quantity, quality, or timing of water available for any lawful use caused by any action taken by another person if: (a) the social utility† of the injured use is greater than the social utility of the action causing the injury; or (b) the cost of avoiding or mitigating the injury is materially less than the costs imposed by the injury." A state agency ordinarily will determine reasonableness, rather than a court under the code.

11.10 LIABILITY FOR DAMAGES BY CAPTURED WATER

Special rules often apply to waters that are trapped and captured.‡ The liability for damage caused by such waters to the property of others is higher than for redirected runoff.§ Some jurisdictions even apply strict liability,¶ meaning the owner is liable for all damages caused by virtue of assuming the risk. Many other jurisdictions require proof of negligence.** Some jurisdictions have used the

---

* *Braham v. Fuller*, 728 P.2d 641 (Alaska 1986) (permafrost is a form of groundwater; same rules as surface-water drainage).
† See § 12.2 "Utility Theory."
‡ See 78 Am Jur 2d, Waters §§ 211–12, 216; most jurisdictions, with no statute on the matter, seem to require at least some negligence, and almost all exempt owners from liability for "acts of God." To constitute an act of God, an event must be unprecedented and completely unforeseeable. See *Grossner v. Utah Power & Light*, 612 P.2d 337 (Utah 1980) (defendant must use reasonable care but not liable for acts of God). The doctrine of absolute liability is often termed the *Rylands v. Fletcher* doctrine after the leading case from England, LR 3, HL 330 (1868). For more on acts of God, see 6 Am Jur Proof of Facts 3d §§ 1–57, pp. 319–456 (1989). For liability related to dam failure due to improper design, see 19 POF2d Dam Failure as Result of Negligent Design or Maintenance, 75–161 (1979) with 1989 update. See also 17 POF2d *Failure to Manage Dam or Reservoir to Prevent Flooding,* 133–90 (1978) with 1989 update.
§ See *Barr v. Game, Fish and Parks Comm'n*, 497 P.2d 340 (Colo. App. 1972) (rejecting the act of God defense where a dam was not properly designed to withstand and accommodate foreseeable floods).
¶ *Rylands v. Fletcher*, L.R. 3 H.L. 330 (1868); *Cities Service Co. v. State*, 312 So.2d 799 (Fla. App. 1975). According to Professor Getches, *Water Law in a Nutshell*, 3rd ed. (1997), the majority of states follow this line of thought, that storage of water behind a dam is an ultrahazardous activity and a dam owner may have absolute liability for damages caused by release of water, but see footnote ** regarding requirement of negligence. Some jurisdictions impose strict liability by statute, i.e., Colorado (see Col. Rev. Stat. § 37-87-104) and Nebraska (see Neb. Rev. Stat. § 46-253 (2)). Some statutes impose liability but not in the absence of negligence, ordinarily meaning that "acts of God" and third parties are not enough to create liability, i.e., New Hampshire (see N.H. Rev. Stat. Ann. Title L §§ 482:11 & 39) and Wyoming (see Wyo. Stat. Ann. §§ 41-3-305, 307-317). See 78 Am Jur 2d, Waters § 212 with 1999 update.
** I.e., Arizona, Arkansas, California, Connecticut, Georgia, Idaho, Iowa, Massachusetts, Michigan, Minnesota, Montana, Nebraska, Nevada, New Hampshire, New York, North Carolina, Ohio, Oregon, Texas, Washington, and Wyoming. See 78 Am Jur 2d, Waters § 211 fn 33 with 1999 update.

doctrine of *res ipsa loquitur*, meaning that negligence is implied by the fact that an incident occurred.* This shifts the burden of proof to the defendant to show some other cause than negligence.

11.11 PUBLIC UTILITIES: PROPRIETARY AND GOVERNMENTAL FUNCTIONS

The operation of public utilities has been found to be a proprietary interest in many cases.† Municipalities are usually held under no obligation to provide sewers‡ and drains.§ They are not ordinarily liable for damage caused by the failure to do so but once they choose to provide such services they incur liability for unreasonable

---

* 78 Am Jur 2d, Waters § 220 with 1999 update. *See Weaver Mercantile Co. v. Thurmond,* 68 W. Va. 530, 70 SE 126 (bursting of tank storing water to injury of neighbors' land raises presumption of negligence). *Res Ipsa* is not appropriate when there is an intervening act by a third party, causing the harm. See *Hamilton v. State,* 42 Or. App. 821, 601 P.2d 882 (1979).

† This includes waterworks. See Osborne M. Reynolds, *Local Government Law,* Cp. 30, § 193 (1982). *City of Birmingham v. Lake,* 243 Ala. 367, 10 So.2d 24 (1942); *Richmond v. City of Norwich,* 96 Conn. 582, 115 A. 11 (1921); *City of Elberton v. J.C. Pool Realty Co.,* 111 Ga. App. 765, 143 S.E.2d 407 (1965); *McGinley v. City of Cherryvale,* 141 Kan. 155, 40 P.2d 377 (1935); *Harvard Furniture Co. v. City of Cambridge,* 320 Mass. 227, 68 N.E.2d 684 (1946); *Carlisi v. City of Marysville,* 373 Mich. 198, 128 N.W.2d 477 (1964); *Henderson v. Kansas City,* 177 Mo. 477, 76 S.W. 1045 (1903); *Roberts Realty Corp. v. City of Great Falls,* 160 Mont. 144, 500 P.2d 956 (1972); *Faw v. Town of North Wilkesboro,* 253 N.C. 406, 117 S.E.2d 14 (1960); *Helz v. City of Pittsburgh,* 387 Pa. 169, 127 A.2d 89 (1956); *City of Waco v. Busby,* 396 S.W.2d 469 (Tex. Civ. App. 1965) refused n.r.e. See *City of Boston,* 334 Mass. 401, 135 N.E.2d 658 (1956) (action for flooding of property owner's basement when city water main broke; question of fact whether city was negligent); *Oklahoma City v. Moore,* 491 P.2d 273 (Okl. 1971) (plaintiff fell into water-meter box installed and maintained by city; held proprietary function). Cf. *Foust v. City of Durham,* 239 N.C. 306, 79 S.E.2d 519 (1954) (water main burst, allegedly due to city's negligence; cause of action held stated). According to Professor Reynolds, most places have found cities potentially liable even when water is used for largely governmental functions, such as fire fighting and street cleaning. See *Cole Drug Co. v. City of Boston,* 326 Mass. 199, 93 N.E.2d 556 (1950); *Koch Brothers Bag Co. v. Kansas City,* 315 S.W.2d 743 (Mo. 1958); *Bowling v. City of Oxford,* 267 N.C. 552, 148 S.E.2d 624 (1966); *Russell v. City of Grandview,* 39 Wn.2d 551, 236 P.2d 1061 (1951). But see *Cross v. Kansas City,* 230 Kan. 545, 638 P.2d 933 (1982) (furnishing water for fire department is governmental function). A Utah case held that a water storage tank was not uniquely governmental and not immune from liability. See *Bennett v. Bow Valley Dev. Corp.,* 797 P.2d 419 (Utah 1990).

‡ For more on sewers, see 8 POF2d *Municipality*'s Failure to Maintain Sewers Properly, 101–44 (1976) with 1989 update.

§ See *Martinez v. Cook,* 56 N.M. 343, 244 P.2d 134 (1952); *McCutchen v. Village of Peekskill,* 167 Misc. 460, 3 N.Y.S.2d 277 (1938); *Boone v. City of Akron,* 69 Ohio App. 95, 43 N.E.2d 315 (1942); *Strach v. City of Scranton,* 353 Pa. 10, 44 A.2d 258 (1945). For no liability for failure to provide, see *Daniels v. City of Denver,* 2 Colo. 669 (1875); *Beals v. Inhabitants of Brookline,* 174 Mass. 1, 54 N.E. 339 (1899). For no obligation to provide drainage for public streets, see *Randle v. City of Rome,* 23 Misc.2d 436, 195 N.Y.S.2d 373 (1960). Local municipalities may be liable to property owners when they divert water onto private property. See *City of McComb v. Rodgers,* 246 So.2d 913 (Miss. 1971). Logically, this could be considered a taking of private land for a public function and should require compensation for the extent of the taking. *LaForm v. Bethlehem Township,* 346 Pa. Super. 512, 499 A.2d 1373 (1985) (City not liable for death of motorist who drowned when car stalled on surface of floodwaters); *Wilber Dev. Corp. v Les Rowland Constr. Inc.,* 83 Wash.2d 871, 523 P.2d 186 (1974) (city not liable for changing flow onto private land if damage results from opening of streets and other ordinary municipal growth); *Patterson v. City of Bellevue,* 37 Wash. App. 535. 681 P.2d 266 (1984) (city not liable for increasing rate of flow if no new water added to drainway and natural drainage not disturbed). See § 3.10.

activity, such as negligent planning* or construction.† Most jurisdictions also hold the municipality liable for negligent maintenance.‡ Utility functions may be found to be largely governmental, meaning liability is unlikely to attach, or proprietary, which holds a higher liability for the utility company.

11.12 INTERFERENCE WITH WATER RIGHT AND TRESPASS§ (SEE 11.15)

Any wrongful interference with the rights of others to water is considered a trespass¶ or nuisance, within tort law. The usual remedy is damages,** though

---

* *Bowman v. Town of Chenango*, 227 N.Y. 459, 125 N.E. 809 (1920); *McCormick v. Town of Thermopolis*, 478 P.2d 67 (Wyo. 1970). For cases holding municipality not liable for overflow due to extraordinary rainfall. See *Savage v. Town of Lander*, 77 Wyo. 157, 309 P.2d 152 (1957) (may be liable only if was so manifestly negligent that liability found as matter of law). But see *Carlo v. City of Pittsfield*, 339 Mass. 624, 161 N.E.2d 757 (1959) (finding no liability for inadequate planning but could be liable for negligent maintenance).

† *Durante v. City of Oakland*, 19 Cal. App.2d 543, 65 P.2d 1326 (Dist. Ct. 1937); *Galluzzi v. City of Beverly*, 309 Mass. 135, 34 N.E.2d 492 (1941) (negligence in connecting property with sewer); *Bean v. City of Moberly*, 350 Mo. 975, 169 S.W.2d 393 (1943); *Stoddard v. Village of Saratoga Springs*, 127 N.Y. 261, 27 N.E. 1030 (1891); *Accurate Die Casting Co. v. City of Cleveland*, 68 O.L.A. 230, 113 N.E.2d 401 (1953); *Rikansrud v. City of Canton*, 79 S.D. 592, 116 N.W.2d 234 (1962); *Hatten v. Mason Realty Co.*, 148 W.Va. 380, 135 S.E.2d 236 (1964). A few cases have held the construction of sewers governmental and found no liability. See *Register v. Burton Elliott, Inc.*, 229 A.2d 488 (Del. 1967); *Elledge v. City of Des Moines*, 259 Iowa 284, 144 N.W.2d 283 (1966). Texas has treated storm sewers as proprietary but sanitary sewers as governmental. See *Chambers v. City of Dayton*, 447 S.W.2d 425 (Tex. Civ. App. 1969).

‡ See *Mulloy v. Sharp Park Sanitary District*, 164 Cal.App.2d 438, 330 P.2d 441 (Dist. Ct. 1958); *Kellom v. City of Ecorse*, 329 Mich. 303, 45 N.W.2d 293 (1951); *Allen v. Town of Hampton*, 107 N.H. 377, 222 A.2d 833 (1966); *Ulibarri v. Village of Las Lunas*, 79 N.M. 421, 444 P.2d 606 (1968); *Barker v. City of Santa Fe*, 47 N.M. 85, 136 P.2d 480 (1943); *Milner Hotels v. City of Raleigh*, 268 N.C. 535, 151 S.E.2d 35 (1966); *Stone v. City of Ashland*, 30 O.L.A. 367, 32 N.E.2d 560 (1939); *Oklahoma City v. Romano*, 433 P.2d 924 (Okl. 1967); *Hayes v. City of Vancouver*, 61 Wash. 536, 112 P. 498 (1911). *The Lobster Pot v. City of Lowell*, 333 Mass. 31, 127 N.E.2d 659 (1955); *Dize Awning & Tent Co. v. City of Winston-Salem*, 271 N.C. 715, 157 S.E.2d 577 (1967); *Ball v. Village of Reynoldsburg*, 86 O.L.A. 293, 176 N.E.2d 739 (1960). But see *Kershner v. Town of Walden*, 144 Colo. 67, 355 P.2d 77 (1960) (finding duty of only reasonable care).

§ See 78 Am Jur 2d, Waters §§ 164–69 (interference with source of supply), 204 (interference with dams or facilities of water), 239 (interference in possession of water) with 1999 update.

¶ *Fall River Valley Irr. Dist. v. Mt. Shasta Power Corp.*, 202 Cal. 56, 259 P. 444, 56 A.L.R. 264 (1927); *McNabb v. Houser*, 171 Ga. 744, 156 S.E. 595, 74 A.L.R. 1122 (1931). See also *Collens v. New Canaan Water Co.*, 155 Conn. 477, 234 A.2d 825 (1967) (interference is a tort). Note that a public utility may still be liable for damages even if its actions were necessary to carry out public duty. *Taylor v. Indiana & Michigan Elec. Co.*, 184 Mich. 578, 151 N.W. 739 (1915). But see *Finley v. Teeter Stone, Inc.*, 251 Md. 428, 248 A.2d 106 (1968) (holding that interference with support by water Is not the same as soil and other elements of land); *Taft v. Bridgeton Worsted Co.*, 237 Mass. 385, 130 N.E. 48, 13 A.L.R. 928 (1921) (interference with ice field).

** *Price v. High Shoals Mfg. Co.*, 132 Ga. 246, 64 S.E. 87 (1909); *Wagner v. Purity Water Co.*, 241 Pa. 328, 88 A. 484 (1913); *Pickens v. Coal River Boom & Timber Co.*, 58 W. Va. 11, 50 S.E. 872 (1905) (economic damages established with reasonable certainty); *Maynard v. Nemaha Valley Drainage Dist. No. 2*, 94 Neb. 610, 143 N.W. 927 (1913) (speculative damages not recoverable; *Gerlach Livestock Co. v. U.S.*, 111 Ct. Cl. 1, 76 F. Supp. 87 (1948), cert. granted, 335 U.S. 883, 69 S. Ct. 234, 93 L. Ed. 422 (1948) and cert. granted, 335 U.S. 883, 69 S. Ct. 234, 93 L. Ed. 422 (1948) and cert. granted, 335 U.S. 883, 69 S. Ct. 235, 93 L. Ed. 422 (1948) and judgment aff'd, 339 U.S. 725, 70 S. Ct. 955, 94 L. Ed. 1231, 20 A.L.R.2d 633 (1950) (allowing recovery of interest on damages). Note some jurisdictions allow interest on damages under some circumstances; *Foster v. City of Augusta*, 174 Kan. 324, 256 P.2d 121 (1953); *McKain v. Platte Val. Public Power & Irrigation Dist.*, 151 Neb. 497, 37 N.W.2d 923 (1949) (no interest recovery allowed).

an injunction* may be possible. Recently, the Birmingham Water Works began a new push to prosecute water thieves who have hooked up illegally to the city's water system or tampered with meters. Birmingham estimates that 26 percent of its water is lost to theft.† The national average is 15 percent.‡ This type of theft from the city water system will almost certainly qualify for criminal penalties§ beyond the damages available through tort law. Whether charges are filed or not may depend on a balance between the cost and benefits of prosecuting these cases criminally, the dollar value of the theft, and the sufficiency of other alternatives to criminal prosecution.

11.13   BREACH OF CONTRACT¶

Breach of contract is a separate set of law and rules from tort law.** Contract principles are generally the same for water rights as for other areas of contract law. The usual remedy for breach of contract is damages.†† Where no other remedy is sufficient, a court may require specific performance.

11.14   STANDING‡‡

The requirements for Constitutional Standing§§ are 1) injury in fact,¶¶ 2) causation by the defendant's alleged action, and 3) that the injury can be redressed by a favorable outcome. Injury in fact requires a live case or controversy.*** If the case is speculative††† in nature about some presumed future harm, or becomes moot‡‡‡ at some point, then it fails to meet this requirement.§§§ A handful of states issue advisory opinions,

---

* *Atchison v. Peterson*, 87 U.S. 507, 22 L. Ed. 414 (1874); *Walton v. Town of New Hartford*, 223 Conn. 155, 612 A.2d 1153 (1992); *Robertson v. Arnold*, 182 Ga. 664, 186 S.E. 806, 106 A.L.R. 681 (1936); *Woody v. Durham*, 267 S.W.2d 219 (Tex. Civ. App. Fort Worth 1954), writ refused.

† Available online at http://www.uswaternews.com/archives/arcrights/7birmwate9.html, September 15, 2007.

‡ *Id.*

§ See R.C.W.A. 90.03.400–20, *Crimes against water code,* (400) unauthorized use; (410) interference with works, wrongful use of water, property destruction; (420) obstruction of right-of-way.

¶ 78 Am Jur 2d, Waters §§ 189–94 with 1999 update (well drillers' breach and liability).

** *Taylor v. Holter*, 1 Mont. 688, 1872 WL 7229 (1872) (extent of damages available).

†† See § 11.15.

‡‡ The reader should be advised that federal standing is quite complicated, and as this is not a book on Constitutional law of standing doctrines, coverage here is very limited and basic.

§§ Many environmental laws authorize citizen standing to sue for compliance. This is ordinarily allowed where citizens have at least an aesthetic interest at stake. *Lujan v. Defenders of Wildlife*, 504 U.S. 555, 112 S.Ct. 2130, 119 L.Ed.2d 351 (1992).

¶¶ Injury in fact is usually not a high hurdle. It need not be economic. *United States v. S.C.R.A.P.*, 412 U.S. 669, 93 S.Ct. 2405, 37 L.Ed.2d 254 (1973). Note that this case represented the extreme in finding standing, and that there is some indication that modern standing doctrine no longer stretches this far. Regardless, standing is not a high hurdle. See *Warth v. Seldin*, 422 U.S. 490, 95 S.Ct. 2197, 45 L.Ed.2d 343 (1975) (no standing).

*** United States Constitution, Article III, Section 2.

††† Cases must be ripe for decision. This means that they must not be brought too early. *United Public Workers v. Mitchell*, 330 U.S. 75, 67 S.Ct. 556, 91 L.Ed. 754 (1947).

‡‡‡ *Ex parte Baez*, 177 U.S. 378, 20 S.Ct. 673, 44 L.Ed. 813 (1900).

§§§ Certain exceptions to this doctrine do exist where cases are not likely to be heard at all because of the short life of the matter. Some matters are considered by the Court to be of enough importance that they choose to hear the case anyway. *Roe v. Wade*, 410 U.S. 113, 93 S.Ct. 705, 35 L.Ed.2d 147 (1973).

but the United States Supreme Court does not.* Many federal agencies, acting under the Administrative Procedure Act,† must abide by the zone of interest test, which asks whether the plaintiff's injury within the zone of interest is intended to be protected by the law. Third-party standing is ordinarily not allowed.‡

11.15 REMEDIES§ (SEE 11.12)

The most common remedies for injury to one's rights are damages¶ and injunction.** Damages ordinarily consist of actual proven costs.†† One may recover for lost opportunities but must show with reasonable certainty‡‡ that the losses were real and not speculative.§§ In certain circumstances special remedies may apply, such as specific performance. Courts may issue injunctions¶¶ against harmful activity to prevent further harm. Individuals are not responsible for remote damages that are not proximately caused by the defendant's action.*** In criminal cases,††† the remedy may

---

* *Muskrat v. United States*, 219 U.S. 346, 31 S.Ct. 250, 55 L.Ed. 246 (1911).
† 5 U.S.C.A. 551–59.
‡ *McNabb v. Houser*, 171 Ga. 744, 156 S.E. 595, 74 A.L.R. 1122 (1931); *Humphreys-Mexia Co. v. Arseneaux*, 116 Tex. 603, 297 S.W. 225, 53 A.L.R. 1147 (1927) (nonriparian may not ordinarily sue for damages to riparian rights). Note that the most likely exceptions would be organizations representing the interests of their members, or those acting as legal guardians or representatives for individuals who lack capacity to represent themselves. See *Combs v. Farmers Highline Canal & Reservoir Co.*, 88 P. 396 (Colo. 1907) (finding that the corporation is the representative of its members in a dispute with members of another company).
§ For discussion of remedies in water law, see 78 Am Jur 2d, Waters §§ 32–38 (riparian), 58 (lakes and ponds), 108–16 (navigation), 140–45 (surface water), 154 (underground stream), 170–75 (percolating waters), 181 (springs), 187 (injury to wells), 219 (artificial water), 256 (interference with water rights), 297–303 (stream diversion), 365–72 (injury to property), 374 (general nuisances).
¶ *Snyder v. Callaghan*, 168 W. Va. 265, 284 S.E.2d 241 (1981).
** *Wisniewski v. Gemmill*, 123 N.H. 701, 465 A.2d 875 (1983); *Snyder v. Callaghan*, 168 W. Va. 265, 284 S.E.2d 241 (1981); *Harris v. Brooks*, 225 Ark. 436, 283 S.W.2d 129, 54 A.L.R.2d 1440 (1955). See also *Stock v. Jefferson Tp.*, 114 Mich. 357, 72 N.W. 132 (1897) (injunction not refused to one who acted promptly regardless of minor impact compared to injury to diverting party). But see *Ulbricht v. Eufaula Water Co.*, 86 Ala. 587, 6 So. 78 (1889) (no injunction when plaintiff made no use of the water). A court of equity will not enjoin an action to divert water by a riparian unless the one bringing suit suffered an injury by the act. *Humphreys-Mexia Co. v. Arseneaux*, 116 Tex. 603, 297 S.W. 225, 53 A.L.R. 1147 (1927); *State of New York v. State of Illinois*, 274 U.S. 488, 47 S. Ct. 661, 71 L. Ed. 1164 (1927) (where no use of water was contemplated by plaintiff, injunction was denied).
†† The usual recovery for a permanent taking would be the difference in the value of the property before and after the taking. *Seneca Consol. Gold Mines Co. v. Great Western Power Co. of California*, 209 Cal. 206, 287 P. 93, 70 A.L.R. 210 (1930).
‡‡ *Pickens v. Coal River Boom & Timber Co.*, 58 W. Va. 11, 50 S.E. 872 (1905) (economic damages established with reasonable certainty); *Ulbricht v. Eufaula Water Co.*, 86 Ala. 587, 6 So. 78 (1889); *Clark v. Pennsylvania R. Co.*, 145 Pa. 438, 22 A. 989 (1891) (recovery for no more than nominal damages unless special damages proven); *Auger & Simon Silk Dyeing Co. v. East Jersey Water Co.*, 88 N.J.L. 273, 96 A. 60 (N.J. Ct. Err. & App. 1915) (recovery may vary based on type of use plaintiff deprived of).
§§ *Parke v. Bell*, 97 Idaho 67, 539 P.2d 995 (1975).
¶¶ See 78 Am Jur 2d, Waters §§ 33, 298, 313.
*** *Woodstock Iron Works v. Stockdale*, 142 Ala. 550, 39 So. 335 (1905) (consequence must be foreseeable and proximately caused by defendant to be recoverable).
††† I.e., R.C.W.A. §§ 90.03.400, 90.03.410, 90.03.420 (Washington crimes against water code). Such provisions provide for remedy against willful or negligent waste to detriment of another, use without legal right, storage or diversion of water without permit, willful interference with dam, dike, etc. of another, interference with measuring devices such as meters, and interference with lawful right-of-way.

include confinement or fines.* Many environmental statutes, such as CERCLA† or the Clean Water Act‡ include provisions for fines, when individuals or groups are not in compliance with the provisions of the act. Some remedies also include loss of permits or licenses. Plaintiffs may be barred from recovery, for unreasonable delay in bringing suit, by what is known as laches.§ Many actions also have statutes of limitations,¶ which state the time in which an action must be brought if the potential plaintiff wishes to have his or her grievance heard.

The two basic measures of damage in injury cases to land are diminution and restoration. The diminution method has been widely used where water is permanently ruined or cut off from the land.** Where the damage or loss is reversible, the restoration cost is often used.††

---

* For example of civil penalties, see R.C.W.A. § 90.03.600.
† 42 U.S.C.A. § 9601–15, 9631–57.
‡ 33 U.S.C.A. § 1311.
§ *State of Washington v. State of Oregon*, 297 U.S. 517, 56 S. Ct. 540, 80 L. Ed. 837 (1936).
¶ The statute runs from time of injury. If the injury is temporary, then the statute runs with each successive injury. *Wong Nin v. City and County of Honolulu,* 33 Haw. 379, 1935 WL 3373 (1935); *Wagner v. Purity Water Co.,* 241 Pa. 328, 88 A. 484 (1913).
** See Dan B. Dobbs, *Remedies: Damages, Equity, Restitution* at 312 (1973). *Mid-Continent Pipe Line Co. v. Eberwein,* 333 P.2d 561 (Okl. 1958); *General Crude Oil Co. v. Aiken,* 162 Tex. 104, 344 S.W.2d 668 (1961).
†† *Ward v. Chevallier Ranch Co.,* 138 Mont. 144, 354 P.2d 1031 (1960).

12 Social Theory*

12.1 GENERALLY

It is important that the student of water rights understand not only what the two major practices in water rights are (meaning prior appropriation and riparian), but also why these practices exist as they do and how they came into being. It is equally important to realize that as times change so, too, do the needs of society. Those who work in the realm of policy making or implementation must understand the policy as it exists and the reasons for that policy, and be able to anticipate public reaction and needs in the future. Sometimes it is necessary to change course if we are to navigate clear of the ever-increasing number of new obstacles that may come our way.

One of the important means of resolving any problem is to accurately identify the root of the problem and not merely some effect manifesting itself on the surface. The study of social theory can help organize and identify potential pros and cons of issues and problems. For this reason, I have included a very brief section here to help the reader organize his or her ideas and better navigate these issues. The discussion of social theory in this volume represents only a fraction of the applicable information on the subject and is not intended as a complete guide to social thinking.

12.2 UTILITY THEORY

The works of individuals like Jeremy Bentham, James Mill, and John Stuart Mill were the basis for the creation of utilitarian theory. The basic premise of this line of thought is that individuals have rational preferences based on self interest. Each potential choice is assigned a fictional value, known as utility. Higher values are assigned for those ends most desired by the individual and lower values for those least desired. We can predict that a rational person will pursue those ends with the highest overall utility. Utility theory is the basis for much of modern social theory, including game theory, rational choice, and much of the study of economics.

12.3 RATIONAL CHOICE AND GAME THEORY

This is a way of analyzing human behavior on the basis of a set of individual beliefs and goals. The premise is that behavior is goal oriented. Individual actors are assumed to have a set of preferred outcomes and interests. These preferences are evaluated in light of alternative choices. The actor will then weigh the pros and cons of each choice and choose the path most likely to get them closest to their preferred outcome.

---

* See also §§ 2.1, 8.11, 8.12.

12.4   NASH EQUILIBRIUM

John Nash first proposed the Nash Equilibrium in 1950 as a solution of sorts to game theory. Game theory involves two or more players and predicts that rational individuals within the game will have an incentive not to change their choice when acting unilaterally. This theory is best illustrated with an example known as prisoner's dilemma, where two similarly situated individuals accused of a crime must decide whether to turn state's evidence and receive a lighter sentence or hold out in the hope that the other will also hold out, with possible outcomes as follows. If player 1 holds out and player 2 turns state's evidence, then player 2 gets off with a one-year sentence while player 1 gets ten years. If player 1 turns state's evidence while player 2 holds out, then player 1 gets one year and player 2 gets ten. If both turn state's evidence, then they each get five years, and if neither turn state's evidence, they each get two years. Regardless of what player 2 decides to do, player 1 has an incentive to turn state's evidence. Player 1 will always be better off with that choice. The Nash Equilibrium predicts that both self-interested players will choose to turn state's evidence and receive five years.

Player 1

	Cooperate	Not Cooperate	
A		**B**	**Player 2**
	1-10	2-2	**Not Cooperate**
C		**D**	
	5-5	10-1	**Cooperate**

Square C in the above chart represents the Nash Equilibrium, or predicted outcome based on each player's individual incentive to cooperate. Some games are more complex and may involve multiple Nash Equilibrium in the same game.

In order to counter the problems of prisoner's dilemma, players are encouraged to communicate, and various incentives and disincentives have been proposed over the years.

12.5   TIT FOR TAT

One of the more successful strategies to encourage cooperation is known as tit for tat. According to this strategy a player is encouraged to cooperate until the other player ceases cooperation. At that point the strategy is to retaliate. A variation of this strategy is to retaliate once, and then begin cooperation again. This is known as an iterated tit for tat strategy. It is intended to escape the escalating retaliation of tit for tat. While this strategy is not ideal for a onetime, one-on-one competition, it performs remarkably well in repeated games with many players.

12.6 COST-BENEFIT ANALYSIS

Cost-benefit analysis is used to weigh the costs of proposed actions with the benefits in order to weigh economic advantages and disadvantages. Cost-benefit analysis is closely related to the above theories, especially utility theory. It is an attempt to weigh the social utility of a proposed action.

12.7 SOCIAL THEORY AND WATER RIGHTS (FOR A DISCUSSION OF TRAGEDY OF THE COMMONS AND OTHER COLLECTIVE ACTION ISSUES RELATING TO GROUNDWATER, SEE 8.12 "COLLECTIVE ACTION THEORY")

Like all scarce resources, demand for water often exceeds supply. This encourages competition among interested parties and often results in waste. The interests of all individual actors, if pursued, usually lead to a less desirable outcome, collectively. Just as in the illustration above for prisoner's dilemma, the best overall results are seldom achieved when all individuals pursue their own interests unilaterally.

Water law exists, in no small part, to alleviate the collective action problems that arise with limited resources and high demand. Were water not regulated, its use would all too often be wasteful and diversions would exist as far upstream as individuals could find to divert the water. The law helps prevent physical conflicts and provides a fairer division of resources throughout the nation.

In an area of law where competing interested parties often distrust each other and routinely compete for scare supplies, waste is inevitable. Absent some outside incentive to cooperate and some mode of control capable of keeping the competing parties honest, all participants are likely to use as much water as they can, as rapidly as possible. The incentive is to use it before someone else can. The participant who attempts to play fair stands to lose his or her own water supply to a more aggressive party. To resolve this problem, the federal government and various states, either through their legislatures or their courts, have created laws and rules to govern and manage water resources. These rules of the game are intended to counter the effects of Nash Equilibrium in a typical prisoner's dilemma scenario. As it is, each year sees numerous *State v. State* lawsuits over water resources.* Some have been ongoing for generations and still have no end in sight. Many of the suits involve two very different, self-serving readings of the same rule, compact clause, or law.† It is difficult to imagine that all of these debates are raised in good faith, and from the language of the arguments and cases it seems apparent that the various state actors are self-interested and often even underhanded in their dealings. The greatest problems seem to occur during times of shortage, when it must be decided who stands to lose what amount of their allotted supply.

---

* For a review of interstate competition for water resources, including cases and rules, see Chapter 2.
† See numerous examples of self-serving interpretation in the footnotes to Chapter 2.

Unregulated access to water can lead to problems not unlike those, which exist elsewhere in the world. Unregulated dumping and other uses have historically been associated with great outbreaks of disease, serious illness, and even death. Many of the world's greatest plagues have been furthered by polluted waters, often containing sewage, chemicals, and other unnatural contents. One of the major problems today has to do with agricultural runoff and return flows. These often contain fertilizers and other substances, which deplete the oxygen and kill off fish. With so many powerful lobbies, it is often difficult to take immediate action. Progress tends to be slow and often accidental. Everyone seems to want change but no one wants to change.

12.8   DISINCENTIVES, OVERAPPROPRIATION, AND INEFFICIENCY

There is ordinarily a lack of incentive for appropriators to conserve water, since any water saved is generally placed back into the pool for the next appropriator to use.* While technology is available to conserve a considerable amount of water, the fact that it would be lost to the appropriator, who invests in such conservation, often serves as a disincentive to conserve. This is a classic example of a collective action problem that can only be solved by understanding human motivation, instinct, and incentives. In order to break the cycle, states must offer incentives to users,† such as allowing them to sell back at least part of what they conserve, or authorizing them to retain the full appropriation for some reasonable period of time.‡

Lax enforcement and oversight often magnify the problems with the appropriation system. In some cases unused water rights, which should have been forfeit years earlier are still on the books.§ Speculators are all too often allowed to hold rights for fictitious projects that never occur. This drives up the cost of water and limits the available supply for beneficial use.

---

* See *Southeastern Colorado Water Conservancy District v. Shelton Farms, Inc.*, 187 Colo. 181, 529 P.2d 1321(1974) (holding that salvaged water returns to the available supply for the next senior appropriator); codified by Colo. Rev. Stat. §§ 37-90-103, 37-92-103(9); *Salt River Valley Water Users'* Association v. Kovacovich, 3 Ariz. App. 28, 411 P.2d 201 (1966) (holding that there is no right to salvaged water by user who prevents waste). See also *Basin Elec. Power Coop. v. State Bd. of Control*, 578 P.2d 557 (Wyo. 1978) (statute prevents sale of more than consumptive use, wasted water cannot be sold).

† Some states allow juniors to pay the expense of seniors to upgrade their systems and improve efficiency so that the junior's right might be met. See *Morrison v. Higbee*, 204 Mont. 501, 668 P.2d 1029 (1983); *Alamosa-La Jara Water Users Protect. Ass'n v. Gould*, 674 P.2d 914, 935 (Colo. 1983); *Big Cottonwood Tanner Ditch Co. v. Shurtliff*, 56 Utah 196, 189 P. 587 (1919); *Tonkin v. Winzell*, 27 Nev. 88, 73 P. 593 (1903). See also Mont. Code Ann. §§ 85-2-103(14), 1-402(2)(e), 1-419; Cal Water Code §§ 109, 1011.

‡ See § 14.8. See also *Basinger v. Taylor*, 36 Idaho 591, 211 P. 1085 (1922); *Reno v. Richards*, 32 Idaho 1, 178 P. 81 (1918) (holding that one who salvages water by more efficient means has right to it); *East Bench Irr. Co. v. Deseret Irr. Co.*, 2 Utah 2d 170, 271 P.2d 449 (1954) (holding that salvager has right to water so long as no junior is harmed).

§ See Joseph L. Sax, Robert H. Abrams, Barton H. Thompson Jr., John D. Leshy, *Legal Control of Water Resources: Cases and Materials,* 185–86 (3rd ed., 2000), citing Jackson Battle, *Paper Clouds Over the Waters: Shelf Filings and Hyperextended Permits in Wyoming,* 22 Land & Water L. Rev. 673, 680 (1987).

13 Louisiana and Hawaii Water Law*

13.1 LOUISIANA CIVIL CODE

Louisiana is listed as a riparian state but its system of water law is based on the old French and Spanish law and warrants some separate mention. Louisiana has a civil code system for settling disputes over water rights. Courts are allowed to refer to common law when applicable and are not bound by stare decisis (case law precedent).

13.2 ARTICLE 657†

This article states that an owner of an estate bordering running water may use the water to water his or her estate or for other purposes.

13.3 ARTICLE 658

This article allows an owner of an estate over which water runs to make use of it while it runs over his or her lands. The owner cannot stop the water flow or change its course and is bound to return it to its original channel where it leaves his estate.

13.4 LOUISIANA GROUNDWATER (SEE 8.14 "ABSOLUTE OWNERSHIP DOCTRINE")

Article 490 states, "Unless otherwise provided by law, the ownership of a tract of land carries with it the ownership of everything that is directly above or under it." Regulation of groundwater in Louisiana is a relatively recent thing.‡

13.5 TRADITIONAL HAWAIIAN WATER LAW

Anciently, water rights in Hawaii came from grants given by the chiefs. Typically, the grants in land would be wedge-shaped beginning at the top of the mountain and

---

* See Chapter 6 on riparian uses
† This article is almost verbatim from Napoleon's French Civil Code. It is not clear if the provision expresses a natural flow or reasonable use rule. See *Long v. Louisiana Creosoting Co.*, 137 La. 861, 69 So. 281 (La. 1915) (applying the reasonable use rule to Article 657).
‡ See *Adams v. Grigsby*, 152 So.2d 619 (La. App. 2 Cir. 1963) (ruling that water is not owned until reduced to possession by pumping). In response to this ruling, the legislature authorized the Department of Public Works to regulate wells that produce more than 50,000 gallons per day. This regulation was to alleviate the concern that the ruling might result in waste.

growing wider as the land came nearer to the sea. These grants generally followed watershed lines. Peasants worked the lands in exchange for a part of their crops that went to the chief. Water rights were considered appurtenant to the land, in sufficient amounts to cultivate food and meet the basic needs of the people living there. Water rights eventually became alienable with the land after westerners arrived on the islands.

13.6   EARLY TRANSFERS OF WATER RIGHTS

Early cases dealing with the sugar plantations permitted transfers of appurtenant water rights but only after a showing that no harm would be done to other rights holders.*

13.7   HAWAII PRESCRIPTIVE RIGHTS (SEE 3.7 AND 7.10)

Appurtenant rights were once available by prescription in Hawaii.† Prescriptive rights are not allowed against state-owned land, since adverse possession does not run against the government. In 1973 the courts held that appurtenant rights could not be acquired by prescription.‡ This is because landowners do not own the rights, since water is held in trust by the state for the benefit of the whole and prescription against the state is not possible.

13.8   APPLICATION OF RIPARIAN PRINCIPLES

In 1973, the Hawaii Supreme Court applied riparian ideas to forbid the sale or transfer of surplus water rights to any land except the riparian parcel. The court stated that appurtenant rights by their nature could not be used on other lands.§ It also accepted the "natural flow doctrine."¶ The court later changed its view some, holding that reasonable use applies to riparian rights.**

13.9   NEW LAW AND HAWAII PERMIT SYSTEM

In 1987, a new water law was adopted for Hawaii. The law establishes a comprehensive water plan and designates water-management areas. It requires all existing and new rights to acquire a permit. Water use is no longer restricted to riparian or appurtenant lands.

---

* *Kahookiekie v. Keanini*, 8 Haw. 310 (1891); Peck v. Baily, 8 Haw. 658 (1867) (scrutinizing change in place of use); *Carter v. Territory* 24 Haw. 47 (1917) (scrutinizing change in point of diversion).
† *Lonoaea v. Wailuku Sugar Co.*, 9 Haw. 651 (1895).
‡ *McBryde Sugar Co. v. Robinson*, 54 Haw. 174, 504 P.2d 1330 (1973).
§ *McBryde Sugar Co. v. Robinson*, 54 Haw. 174, 504 P.2d 1330 (1973). The *McBryde* court stated that riparian law in Hawaii is rooted in Hawaiian custom, giving the state greater authority to allocate, since it is the custodian of the rights. This is a significant variance from other riparian states. The state is obligated to make a fair distribution of waters among all who put it to productive use. The state constitution was revised around this time to recognize the state's obligation in trust. See § 13.9 for 1987 statutory change in the law.
¶ See §§ 6.2, 6.18, 6.19.
** *Reppun v. Board of Water Supply*, 65 Haw. 531, 656 P.2d 57 (1982).

14 The Future of Water Resources in the United States

14.1 INTRODUCTION

States are said to have a duty to preserve water resources for future use and ensure its continued availability.* Demand for fresh water is continually increasing, and supplies are dwindling.† As we move into the future, we are likely to see a shift toward new sources. Some of these sources are more promising than others.‡ Even so, we have a limited supply of suitable fresh water, and it is not evenly dispersed among the several states. This creates problems for planners and likely means that we will eventually be forced to look to the oceans in order to meet future needs.§ Unfortunately, the saline content of the ocean's water makes it necessary to treat the water first in order to make it suitable for use. This can be expensive and time consuming.

14.2 CLOUD SEEDING (SEE 7.18, "ARTIFICIALLY INDUCED PRECIPITATION")

Cloud seeding involves introducing substances into the sky to cause water droplets to congeal and fall. There is limited evidence that cloud seeding can work. Even so, for the process to work, there must be moisture in the air. Cloud seeding has shown only limited promise and is not a likely solution for future needs.¶

---

* In re Water Use Permit Applications, 94 Haw. 97, 9 P.3d 409 (2000).
† This is a worldwide problem that affects many parts of the world far more than the United States. See slide show at http://www.time.com/time/photogallery/0,29307,1724375_1552659,00.html, (accessed March 24, 2008).
‡ Sixteen eastern states had legislation addressing water conservation as of the early 1990s: Connecticut, Delaware, Florida, Illinois, Indiana, Kentucky, Maine, Massachusetts, New Hampshire, New Jersey, New York, North Carolina, Ohio, South Carolina, Virginia, and Wisconsin. Such legislation was proposed in Mississippi, Pennsylvania, West Virginia, and Vermont.
§ See § 14.3 on desalinization.
¶ For more information on cloud seeding, see the Denver Water Web page, online at http://www.water. denver.co.gov/cloud_seeding.html. For information on liability for flood damage related to cloud seeding, see 10 POF *Rain and Other Weather Phenomena*, 49–136 (1961) with 1989 update.

14.3 DESALINIZATION* (SEE 8.26, 14.6)

The process of desalinization removes the salt from seawater. This process is fairly expensive and will likely only be used after other sources of fresh water are depleted. Some parts of the world already use large desalinization plants to create fresh water.†

14.4 RECYCLED WATER (SEE 7.9, 8.24)

One likely source of water for future use is recycled water, often referred to as gray water. This is water used for one purpose, then processed in a manner to make it suitable for other uses. One such example is the use of treated sewage water to irrigate fields. Treated sewage is often injected back into the ground and stored for such uses. This process is called recharge. Many forms of pollution and quality control help make used water suitable for reuse. It is almost certain that water will be put to more efficient uses in the future, and we are already seeing a significant movement in this direction.‡

14.5 IMPROVED TRANSPORT AND STORAGE OF WATER RESOURCES (SEE 7.17 "PHREATOPHYTES AND DEVELOPED WATER")

Large amounts of surface water are lost to evaporation. When water is transported in rivers and streams, it is subject to loss by evaporation and seepage into the soil. More efficient methods of storage and transportation can reduce evaporation and seepage and make larger quantities of water available for use. Improved irrigation methods are making more efficient use of water a reality. It is almost certain that we will continue to see more pressure on water users to make efficient use and avoid waste. This will eventually require upgrades to many present systems. Some possible methods of improvement include use of pipes to transport water, use of certain trees and plants for shading of water sources, covering water-storage areas, and more-efficient watering systems for irrigation. Some of the major developments in water transport include better measuring devices, construction of impervious ditches in place of porous soil ditches, the use of regulating ponds and other "check" structures, and the use of pipelines. Developments in water use include better sprinklers and micro-irrigation systems designed to decrease surface runoff.§

---

* See Figures 14–19.
† For more information on desalinization of seawater, see the California Coastal Commission's page online, http://www.coastal.ca.gov/desalrpt/dchap1.html. It is estimated that desalinization in Spain produces at least 800 million gallons of freshwater, per year, from 700 plants. See http://www.time.com/time/photogallery/0,29307,1724375_1552684,00.html (accessed March 24, 2008).
‡ For more on the benefits of recycled water, see the Environmental Protection Agency page online, http://www.epa.gov/region09/water/recycling/.
§ In the West approximately 90 percent of water is used for irrigation. See Dan Tarlock, David H. Getches, James N. Corbridge, *Water Resource Management: A Casebook in Law and Public Policy,* 221 (4th ed., 1993). It is estimated that installation of sprinklers instead of flooding fields can reduce waste by 47 to 29 percent. *Id.* One irrigation technology that is being implemented in many places is the drip irrigation system.

14.6 IMPROVED FILTERING SYSTEMS FOR PUBLIC WATER SUPPLIES (SEE ALSO 14.3 "DESALINIZATION")

Pollutants are increasing in the waters, and with new federal regulations many groundwaters are no longer deemed suitable for human consumption due to the presence of harmful substances, such as arsenic, which often exist naturally in the water. With the onset of acid rain, many of our waters, including the oceans, are growing increasingly acidic. Technology is making it possible to improve the condition of these waters. As demand grows, many sources deemed less suitable for consumption may be improved for such uses.

14.7 NEGATIVE REINFORCEMENT: USE OF FINES AND CRIMINAL SANCTIONS (SEE CHAPTER 11 ON NUISANCE)

Sanctions are already commonplace in the United States. As water demand increases, so likely will the attention we pay to abuses. We will probably see an increase in fines and prosecutions for abuses such as pollution, theft, waste, and other harms. Also, new building codes in some areas, such as Albuquerque,* require use of water-saving appliances and toilets and collection devices for rainwater. Violations of these codes often lead to denial of permits, licenses, and other privileges. Sanctions may be sought, and there is often greater liability on builders who do not build to code. It is likely that as water supplies continue to dwindle, cities will continue to require more-efficient construction and appliances in order to conserve what water they can.

14.8 POSITIVE REINFORCEMENT: TAX INCENTIVES AND BENEFITS† (SEE CHAPTER 16, "RECLAMATION;" SEE ALSO 12.8)

Government often resorts to creating incentives to promote efficient use of resources or to promote desirable behaviors. These methods may include tax credits, grants, or other benefits. The city of Albuquerque, New Mexico,‡ for example, offers incentives to individuals§ who purchase front-loading washers and high-efficiency toilets§ for use in their homes, because such washers and toilets¶ use only a fraction of the water consumed by traditional items.**

---

* See http://www.uswaternews.com/archives/arcconserv/8waterule2.html (accessed February 2008).

† An example of such incentives is the use of rebates for replacing grass with "water smart landscaping." Such a program was applied by the Southern Nevada Water Authority. The program allegedly helped save 18 billion gallons of water in 2006. See http://www.snwa.com/html/cons_wsl.html (accessed January 2008).

‡ http://www.abcwua.org/waterconservation/washingmach.html (accessed) February, 2008

§ An Australian company is currently developing a high efficiency toilet, which uses less than three-quarters of a gallon of water per flush. Most high-efficiency toilets today use a gallon and a half. See http://www.time.com/time/photogallery/0,29307,1724375_1552690,00.html (accessed March 24, 2008).

¶ For more on high-efficiency toilets, see http://www.epa.gov/watersense/pp/het.htm (accessed February 2008). A low-flow toilet uses less than 1.6 gallons per flush, while high-efficiency toilets use less than 1 gallon per flush. See http://www.abcwua.org/waterconservation/opflow.html.

** For list of other utilities offering rebates and credits for front-loaders, see http://www.thorappliances.com/features/rebates.php (available February 2008).

14.9   ICE AND OTHER IMPORTS OF FRESHWATER FROM THE NORTH

Most of Earth's freshwater is stored in the form of ice within ice caps and glaciers. Someday it may become commonplace to pump water from the Alaskan and the Canadian North to drier areas of the country much as oil is now pumped using the Alaska pipeline.* Pipelines of this magnitude are expensive to build and maintain, and water from ice must first be melted for transport. Freshwater beneath the ice may also prove difficult to access. The extreme north is not terribly hospitable, making repairs difficult. Since water expands when frozen, there is a significant risk of burst pipes. At present such pipelines appear to be cost-prohibitive.

Proposals have been considered to transport icebergs from the north in order to supplement water supplies. Ordinarily, such proposals are unfeasible because of the difficulty in transporting the large icebergs and losses due to melt during transport.

---

* See http://www.alyeska-pipe.com/Default.asp for information on the Alaska pipeline (available February 2008).

15 Water Corporations

15.1 GENERAL PURPOSE*

Water corporations exist for many different reasons, and their powers often vary based on their purpose. Some corporations are created by statute to serve some public function or benefit, while others are created to resolve disputes, provide easier access to water, or allocate water rights in the possession of the corporation in accordance with shares of corporate stock. It sometimes occurs that existing rights are deeded to the corporation in exchange for shares. Some farmers who already possess water rights may go to the corporation in order to change the mode of delivery. In such a case they will usually retain their original rights in the form of shares, and the corporation would construct ditches and other structures to promote greater efficiency. There is no set formula for water corporations, and most are defined either by statute or by their charter.

15.2 THE CAREY ACT†

Very few corporations still exist for purposes under this act, but some are still around. The purpose of these corporations was usually to promote settlement of arid lands and provide water to make those lands usable. The federal government under the bill authorized the donation of one million acres of public lands to each of several western states, provided that the states see to the land's reclamation and settlement. Most of the states involved elected to contract with private companies to perform the reclamation work necessary. As the state sold the lands to settlers, the settlers would buy water rights from companies in the form of stock in the water corporation. This usually occurred under contract, secured by a mortgage in favor of the construction company. Many of these original Carey Act Corporations have since been reorganized into irrigation districts.

---

* For a general discussion of water corporations, see *Jacobucci v. District Court*, 189 Colo. 380, 541 P.2d 667 (1975).
† 28 Stat 372 (1894), 43 U.S.C.A. § 641.

15.3   WATER USERS ASSOCIATION

These were corporations formed to contract with the Secretary of Interior in order to repay the government the costs of a federal reclamation project. Many of these associations have changed their form due to a preference on the part of the Bureau of Reclamation for dealing with irrigation districts. These associations are still widely used in Arizona and Utah.

15.4   MUTUAL WATER CORPORATE SHARES

Shares are usually issued based on acreage to be irrigated, one share for each acre. A less common method entitles the holder to a specific share of water, regardless or acreage.*

15.5   SEPARATE APPROPRIATIONS AND CLASSES OF STOCK

Some corporations may acquire different sources of appropriation as they expand. When this occurs, they may elect to create different classes of stock. Ordinarily, the rights of holders to shares of the same class of stock are equal.† Courts have differed as to whether a corporation may sell more shares than its water supply might reasonably support.‡

15.6   ASSESSMENTS AGAINST SHAREHOLDERS

Mutual Corporations are usually authorized to make assessments against shareholders for their interests in order to raise funds to continue operations. The corporation may deny water to holders who refuse to pay.§ Assessments for the same class of shares are ordinarily determined pro rata. It is possible for assessments to vary among different classes when considering reliability of the source, use, and other factors, such that one class may be assessed at a higher rate than another class.¶

---

* It should be noted that some corporations contract to provide a mode of transport and not the water itself. This is different from a mutual corporation. Also, if the company charter expires, the water rights ordinarily survive in the consumers. See *St. George City v. Kirkland*, 409 P.2d 970 (Utah 1966). But see *Combs v. Farmers Highline Canal & Reservoir Co.*, 88 P. 396 (Colo. 1907) (finding that the corporation is the representative of its members in a dispute with members of another company).

† See Idaho Code § 42-904; but see *Brose v. Nampa & Meridian Irrigation Dist.*, 118 P. 504 (Idaho 1911) (ruling that statutory equality of distribution, within a class, only exists until landowners can get individual priorities adjudicated). Many bylaws and corporate articles provide for a pro rata sharing during times of shortage so that there are no individual priorities among shareholders of the same class. See *Sanderson v. Salmon River Canal Co.*, 200 P. 341 (Idaho 1921).

‡ See *Anderson v. Wyoming Dev. Co.*, 154 P.2d 318(Wyo. 1944) (holding that company is not liable); but see *Laramie Rivers Co. v. Watson*, 241 P.2d 1080 (Wyo. 1952) (holding that it would be unjust to allow a corporation to dilute the present rights of shareholders when the supply is barely enough to support present needs). Equity probably pulls in the direction of holding corporations liable under such circumstances, where it is clear that the corporation is overselling shares when they do so.

§ See *Henderson v. Kirby Ditch Co.*, 373 P.2d 591 (Wyo. 1962).

¶ See *Robinson v. Booth-Orchard Grove Ditch Co.*, 31 P.2d 487 (Colo. 1934).

15.7 WATER STOCK APPURTENANT TO THE LAND

Ordinarily, stock in water corporations follows the land benefited.* Division exists among the jurisdictions as to whether a shareholder must receive his or her share of water from the company ditches or may change the point of diversion to some point outside the company's jurisdiction.†

15.8 JOINT, COMMON, AND PARTNERSHIP DITCHES

Joint, common, or partnership ditches are usually a less formal arrangement than mutual corporations. They consist of multiple users agreeing to share a ditch or mode of transportation. Such arrangements are generally governed by the laws of property, contract, or partnership.

15.9 ACEQUIAS

Acequias are a feature unique to New Mexico. This community ditch idea originated with the Moorish Spanish and Indian traditions of the territory. These features usually serviced the public good by bringing water to the community. Under New Mexico law, acequias are political subdivisions of the state.‡ These associations have the power of eminent domain but they do not have power to tax.

15.10 SECONDARY PERMITS (SEE 7.7 "RESERVOIRS AND WATER STORAGE")

Many states use a two-prong permit system, granting a permit to the supplier for storing the water and another to the consumer for using it.§

---

* See *Riverside Land Co. v. Jarvis*, 163 P. 54 (Cal. 1917) (finding reasonable restriction on transfer of stock to be valid and inseverability from the land is reasonable). A Utah provision requires water rights to be transferred by deed the same as real property, but states "except when they are represented by stock in a corporation, in which case the water shall not be deemed to be appurtenant to the land. Utah Code Ann. § 73-1-10. See *Brimm v. Cache Valley Banking Co.*, 269 P.2d 859 (Utah 1954) (construing the statute to establish a rebuttable presumption that water rights represented by shares in a corporation did not pass to the grantee as an appurtenance). See also *In re Bear River Drainage Area*, 2 Utah 2d 208, 271 P.2d 846 (1954) (holding that rights in water reflect an interest in real property). See also *Salt Lake City v. Cahoon & Maxfield Irrigation Company*, 879 P.2d (Utah 1994) (declaring that water stock is not a security interest under the state's commercial code, but is rather an interest in real property subject to the bylaws of the corporation as terms of a contract). Courts often will uphold terms so far as their application is not arbitrary or capricious. See *In re Water Rights of Fort Lyon Canal Co.*, 762 P.2d 1375 (Colo. 1988). States often have special statutes governing the transfer of water rights. In Utah only a mutual company may apply for a transfer of diversion point in its shares. See Utah Code Ann. § 73-3-3 (2). In Idaho, no change will be allowed in the point of diversion, if the water right is in the form of shares in a corporation, unless the corporation first consents. See Idaho Code § 42-108.
† See *Consolidated People's Ditch Co. v. Foorhill Ditch Co.*, 269 P. 915 (Cal. 1928) (finding that the diversion could not be made upstream from the companies diversion ditches); but see *Wadsworth Ditch Co. v. Brown*, 88 P. 1060 (Colo. 1907) (allowing the diversion outside the company's service area).
‡ http://www.seo.state.nm.us/isc_acequias.html.
§ See Ariz. Rev. Stat. Ann. § 45-161; Nev. Rev. Stat. § 533.440; Or. Rev. Stat. § 537.400; Wash. Rev. Code Ann. § 90.03.370; Wyo. Stat. §§ 41-3-301, -302.

15.11   IRRIGATION DISTRICTS

Irrigation districts are public agencies established to provide water to lands within the district. All of the western states have laws creating or regulating such districts. The formation of a district is typically presented to the county commissioners and presented for a vote to the citizens in the area to be serviced.* Districts are often funded by issuing assessments against the lands they service.† Some states allow districts to sell excess water for use outside the district and others allow members to transfer their water rights with permission of the district.‡

15.12   REGULATORY TAKINGS (SEE 3.10 AND 3.11 ON DEALING WITH REGULATORY TAKINGS)

The courts, generally do not consider an ad valorem tax§ (one based on the value of the land) to be a taking under the Constitution, requiring compensation. The tax power is rare among water districts. Most districts created outside incorporated areas are authorized to charge for water actually used in order to pay off debts and cover overhead costs.¶ When a city annexes an area serviced by a district, the district may be protected from encroachment on its services by the city.**

15.13   BASIC CONTRACT RULES FOR WATER PROVISION

Ordinarily, when a company or individual agrees to provide water for a fee, the risk of shortage falls on the provider.†† It remains possible that the terms of the contract can specify a different result and the provider may not assume the risk. If the agreement is dependent on a set of conditions being met, such as water being available, then absent those conditions the company would not ordinarily be held liable for

---

* On a side note, the United States Supreme Court has found that voting schemes granting voting powers according to land owned within a district is a valid scheme under the Constitution and an exception to the rule of one person one vote. *Ball v. James*, 451 U.S. 355, 101 S.Ct. 1811, 68 L.Ed.2d 150 (1981).

† See *Wheatland Irrigation Dist. v. Short*, 339 P.2d 403 (Wyo. 1959) (dealing with assessments against land based on benefit received); *Reed v. Oroville-Wyandotte Irrigation Dist.*, 304 P.2d 731 (Cal. Dist. Ct. App. 1956) (dealing with assessments based on ad valorem basis, based on the value of the land). Some states also allow assessments against all land equally. Other states authorize tolls for actual water delivered. These tolls often exist in addition to assessments.

‡ For statutes permitting sale of water by district, see Colo. Rev. Stat. Ann. § 37-42-135; Mont. Code Ann. § 85-7-1911(3); Utah Code Ann. §§ 17A-2-711. For those authorizing transfer of member rights, see Ariz. Rev. Stat. Ann. § 45-172 (4); Idaho Code § 42-108; Utah Code Ann. § 17A-2-711.

§ See *Millis v. Board of County Commissioners*, 626 P.2d 652 (Colo. 1981).

¶ The court in *Lockary v. Kayfetz*, 917 F.2d 1150 (9th Cir. 1990), discussed the constitutionality of moratoriums on hookups to water supplies in a district. The court found such moratoriums may be a regulatory taking, if they are arbitrary and remove all economic value of the property for its owner. It stated that if the moratorium is generally due to water shortages it does not constitute taking, though.

** See 7 U.S.C.A. § 1926 (a), providing for low-interest rate loans to many rural districts and protecting those districts covered by the loans from encroachment by cities during the term of the loan.

†† See Arthur L. Corbin, *Corbin On Contracts: One Volume Edition*, § 1328 (1951).

failure to provide water.* The same rules generally apply to contracts for water rights as to other contracts.†

15.14 CLASSIFICATION OF AGENCIES BY SERVICES PROVIDED

Classification of this sort is difficult and often misleading. Some companies listed as providing two services may in fact only provide one, while some listed as providing one may also provide another. The list below is only a rough distinction.‡

A. Types of Agencies Predominantly or Entirely Engaged in Supplying Irrigation Water:
   1. Irrigation districts
   2. California water districts
B. Types of Agencies Engaged to a Significant Degree in Supplying Both Irrigation Water and Urban Water:
   1. Individual proprietorships
   2. Mutual water companies
   3. Privately owned public utilities
   4. Water-storage districts
   5. Water-conservation districts
   6. Water storage and conservation districts
   7. County water districts
   8. Flood control and water conservation districts
   9. County flood control and water conservation districts
   10. County water authorities, and county water agencies
C. Types of Agencies Predominantly or Entirely Engaged in Supplying Urban Water:
   1. Municipal water departments
   2. County waterworks districts
   3. Public utility districts
   4. Community service districts
   5. Municipal water districts
   6 Municipal utility districts
   7. Metropolitan water districts

In general, a public utility has less discretion to provide services or not than a municipality, though many municipalities have been treated as public utilities where they are the exclusive provider of services in an area.§

---

* See Arthur L. Corbin, *Corbin On Contracts: One Volume Edition,* § 1339 (1951); See *Souther v. San Diego Flume Co.*, 121 F. 347 (9th Cir.1903) (liability avoided through clause in the contract).
† *Consolidated Canal Co. v. Mesa Canal Co.*, 177 U.S. 296, 20 S. Ct. 628, 44 L. Ed. 777 (1900); *Faught v. Platte Val. Public Power & Irr. Dist.*, 155 Neb. 141, 51 N.W.2d 253 (1952); *Gold Ridge Min. Co. v. Tallmadge*, 44 Or. 34, 74 P. 325 (1903) (water must be fit for intended purpose of contract).
‡ Dan Tarlock, James N. Corbridge Jr., and David H. Getches, *Water Resource Management: A Casebook in Law and Public Policy,* 593 (University Casebook Series 4th ed., 1993).
§ See *Zepp v. Mayor & Council of Athens*, 348 S.E.2d 673 (Ga. App. 1986) (no duty to provide outside boundaries); but see *Robinson v. Boulder*, 190 Colo. 357, 547 P.2d 228 (1976) (finding duty as exclusive provider in area).

16 Reclamation*

16.1 HISTORY

With large federal land holdings in the sparsely populated, arid West, Congress sought to encourage settlement. The Reclamation Act provided free and inexpensive land to individuals willing to settle. Unfortunately, the arid climate made much of the land virtually useless for settlement without water. Since would-be settlers could not readily afford to bring water to the region, Congress passed the Reclamation Act in 1902. The act established the Bureau of Reclamation within the Interior Department. The act was intended to provide water and make the land profitable without fueling speculation.

16.2 GOVERNMENT EXPENSES

The Reclamation Act was intended to be self-sustaining. Most of the settlers lacked the funds to repay the costs, which forced a revision of the program. This led to substantial federal subsidies.†

16.3 SPECIFIC PROJECTS

Many construction projects were enacted following the Reclamation Act. Among these were the Boulder Canyon Project and the Colorado River Storage Act,‡ which provided for construction of dams (the most famous of which is probably the Hoover Dam).§

16.4 THE SMALL PROJECTS ACT¶

This act was intended to speed up construction of small reclamation projects with some federal funding. It required that local governments secure water rights, easements, and land.

---

* For information on current reclamation activities, see Bureau of Reclamation Web page at http://www. usbr.gov.
† The federal projects cost between $43 and $163 per acre served, for an average of $85. Private enterprises would not have undertaken such expenses. Less costly projects, ranging between $15 to $20 per acre, had already been built by private groups. Reclamation projects generated electricity valued at $648 million in 1990 and provided water for 19 percent of the nation's irrigated land, approximately 9.3 million acres, at a cost of about $9.9 billion as of that time. Debate still rages today over the actual value of the reclamation projects and whether they are worthwhile. The Bureau of Reclamation issued a report in the late eighties stating that the mission of the bureau should change from one of construction to one of resource management.
‡ 43 U.S.C.A. § 620.
§ 43 U.S.C.A. § 617.
¶ 43 U.S.C.A. § 422.

16.5 BASIS FOR CONGRESSIONAL POWER

The two major bases for congressional authority to enact the Reclamation Act and subsequent acts were the "Property Clause"* and the "General Welfare Clause."†

16.6 THE RECLAMATION PROJECT ACT OF 1939‡

The purpose of this act was to authorize federal expenditure for large-scale reclamation projects. It made the funds available without the need for reimbursement.

16.7 BENEFIT LIMITATIONS

Congress was concerned about land speculators and placed limitations on the program benefits in 1902. No single owner could receive water for more than 160 acres.§ The owner also had to reside on or near the land to obtain the benefit. The original act required beneficiaries to pay back part of the cost. Later amendments authorized extensions or forgave most of the debts. The 1983 Reclamation Reform Act¶ increased the minimum acreage from 160 to 960, but also increased the charges for water. It addressed leasing by setting an overall limit of 2,080 acres. Excess lands are subject to full cost for waters. Later increases in fees for water were found constitutional on the basis that federal sovereignty was never surrendered in the contracts and settlers operated subject to such increases. The court denied challenges for violations of due process and takings.**

---

* United States Const. Art. 1 § 8 clause 17; *Kansas v. Colorado*, 206 U.S. 46, 27 S.Ct. 655, 51 L.Ed. 956 (1907). See § 2.2.

† United States Const. Art. I, sec. 8: "The Congress shall have Power to lay and collect Taxes, Duties, Imposts and Excises, to pay the Debts and provide for the common Defense and general Welfare of the United States; but all Duties, Imposts and Excises shall be uniform throughout the United States; . . ." See § 1.14. See *United States v. Gerlach Live Stock Co.*, 339 U.S. 725, 70 S.Ct. 955, 94 L.Ed. 1231 (1950). See §§ 3.11 and 6.13.

‡ 43 U.S.C.A. §§ 485-485k.

§ 43 U.S.C.A. § 381, Section 5 of the Reclamation Act. A 1926 Act, The Omnibus Adjustment Act, 43 U.S.C.A. 423e, provided that owners must enter into a recordable contract with the Secretary of Interior, agreeing to sell all excess land (that is, above the 160 acres). This was in order to counter attempts by speculators to subvert earlier precautions. See *Bryant v. Yellen*, 447 U.S. 352 (1980). Note: the Department of the Interior took the position that the Section 5 residency requirement was superseded by the contract provision of the Omnibus Adjustment Act. Subsequently, the residency requirement was seldom enforced. It was appealed in the Reclamation Reform Act.

¶ 43 U.S.C.A. §§ 390aa – 390zz-1, Pub. L. No. 97-293, 96 Stat. 1261 (primarily addressing concerns about the acreage limits). For certain nonlease exceptions, see 43 C.F.R. 426.

** See *Peterson v. U.S. Dep't* of Interior, 899 F.2d 799 (9th Cir. 1990), cert. Denied, 498 U.S. 1003, 111 S.Ct. 567, 112 L.Ed.2d 574 (1990). The challenged provision, known as the "hammer clause," required districts to amend their contracts by 1987 in order to be consistent with the new law. The law offered a higher acreage limit at higher cost per acre (they lost the right to pay less than cost). It dealt primarily with leased lands. Section 9 of the 1939 act, 43 U.S.C.A. § 485h(d), requires districts to enter a contract before water may be delivered. Under the contract the irrigators obtain a vested right to water, appurtenant to their land. See *Ickes v. Fox*, 300 U.S. 82, 57 S.Ct. 412, 81 L.Ed. 525 (1937). In *Ickes*, the costs were paid off already. *Nevada v. United States*, 463 U.S. 110, 103 S.Ct. 2906, 77L.Ed.2d 509(1983), the court found that once lands were acquired, the United States interest in water rights was nominal at best.

16.8   SECTION 9(E) CONTRACTS*

Under this section, the secretary of the interior is authorized to enter into contracts for up to forty years, with what is termed "appropriate share" cost recovery. This allows the secretary to fix rates at less than full cost.†

16.9   ROLE OF STATE LAW

To the extent possible, water is to be distributed according to state law, so long as that law does not conflict with the reclamation law.‡ Many states established their own laws regarding reclamation to assist the federal government.§

16.10   THE FLOOD CONTROL ACT OF 1944¶

This act allows the Army to supply the Bureau of Reclamation with water from its dams for use in irrigation. The Bureau must get approval from the Army Secretary before acquiring any water stored at Army Corps sites.**

---

* 43 U.S.C.A. § 485h(e)
† These contracts were widely used in California's Central Valley Project. The Warren Act, 43 U.S.C.A. §§ 523, 524, provides for a less permanent water right for users outside the project area.
‡ *California v. United States*, 438 U.S.645, 98 S.Ct. 2985, 57 L.Ed.2d 1018 (1978). This decision did not claim to overrule earlier decisions holding the federal government has extensive preemptive authority in reclamation matters. See § 1.3.
§ See R.C.W.A. 89.12.010, stating that "it is the policy of the state of Washington in connection with lands within the scope of this chapter which may be irrigated through works of federal reclamation projects, to assist the United States in the reduction or prevention of speculation in such lands and in limiting the size of the holdings thereof entitled to receive water by means of the works of such projects and otherwise to cooperate with the United States with respect thereto." Another provision of Washington State law states, "The object of this chapter is to provide for the reclamation and development of such lands in the state of Washington as shall be determined to be suitable and economically available for reclamation and development as agricultural lands, and the state of Washington in the exercise of its sovereign and police powers declares the reclamation of such lands to be a state purpose and necessary to the public health, safety and welfare of its people." R.C.W.A. § 89.16.010.
¶ 43 U.S.C.A. § 390b. See 33 U.S.C.A. §§ 701a, 701a-1 (Corps authority for improvements).
** *ETSI Pipeline Project v. Missouri*, 484 U.S. 495, 108 S.Ct. 805, 98 L.Ed.2d 898 (1988). 78 Am Jur 2d, Waters § 75 with 1999 update.

17 Western State Water Systems and Fees*

17.1 INTRODUCTION†

This chapter is intended to offer a brief contrast of western state water systems. As such, the information here is greatly abbreviated. For further information the reader should consult the appropriate Web sites listed near the end of this book for the various state water agencies and codes. This chapter covers most of the major fees for water rights and some basic procedures. Applications and other fees are available through state water agencies on their Web pages. Most of these Web pages provide further information on water rights in the state. Some even offer links to their relevant rules and water codes.‡ Since not all fees and procedures are listed here, the reader should consult the appropriate Web pages for further instructions.

17.2 ALASKA§

Under the Alaska Constitution, water is a public resource. The state is obliged to maximize public utility. All uses must be consistent with public interests. Alaska Stat. 46.15 defines a water right as a legal right to use surface or groundwater. The right becomes appurtenant to the land where the water is used, for as long as beneficial use continues. The Department of Natural Resources must approve a separation of the right from the land when the land is sold, or else the right follows the land. Alaska does not recognize the rights of landowners to use water solely based on ownership of the land (riparian rights). Using water without a permit gives no right to the water. Alaska does not recognize prescriptive rights. Water-permit applications require a filing fee in Alaska, which varies based on the proposed use. A single-family home must pay a fee of $100. Other fees vary upward, beginning with $200 for a use of 5,000 gallons or less not listed among other uses and ranging through $900 for uses of more than 100,000 gallons. The fees range upward to $1,200 for activities associated with oil and gas. Other listed uses include hydroelectric power, mineral mining,

---

* See Appendix 4 for BLM fact sheets of 11 western states.
† See "State Water Codes and Western State Water Departments and Permitting Agencies" at end of appendices. See also listing of "Case Law by State."
‡ Web sites for the state water codes are listed in the appendices. For BLM fact sheets on western state water law, see Appendix 4, covering Alaska, Arizona, California, Colorado, Idaho, Montana, Nevada, New Mexico, Oregon, Utah, and Wyoming.
§ See Appendix 4, BLM Fact Sheet.

and hydrologic unit removal, all of which have varying fees. Alaska uses a single application form for both surface and groundwater uses.

Alaska has a notification requirement for uses over 5,000 gallons per day and users may be required to pay for advertising in at least one local paper. Under certain circumstances notice may be required for uses less than 5,000 gallons per day. There is an annual $50 service fee for the use certificate. Domestic uses and those for less than 500 gallons per day are exempt from the annual fee. It is a misdemeanor to withdraw significant amounts of water without a permit. Alaska allows some rights to in-stream flows for fish, wildlife, and recreational uses. A separate application is available for such uses, titled an application for reservation of water. A separate application exists for temporary water use, which must be accompanied by a fee of $350.

17.3   ARIZONA*

Arizona's surface water law was enacted in 1919. The basis for water appropriation and use in AZ is beneficial use. Notice is required at the point of diversion. Arizona has a separate set of laws governing groundwater but uses the same application for appropriation of surface and groundwater. Arizona requires separate permits for construction of a reservoir to store water and for water use. When land is transferred with water rights, a filing, titled a request for assignment, is required.† Ordinarily, a person must complete an application to transfer a water right to new land. This is termed a severance and transfer. The director publishes notice of the proposed transfer once a week for three weeks. No transfers are allowed, which will negatively affect vested rights, or where the water right to be transferred was forfeited or abandoned.

Temporary rules for groundwater were adopted in 1983. For various reasons, no permanent rules were adopted until the initiation in 2004 of such rules. Permanent rules were still in development as of the time of this record. Arizona requires a $50 application fee for less than 50 acre-feet or $75 for more than 50 acre-feet. A fee of $50 is required for a certificate of water right as well. Applications are reviewed within 450 days. Water rights may be denied if they conflict with vested rights, menace public safety, or are against public interests. Under some circumstances they may be approved for less than the amount applied for. Arizona gives preferential treatment to municipal uses and may grant a municipal right to the exclusion of all subsequent applications if circumstances demand.

17.4   CALIFORNIA‡

California is one of the few western states that recognize new riparian rights under some circumstances. California has one of the most extensive water codes in the country. The state offers some incentives for conservation of water by allowing the sale of some rights and application of conserved water to other lands or uses. California first recognized water rights prior to 1872, for just putting the water to

---

* See Appendix 4, BLM Fact Sheet.
† A.R.S. § 45-164.
‡ See Appendix 4 BLM Fact Sheet.

beneficial use. In 1872 the first permitting system in state was established. Priority has been established by publication of notice at the location of diversion and recording with the county recorder. The right is fixed by actual beneficial use and not the application. Abandonment requires both act and intent and is always voluntary. Rights are forfeit after five years of consecutive nonuse. Because the state needs to communicate with rights holders, any change in ownership must be communicated to the agency.

Riparian owners have a correlative right to the natural flow of water passing the landowner's land. These rights include reasonable beneficial use, and do not require a permit. Under state policy, natural flow does not include return flows. Riparian rights do not exist to use water on land not abutting the stream or lake and land loses its riparian rights when severed from land bordering the stream. There is no priority of use based on first in time regarding riparian rights and uses. Riparian rights exist regardless of use or nonuse but may be lost by prescription. Riparian rights cannot be transferred to a nonriparian parcel. Riparian rights apply only to natural streams and not foreign water. Water cannot be stored for later use under claim of riparian rights.

Appropriative rights in California are subject to all prior uses of water, whether riparian or appropriation. When a prior right holder fails to take legal action to protect his rights during a five-year period of continuous adverse use, the holder is barred from taking action to protect his right. Prescriptive uses must be open, notorious, hostile, and exclusive, under a claim of right, continuous without interruption and an invasion of the prior owners' rights. This must give the prior owner an opportunity to take legal action and the adverse user must pay necessary taxes. If any of these elements are missing, then the prescriptive right fails. Prescriptive rights only flow against downstream users, as upstream users cannot be harmed by downstream uses after the water has passed their land. Ordinarily, adverse users must acquire a permit to have their rights protected. In this regard, a junior appropriator may gain rights against seniors but one without a permit has no rights. Priority of any right relates back to the filing for a permit and not first use.

Water rights are property rights and are protected in the same manner as other property rights. Illegal diversion is actionable as a trespass. California is a public trust state and holds that the state took title to nontidal navigable waters on admission to the union, in trust for the public. Public trust has been used to protect fishing, hunting, navigation, commerce, swimming, boating, wading, and operation of fisheries. Navigable is interpreted broadly in California, and includes rafting and kayaking. Recently, the courts have recognized environmental protections as part of the public trust. All uses, including the public trust, must conform to reasonable use.

Appropriation of underground water is limited to underground streams known to flow in definite channels. A statement of diversion should be filed for use of water on overlying land. If the use is on nonoverlying land, then the user must apply for a permit. Nonflowing underground water is not part of the appropriation system and overlying landowners have the first right to make reasonable beneficial use of such waters under the correlative rights doctrine. Surplus waters may be appropriated subject to future reasonable uses on overlying lands. A notice of intent must be filed for any well that is to be dug, drilled, or expanded. The doctrine of "correlative rights" applies to springwater that enters a stream; however, a spring with no natural outlet may be used by the landowner without a permit.

Owners of dams must allow sufficient water to provide for fish downstream. All diverters of water must file a statement of use whether riparian or appropriation, with some exceptions noted usually where water is not subject to the permit process, or where other documents on file serve the primary purpose.

17.5   COLORADO* (SEE 7.23, COLORADO APPROPRIATION)

Colorado was the first state to use public officials to distribute water, more than 120 years ago. Colorado is unique in that it is the only western state not to use a permit system. Colorado still reaches the same effect through courts, which protect water rights based on first in time, first in right. Water rights are registered by court decree verifying the user's priority. Due to overappropriation, very few new groundwater appropriations are allowed in Colorado without some plan to augment the supply. Some surface rights may still be acquired, but water users must be capable of shutting down the use if they receive a call from a senior rights holder. Prior to 1957 groundwater was not managed in the state. Now well construction requires a permit from the state engineer. A few exemptions exist for some small domestic wells. The State Engineer's Office was officially renamed the Division of Water Resources in 1969, at the same time the state began administering ground and surface waters together.

An application for a surface or groundwater right requires a $136 filing fee. A change in water right or augmentation application requires a $271 fee. A nonparty's protest to a referee ruling or a statement of opposition requires a $70 filing fee. The initiation of a water right requires a showing of intent to divert water, the location of an intended beneficial use, and a physical demonstration of intent to divert. Once these early requirements are met, a user should file an application with the court in the basin from which the water is diverted. Colorado has seven water divisions. The Colorado Supreme Court appoints judges to these courts. Upon application with the court, the proposed use is published in what is known as a water court resume, to provide public notice. The application is also published in appropriate local newspapers. Courts vary on publishing procedures and sometimes bill the user for costs. Parties opposed to the proposed use have two months from the time of publication to file a "statement of opposition." Absent opposition, the application goes to a water referee for review. The referee makes a ruling and, absent any protest within twenty days, the ruling goes to a judge to become final. If a protest is lodged the mater will go to trial before the judge.

17.6   IDAHO†

All water in Idaho, flowing within natural channels and groundwater is considered public water. Water rights, once established are considered real property in Idaho. A water right is a right to divert public waters and put them to beneficial use. Water

---

* See Appendix 4 BLM, Fact Sheet.
† See Appendix 4 BLM, Fact Sheet.

rights may be lost if not used for a continuous five-year period. Idaho recognizes no riparian rights to divert water. The priority date of a water right is the filing date of the application. Idaho used to recognize rights based on diversion and beneficial use, prior to 1971. The one exception to the permit requirement for water rights is the in-stream right to water livestock. The permit process is the only way to acquire new rights to surface or groundwater in Idaho, today. There is an exception for beneficial domestic use of groundwater, as defined by state law.

Applications are filed and reviewed followed by publication of notice once a week for two weeks. This is to allow other users to be aware and, if necessary, protest the application. Some areas have moratoriums on new permits and uses, and applications for uses of these waters may not be processed. Protests must be received no later than 10 days after the notice is published. It must include the name and address of the protesting party and a nonrefundable $25 fee. A clear explanation of the reasons for objecting and the injury to the protestant must also be given, and a copy must be sent to the applicant. If the problem cannot be worked out privately, either party may request a formal hearing on the matter. The hearing officer will issue an order, and the director will review it then issue a final order. The process can take months or years depending on the complexity of the issues.

Changes in the point of diversion must meet certain conditions, including filing an application. The change must not enlarge an existing right or harm other users. It must also undergo public-interest review. The state may approve the change in whole or in part or may apply conditions. Water rights are appurtenant to the land unless reserved when land is transferred. Permits are issued to all applications that meet the state requirements. This includes proof of beneficial use (a form filled out by the user). A field examination then must be made by either the state or a state certified examiner. The purpose of this examination is to ensure that water is being used in fact as it is described in the permit. Application fees for new water rights range from $100 (up to .20 cfs) to more than $6,610 (500 cfs) and are determined by the cubic feet per second appropriated. Transfer and exchange fees mirror those for new water rights. License examination fees begin at $50 for .20 cubic feet per second and go up to $600 for more than 20.1 cubic feet per second.

17.7 KANSAS

All applications for water rights in Kansas must include at least the following on the prescribed form:

1. Name and mailing address
2. Original signature of the applicant or an authorized representative
3. The source of water supply (surface water or groundwater)
4. Proposed place of use
5. Proposed point of diversion
6. Maximum rate of diversion and quantity of water requested
7. The statutorily required (nonrefundable) filing fee

Fees for new rights begin at $200 for 0 to 100 acre-feet and $300 for up to 320 acre-feet. They increase an additional $20 for each added 100 acre-feet or any part thereof. A storage permit requires a fee of $200 for up to 250 acre-feet and an additional $20 for each added 250 acre-feet. If both use and storage are requested, the greater of the two is used. Power fees are $100 plus $200 for each 100 cubic feet per second. Once construction is completed a $400 inspection fee must be paid.

Applicants may be required to complete a water conservation plan and or install meters and other equipment. Other conditions may also apply, including possible groundwater regulations. Maps, well logs, names and addresses of property owners within half a mile of the proposed diversion, and justification for the amount of water requested may also be required. A use may not impair an existing water right nor unreasonably affect the public interest. An applicant is usually given the balance of the current year plus one year to complete construction. The user must then complete a notice of completion and pay for a field inspection (see fee above). The owner may submit a request for extension if more time is needed. Ordinarily, this would include a fee of $100 and must be submitted prior to the deadline. The owner must show when completion is expected and the reason for noncompletion, as well as the progress made thus far. Other information may be required based on the total time of the delay. More time is given for municipal uses (twenty years) to perfect water rights following construction than other uses (four years). If an application is not approved, an applicant has fifteen days to file for reconsideration.

17.8 MISSISSIPPI

Mississippi uses public-interest review to manage water supplies in the best interest of the citizens of the state. All water of the state, surface and groundwater, is considered property of the state and subject to the water code and permitting process. No one is allowed to use water in the state absent a permit, unless exempted by statute. Surface and groundwater resources are regulated together as interconnected. Permits contrary to public interest may be denied or conditioned. The state may on occasion modify permits to conserve resources should it become necessary to deem an area a water use warning area. This is usually a last-resort solution if the agency cannot find another solution first. Such a warning may be issued when data indicates unacceptable trends in water supply for an area. Other designations are available to speed up results. A fee of $10 must accompany each application and separate applications must be used for each diversion or new well. Applications must be notarized and accompanied by a suitable map of the area.

The state will make a preliminary assessment of all proposed new wells or surface intakes that will be part of the public water system. The state furnishes a notice of intent for publication along with instructions. The user must publish once in a paper of general circulation in the area. The paper must then forward proof of publication to the state. Permits may be denied in whole or in part if they would negatively impact existing rights or fail to propose a beneficial use. Permits may be made conditional. Construction must be initiated within two years of the initial permit for public water-supply wells or one year for all other uses. Some exceptions exist. Failure to meet these requirements means that a new application must be filed. Landowners may show mitigating circumstances. Permits are generally issued for a ten-year period. This period may be longer for some public suppliers. Notice of the need to reapply will be sent to users six months before the expiration. To continue the use the user must then file a new application before the deadline. Permits are automatically extended for the time the reissue application is being reviewed.

Mississippi uses minimum flow levels for many streams and will not allow withdrawals below the minimum flow levels. Some users who return the water back to the stream may be permitted to use water that does not significantly affect the streamflow level or quality of water. Some single-household domestic uses are considered exempt from permitting requirements. Mississippi has a specific prioritization of beneficial uses beginning with the public supply. The state also requires minimum spacing of wells used for withdrawal of groundwater. Free-flowing wells (artesian) must be equipped with stop devices to prevent waste.

17.9   MONTANA*

Montana is a "pure" appropriation state. Montana places sole jurisdiction to resolve disputes over water rights in its Water Court. All rights are required to be filed with the state. Old rights not filed by April 30, 1982, were considered abandoned. Statements filed after that time are considered junior to those filed before 1982. The state may require abstracts and request clarification in assessing older rights, to aid the Water Court in approving new appropriations and current usage. Incomplete information may adversely affect the right.

Montana law provides for state ownership of all land below navigable water bodies. This includes islands that may form in the beds. The state considers those bodies of water navigable for which there is historical documentation of commercial use. The state also owns all waters in any form.

Appropriations of groundwater exceeding 35 gallons per minute or 10 acre-feet per year require an application for beneficial use. A separate application must be filed for each different source. Each application requires a $400 filing fee. If the appropriation is for a use less than that, the fee is only $200. Lesser uses must complete the appropriate addendum to the application. A separate addendum exists for uses in excess of the above amounts. The same form is used for surface diversions. Maps must be included with each application. An affidavit attached to each application must be signed and notarized. A notice of completion of development must also be filed with the state. A separate application exists for water storage. It sets out certain requirements and requires a $50 filing fee, submission of maps, and notarization. The filing fee for a change in a water right is $100 for a replacement well, reservoir, or a stock tank, or $400 for all other changes. Montana also allows applications for extension of time to construct or develop. The fee is $100 for this application and the application must be filed by the end of the year scheduled for completion. There is a $50 fee to transfer ownership of an existing right, only $10 for each additional right, up to $300. An addendum must be filed when only part of an existing right is transferred. Montana also requires a $25 fee to file an objection to a water right.

---

* See Appendix 4 BLM, Fact Sheet.

17.10   NEBRASKA

Nebraska uses a variety of forms for different purposes, including but not limited to appropriations of surface waters; permits for in-stream flows; groundwater recharge; hydro-electric power; transfer of rights; change of diversion; groundwater wells; and so on. Nebraska also uses forms for decommissioning a well or for ceasing a diversion. Maps must either accompany the claim or be filed within six months of approval. As of January 2004, well-registration fees in Nebraska were $70 for those designed to take less than 50 gallons per minute and $110 for wells designed to take more than that. Fees for a domestic surface use are $10. Other fees are as follows: Agricultural use for 0 to –1,000 acres is $200, with $100 for each additional 1,000 acres; The fees are somewhat less for irrigation from a storage facility, $50 for the first 1,000 acres and $25 for each additional 1,000 acres; $10 for manufacturing; $5 for power generation; and $10 for other uses. There is a $10 filing fee for use of in-stream flows and for groundwater recharge uses.

Applicants for rights are required to provide certain information including available water for appropriation, maps, witnesses, impact on senior rights, and public-interest evaluation. Applicants for some rights may also be required to provide other information, listed on the appropriate applications. Surface and ground-water are handled separately, but public interest review exists for all appropriations in state. Specific rules and notices are posted on the Department of Natural Resources Web page, listed below.

17.11   NEW MEXICO*

All ground- and surface water in New Mexico belongs to the public. New Mexico distinguishes between uses of ground water, including domestic, livestock, and temporary uses. Filing fees for domestic uses are $125. The fees for livestock and temporary uses are $5. General appropriations of ground water require a $25 fee and a separate form. Applications to change well locations range from $25 to $50 depending on whether the change also involves a change of use. To change a point of diversion from surface to ground water requires a $50 fee, to change from surface to groundwater. A surface appropriation requires a $25 filing fee. Fees to change from ground- to surface water are $200. Other changes of diversion range from either $100 or $200 depending on the nature of the change. Several supplemental forms may also be required including proof of beneficial use ($25), notice of intention to appropriate ($25), extension of time to perfect ($50), and proof of completion of works ($25).

The Water Resources Allocation Program and the Office of the State Engineer process water rights in New Mexico. The state engineer determines what water is available for allocation, whether a new appropriation will impair existing rights, and that the intended use is not contrary to public interest. Applicants must publish their applications in a public newspaper, and those with legitimate objections must have an opportunity to challenge the proposed use. Objections must be legible and in writing, signed, and contain the objector's full name and address. The objection

---

* See Appendix 4, BLM Fact Sheet.

must set forth all the objections or reasons for the challenge. The objection must be filed in triplicate, within ten days of the date of last publication.

17.12   NEVADA*

Nevada water law is based on the concepts of prior appropriation and beneficial use. The Nevada Division of Water Resources administers water law in the state. All water in Nevada, both underground and surface belongs to the public. Nevada boasts one of the most extensive water codes in the west. Water permits require an application be filed with the State Engineer. Forms are available at the agencies web sight. Applications must include a map made by a water-rights surveyor. The map must include the point of diversion and place of use. Once the application is complete, a notice must be published in a newspaper of general circulation, in the area of use, for thirty days. Objectors have until thirty days from the date of last publication to file a protest. The State Engineer will consider the following before approving or denying the application: available water, effects on existing rights, public interest, and impact on domestic wells. Other criteria may also be weighed depending on the proposed use. The state may request more information or hold a hearing if they wish. Denials may be appealed within thirty days from the time of decision, in the courts of jurisdiction. Water rights once granted have the standing of both real and personal property. They are appurtenant to the land unless specifically withheld in the deed. Water rights are conveyed by deed.

Nevada requires a $250 filing fee for new appropriations. All applications must be mailed in and some require colored paper. Applicants are required to describe all works regarding diversion and may be required to submit plans and other documents. A change in diversion requires a $150 filing fee and $100 for a temporary use permit. Nevada has a water use application for environmental uses and requires a $150 filing fee. Like many other states, Nevada requires a notice of completion (form must be on blue paper) and a $10 filing fee. This document must be notarized. There is also a proof of beneficial use form, which must be accompanied by a $50 filing fee. There is a $100 filing fee for an extension of time application. Nevada has a separate application for an extension of time to prevent forfeiture. This also must be accompanied by a $100 filing fee. Applications must be notarized and signed with an original signature. Individuals wishing to protest an application must fill out a protest form and submit a $25 filing fee. All protests must be filed in duplicate and each must have an original signature. No permit is required for domestic wells in Nevada, but all other wells do require permits. Such wells are forbidden when the land can be supplied from the public water supply.

17.13   NORTH DAKOTA

Application fees in North Dakota include a $500 filing fee for municipal uses over 2500 people, $250 for a municipal use serving a population of less than 2500 people, $200 for irrigation, $250 for minor industrial use (1 cfs or less), $750 for larger

---

* See Appendix 4, BLM Fact Sheet.

industrial uses (more than 1 cfs), $100 for instream uses such as recreation, livestock, or fish, $200 for commercial recreation, and $50 for an amendment to a water right. Where water is to be stored; the amount for use and the amount that will be lost to evaporation must both be identified.

Applications are only valid for a single source of water. Water rights do not exist until actual beneficial use is made. Only a conditional permit to develop water is issued before that time. A perfected permit will be issued after the water is put to beneficial use. A map, prepared by a surveyor must accompany all applications. Examples of the map are given in the application instructions. All maps must be computer generated and not hand written. In addition to the map, the surveyor must submit a certificate stating that the map was made from an actual survey.

Following receipt of the application a notice of application form will be sent out. Instructions are included with the form. Forms must be delivered by certified mail. The receipts must be sent with an affidavit to the state. After fifteen days from the time the notices are sent, a hearing may be scheduled. A notice of hearing is published before the hearing begins. The applicant must pay the publication fees in the paper for two consecutive weeks, prior to the hearing. At the hearing all interested persons may present their views.

17.14   OKLAHOMA

The Oklahoma Water Resource Board is in charge of water management in state. The state issues permits for both ground and surface waters, though domestic uses are exempt from the permit requirement. Domestic uses are defined as those not exceeding five acre-feet per year, and used by a natural person or single-family household. Stream water is considered public property, subject to appropriation. Groundwater is considered private property belonging to the property owner, whose land it is under. Groundwater is still subject to reasonable use regulation by the state. Persons must still submit an application for a permit for non-domestic uses of groundwater. Notice is usually required to landowners within a quarter mile of the well and should be published in the newspapers. The state considers the following to be beneficial uses of water: agriculture, irrigation, water supply, hydroelectric power generation, municipal, industrial, navigation, recreation, and fish and wildlife propagation. During drought, public supply is generally given preference to other uses, though no true priority of uses exists in state.

Four conditions must be met for the board to approve an application for a water right: 1) water must be available, 2) the use must be beneficial, 3) it must not interfere with existing rights, including domestic riparian rights, and 4) it must not interfere with existing or prior proposed uses in the stream or the area's water needs, if the use is outside the area where the water originates. Notice of the application must be published in the county of use and downstream counties and an opportunity given to protest the use. Hearings may be held to determine facts. The board may place conditions on a permit if it thinks them necessary. The normal time allowed for construction is two years. Full use of water for beneficial use must be made at least once every seven-year period.

Oklahoma issues five types of surface water permits, regular (the whole year round), seasonal, temporary, term (for a given number of years), and provisional temporary for up to 90 days (not renewable and not requiring approval by board). Irrigation rights are appurtenant to the lands irrigated; though water rights may generally be transferred or assigned. There are four types of ground water permits, regular, temporary, special, and provisional temporary. Fees for groundwater permits range from $125 for fewer than 320 acre feet, $200 for 321-640 acre feet, and $250 for 641 to –1,500 acre feet, with $100 for every additional 500 acre feet up to $2,000. Applications must be notarized and include a space for a map. There is a $150 filing fee for provisional temporary applications, $200 if expedited. Applications must be submitted in duplicate. The fee for an amendment application to a ground water right is $50 if unpublished and $150 if published. Fees for stream water use mirror those for groundwater use in both cost and acreage. Water transfers for surface rights are $50 subject to a family exception.

17.15   OREGON*

All water in Oregon is publicly owned. Oregon does not recognize riparian rights, absent a permit from the state. Water rights require a three step process in Oregon: 1) an approved application, 2) construction and use, and 3) a survey completed by a certified examiner detailing how and where water is being used. The examiner will then submit maps and other documents to the state. A water certificate will be issued if the use is in accord with the permit. Water may only be used for beneficial purposes without waste. Rights are usually appurtenant to the land and when the land is sold, the right follows. Rights must be used at least once every five years or they are considered forfeited and subject to cancellation. Oregon law does not prefer one use to another and priority will dictate as between all users. If the date of priority is the same, then state law prefers domestic uses and livestock watering. Oregon exempts some uses from the permit requirement. These include: 1) natural springs, which do not form a definite channel nor flow off the property from which it originates at any time, 2) stock watering, 3) salmon-egg incubation, 4) fire control, 5) some forest-management activities (must notify Department of Fish and Wildlife), 6) certain land-management activities, and 7) collection of rainwater.

Exemptions for ground water include domestic uses, noncommercial gardening, small commercial uses for a single industry, watering school grounds, stock watering, and down hole heat exchanges. These uses may still be subject to regulation in times of shortage and must be for beneficial uses. Before any right is granted, the public has a right to protest. Other rights holders may claim the new use would interfere with their use or members of the public can claim the use is contrary to public interest. Maps are required with the application as well as a description of the land involved, names and addresses of anyone who might be affected by the use, if originating on the land of another, permission to access the water, land use information, and if necessary, supplemental forms.

---

* See Appendix 4, BLM Fact Sheet.

Fees are charged for processing an application. There is a base fee of $300 to appropriate ground water and an additional $200 for the first cfs or fraction. Each additional cfs requires another $100. Appropriation of stored water may be made under the same application but requires an additional fee of $20 for the first 10 acre-feet and $1 for each additional acre-foot. The base to appropriate water for storage is $300 with $20 for each fraction up to 10 acre-feet and $1 for each additional acre-foot. Appropriation of stored water requires only a $150 fee with a $15 fee for the first 10 acre-feet and $1 for each additional acre-foot. There is a $250 recording fee once a permit is issued. Other fees apply to specific uses.* A change in the use requires a $350 fee per change. Only a $175 fee is required for a change in place or type of use. The fees are lower for temporary changes or transfers. To substitute surface water for groundwater requires a $250 fee. Leases involving four or more rights or landowners require a $200 fee. All other leases are $100. Examination of water management and conservation plans also require significant fees. An extension of time request carries a $250 fee. Protests also cost $250. Other fees are listed on the Web page for the Water Resource Department.

17.16 SOUTH DAKOTA

All water in South Dakota belongs to the people of the state. Permits are required for all but domestic uses. Even domestic uses may require a permit if they exceed 25,920 gallons per day or a peak pump rate of 25 gallons per minute. Domestic uses include watering a noncommercial garden, stock watering, some uses in schools and parks, drinking, washing, and sanitary uses. Water distribution systems using 18 gallons per minute or less also need not apply for a permit. These include municipalities, rural water systems, suburban housing developments, and so on. Permits are required for construction or dams and other storage facilities.

South Dakota requires with applications, a map showing the diversion and lands for irrigation, prepared by an approved person and a fee of $150 for the first 120 acre-feet to be irrigated, $75 for the second, and $25 for each additional 120 acre-feet. The fee for final inspection of an approved application is $50. The fee for less than 0.10 cfs is $50. A $50 fee is charged to change the point of diversion. Some water sources may require a water-quality survey be conducted. Procedures are laid out in the application. Applications must be notarized. Nonirrigation fees mirror those for irrigation uses but require a different application. Separate applications exist for systems in place before 1955.

Applications must be published in a local paper. Approval takes approximately two months if the application is not contested. An applicant usually has five years to construct and four more to put water to beneficial use. A notice of completion of works must be submitted when construction is finished. The filing fee for a transfer of ownership of a water right is $2.50. Supplemental forms and reports may apply to some construction and use projects. Permits are also available for discharge into either the ground- or surface waters.

---

* http://www1.wrd.state.or.us/pdfs/fees2003.pdf.

17.17 TEXAS

Surface water in Texas is owned by the state and held in trust for the public. Fees are based on the amount of water desired. An application to use less than 100 acre-feet costs $100, for 100 to –5,000 acre-feet is $250, for 5001 to –10,000 acre-feet is $500, for 10,001 to –250,000 acre-feet is $1,000, and more than 250,000 acre-feet is $2,000. Fees to amend are $100 per right amended. There is a recording fee of $1.25 per page. A right for agricultural use is $0.50 per acre of land to be irrigated. Storage requires a fee of $0.50 for each acre-foot of water storage. If the storage is for recreation, then the fee is $1.00 per acre-foot. Other uses are $1.00 per acre-foot based on the maximum amount of diversion. The maximum use fee for temporary applications is $500. The maximum use fee for an extension of time to complete construction is $1,000. Notice of an application must be published in a newspaper, and the applicant must pay that cost. This enables individuals whose interests may be harmed to protest.

The original application must be returned with six copies to the state. This includes six copies of the plans and supporting documents. Maps are required for most projects. The Texas Administrative Code gives the specifications for maps and drawings.* Domestic and livestock uses are exempt from the permit process. Texas recognizes temporary and term permits as well as permits for ordinary use. Texas law recognizes certain riparian rights to make domestic and livestock uses of the water but limits the amount of water available for these uses to no more than an average of 200 acre-feet if storage in any consecutive twelve-month period. Domestic water and water rights may not be sold separately from the land they are used on. These rights are appurtenant to the land, much as other riparian rights. Wildlife management and emergencies are also exempt uses.

Domestic and livestock uses are prioritized ahead of all appropriations in times of shortage. Limited term rights are behind all other rights in line. Protection of water rights in Texas either relies on the honor system, or in some instances a water master, appointed to oversee rights in the basin. This, understandably, creates some problems, as junior holders may not release the water they should or may illegally intercept water released for others. Where full-time water masters are in place, these problems are minor and can be easily dealt with, but in most cases there are no permanent water masters. Water masters monitor the basin and have authority to lock up pumps and take other actions to ensure that water reaches those with priority. Texas does not require permits for groundwater and apparently follows the line of thought that water below the earth belongs to the property owner of the land above. This helps explain the high degree of land subsidence in Texas, particularly around Houston.†

17.18 UTAH‡

Utah was the first state to engage in widespread irrigation in the United States, in the 1840s.§ It was also among the first states to institute the practice of prior appropriation.

---

* Texas Administrative Code 30 (TAC), Sections 295.121–295.126.
† See Figure 20. See also § 8.10.
‡ See Appendix 4, BLM Fact Sheet.
§ This refers to those of Anglo-Saxon descent. Some native tribes did practice large-scale irrigation prior to this date.

Riparian rights are not recognized in Utah. A water right requires actual diversion and application of water to beneficial use. The chief officer over water rights is the state engineer.* All waters in Utah are public property. Rights are established by permit.

Applications should be filled out and submitted to the state engineer. The application is then published and advertised to enable protests. The state engineer evaluates the application and other information and makes a decision based on state law. If the application is approved, the applicant begins construction and files notice with the state. Following application to beneficial use, the state issues a certificate. Maps are prepared by the state to show locations and acreage and must coincide with legal descriptions. A licensed engineer or surveyor must sign on. Instructions are included with the applications and forms. Legal descriptions are included in the application and must be made in conjunction with a competent survey. Detailed instructions are attached to assist in filling out the application. Utah uses an extensive fee schedule for applications. Fees are as follows:†

FEES				
FLOW-CFS		ACRE-FEET		
More Than	Not to Exceed	More Than	Not to Exceed	FEE
0	0.1	0	20	$ 75
0.1	0.5	20	100	$100
0.5	1.0	100	500	$125
1.0	2.0	500	1,000	$150
2.0	3.0	1,000	1,500	$175
3.0	4.0	1,500	2,000	$200
4.0	5.0	2,000	2,500	$215
5.0	6.0	2,500	3,000	$230
6.0	7.0	3,000	3,500	$245
7.0	8.0	3,500	4,000	$260
8.0	9.0	4,000	4,500	$275
9.0	10.0	4,500	5,000	$290
10.0	11.0	5,000	5,500	$305
11.0	12.0	5,500	6,000	$320
12.0	13.0	6,000	6,500	$335
13.0	14.0	6,500	7,000	$350
14.0	15.0	7,000	7,500	$365
15.0	16.0	7,500	8,000	$380
16.0	17.0	8,000	8,500	$395
17.0	18.0	8,500	9,000	$410
18.0	19.0	9,000	9,500	$425
19.0	20.0	9,500	10,000	$440
20.0	21.0	10,000	10,500	$455
21.0	22.0	10,500	11,000	$470
22.0	23.0	11,000	11,500	$485

* The Utah State Engineer's Office was created in 1897. In 1963 the name of the office was changed to the Division of Water Rights.
† Fees taken from http://www.waterrights.utah.gov/wrinfo/forms/fees.asp.

23.0	and above	11,500	and above	$500

Groundwater recharge permits are $2500; requests for extension of time before four-teen years are $25; requests for extension of time after fourteen years are $75; a request for extension of time (fixed time application) is $75; an application to segregate is $25; a report of water right conveyance submission is $25; and sewage efflu-ent reuse applications are $750. For well drillers, the general permit is $50, renewal is $25 if on-time, and $50 if late, Drill rig operator registration is $50 and drill rig operator registration renewal is $25, or $50 if late. Document certification is $4, copies are 25 cents per copy, self-service copies are 10 cents per copy, and there is no fee for fewer than five copies. There is a $2 fee to access computer data; plats and scanned documents run $2 minimum and 25 cents per copy; maps (blueprints) are $3; microfilm is 25 cents per copy; and facsimiles are .25 per copy.

17.19   WASHINGTON*

Waters in Washington State belong to the public and cannot be owned individually. Water rights are granted by the state for beneficial use. Any use of surface water beginning after 1917 requires a permit. Groundwater uses initiated after 1945 also require a permit. Certain groundwater uses are exempt from the permit require-ment but are still considered water rights by the state. There are four exempt uses of groundwater, livestock watering, watering noncommercial lawns and gardens one-half acre in size or less, water to a single home or group of homes requiring 5,000 gallons of water per day or less, and water for industrial purposes (including irrigation) requiring 5,000 gallons per day or less.

The priority date is the date the application was filed or for certain rights includ-ing exempt groundwater rights, the day water was first put to beneficial use. Water rights may be lost after a period of five or more years of nonuse but require an action under due process. The state considers four changes to existing water rights, place of use, point of diversion, additional points of diversion, and purpose of use. Rights existing before 1917 for surface waters and before 1945 for groundwater must still be registered with the state as of 1967. Uses not filed within the time allotted are con-sidered lost. Exempt groundwater users were encouraged to file claims to establish priority dates at this time. Some users were exempt from this requirement including, individuals served by corporations or public utilities, those with valid permits or certificates, other recorded rights issued by court decree, nonconsumptive uses, and livestock watering. The initial registration period ended in 1974. The registration period has been reopened several times since then.

Washington has a cost-reimbursement option for expediting applications. Washington requires a $50 minimum fee for all new water rights and increases the fee $1 per 1/100 cubic feet second. The fee to amend a current right is assessed at

---

* For more information on Washington water law, see http://www.ecy.wa.gov/pubs/0011012.pdf. See also Appendix 4, BLM style Fact Sheet. Note that the Washington fact sheet was created by the author for inclusion in this text and is not available from BLM.

$0.50 per 1/100 cubic feet per second with a minimum fee of $50. The fee for a new water-storage project is assessed at $2 per 1/100 cubic feet per second. The maximum fee to appropriate or store water is set at $25,000. The maximum fee to transfer a right is set at $12,500. The fee for an extension to construct or to file a protest is also $50 per year. There are no fees for donating a right to the state in trust, the acquisition of in-stream flows for public purposes, emergency withdrawal or temporary drought related changes, or agreements under cost-reimbursement agreements under state law.

17.20   WYOMING*

The Wyoming Constitution declares water to be the property of the state. The state engineer manages water rights in four water divisions. Wyoming uses separate application forms for surface and groundwater rights. Wyoming has recently altered its requirements and now requires coordinates to be on its applications. Like other western states, Wyoming requires that water be put to beneficial use before vesting a water right.

Wyoming requires a $25 fee for domestic and stock uses. All other purposes, including some storage, are $50. Stock reservoirs less than 20 acre-feet require a $25 fee. Those between 20 and 100 acre-feet are $50, and those over 100 acre-feet are $125. Temporary applications are $50, petitions to modify a permit or application are $20, and proof of appropriation is $20. Other fees are available on the state fee schedule.†️ The only means of acquiring a water right in Wyoming is by permit from the state. Historic use and adverse possession do not vest rights to water in Wyoming. A licensed surveyor must take surveys and prepare maps for each application. The priority date for rights is as of the date an application arrives in the state engineer's office.

If approved, construction must be completed within the time specified on the permit. Notice of completion must be given on the appropriate forms. An extension may be granted, but requests should be made before the original time lapses. After construction is complete and water has been put to beneficial use, a final proof of appropriation document must be filed. This is then advertised in the local paper and an inspection is made. If there are no problems or protests, the proof goes to the board of control for approval. A certificate is issued once the proof is approved. Once adjudicated, the water right is permanently attached to the specific land. Only the board of control may change this attachment in any way. If water is not used for five consecutive years, it is considered abandoned. Statutory procedure must be followed to bring about legal abandonment.

---

* See Appendix 4, BLM Fact Sheet.
† http://seo.state.wy.us/PDF/FeeSchedule.pdf.

18 Classification of Water Bodies and Types of Precipitation*

18.1 LAKES†

Lakes are distinguished from watercourses, meaning rivers and streams by the fact that they are not flowing. The existence of some current is not ordinarily enough to classify a body as a watercourse instead of a lake.‡ Lakes may be very shallow, such that marsh grass is seen on top.§ A lake is generally defined as "a large body of water surrounded by land."¶ Small mountain lakes are often called tarns. Lakes can be either freshwater or saline. The largest lakes in the United States are the Great Lakes, Lakes Michigan, Superior, Huron, Erie, and Ontario.** The largest saline lake is the Great Salt Lake.††

18.2 PONDS

Like a lake, ponds are distinguished from watercourses by the fact that they are not flowing. Ponds are generally distinguished from lakes by their size and can be defined as a "fairly small body of still water formed naturally or by hollowing or embanking."‡‡

18.3 POOLS

A pool is a "small body of still water."§§ It is generally considered smaller than a pond and may be suitable for such activities as swimming or watering cattle.

---

* For discussion of aquifers and groundwater, see Chapter 8, specifically § 8.4 "Aquifers" and § 8.5 "Artesian Wells."
† Described in Restatemet of Torts (Second) § 842, as reasonably permanent natural body of water at rest in depression in Earth's surface.
‡ *Boardman v. Scott*, 102 Ga. 404, 30 S.E. 982 (Ga.). See 78 Am Jur 2d, Waters § 43 with 1999 update.
§ *Libby, McNeil & Libby v. Roberts*, 110 So.2d 82 (Fla. App.). 78 Am Jur 2d, Waters § 43.
¶ *Oxford Pocket Dictionary and Thesaurus,* American Edition.
** For area covered by individual lakes see *Statistical Abstract of the United States,* 120th ed., no. 385, p. 231 (2000).
†† For area of lakes see *Statistical Abstract of the United States,* 120th ed., No. 385, p. 231 (2000).
‡‡ *Oxford Pocket Dictionary and Thesaurus,* American Edition.
§§ *Oxford Pocket Dictionary and Thesaurus,* American Edition.

18.4   WATERCOURSES* (SEE 7.16)

A watercourse is generally distinguished from lakes and ponds by the fact that it is flowing in a single direction. There are several definitions of watercourse used in different jurisdictions, but most involve some indication of flowing water in a well-defined channel, with currents. Examples of watercourses include rivers and streams. It is usually immaterial whether the water reaches the channel by seepage or from springs, so long as there is a living source "regularly discharged through a well-defined channel made by the force of the water."† A watercourse is said to have a substantial element of permanence and continuity but need not flow continually.‡ Its flow should be regular so that one may rely on it in its season.§

18.5   RIVERS AND STREAMS¶

Rivers are types of watercourses. A river is defined as a "copious natural stream of water flowing to the sea, a lake,"** or another river.†† A stream is defined as a "flowing body of water, especially a small river."‡‡ Brooks are defined as small streams and creeks are synonymous with streams.

18.6   DIFFUSED SURFACE WATERS§§ (SEE 7.34 ON APPROPRIATION AND DIFFUSED SURFACE WATERS AND 6.8 ON DRAINAGE AND UNWANTED RUNOFF)

Diffused surface waters are those that flow across the land with no definite channel or banks.¶¶ They are on the surface because of rainfall, snowmelt, or flooding.*** Most

---

* See 78 Am Jur 2d, Waters § 196 for Waterways and Conduits. Described in Restatement (Second) of Torts § 841 as a natural stream of water with constant or recurring flow in a defined channel on Earth's surface, including lakes, marshes, and springs from which stream originates.

† 78 Am Jur 2d, Waters § 7.

‡ 78 Am Jur 2d, Waters § 8; *United States v. Ide,* 277 F. 373, aff'd 263 U.S. 497, 68 L.Ed. 407, 44 S.Ct. 182; *Rait v. Furrow,* 74 Kan. 101, 85 P. 934; *Collins v. Wickland,* 251 Minn. 419, 88 N.W.2d 83.

§ 78 Am Jur 2d, Waters § 8; *Reynolds v. M'Arthur,* 27 U.S. 417, 7 L.Ed. 470; *Miller & Lux v. Madera Canal & Irrig. Co.,* 155 Cal. 59, 99 P. 502; *Chamberlain v. Hemingway,* 63 Conn. 1, 27 A. 239; *Hutchinson v. Watson Slough Ditch Co.,* 16 Idaho 484, 101 P. 1059; *Card v. Nickerson,* 150 Me. 89, 104 A.2d 427; *Mader v. Mettenbrink,* 159 Neb. 118, 65 N.W.2d 334; *Chicago, R.I. & P.R. Co. v. Groves,* 20 Okla. 101, 93 P. 755; *Wellman v. Kelley,* 197 Or. 553, 252 P.2d 816; *Johnson v. Williams,* 238 S.C. 623, 121 S.E.2d 223; *Hawley v. Sheldon,* 64 Vt. 491, 24 A. 717; *Re Johnson Creek Water Rights,* 159 Wash. 629, 294 P. 566; *McCausland v. Jarrell,* 136 W.Va. 569, 68 S.E.2d 729.

¶ For underground streams, see § 8.7.

** *Oxford Pocket Dictionary and Thesaurus,* American Edition.

†† 78 Am Jur 2d, Waters § 7.

‡‡ *Oxford Pocket Dictionary and Thesaurus,* American Edition.

§§ Restatement (Second) of Torts § 846 refers to surface water as, rain, snowmelt, seepage, springs, or detached floodwater, on surface of the earth but not part of any watercourse.

¶¶ The groundwater equivalent of surface runoff is called seepage, which is water that moves in the pores and crevices of the soil and rock but has no defined channel. Seepage often finds its way into a defined water body.

*** See 78 Am Jur 2d, Waters § 225. Floodwaters become diffused surface waters when separated from the main current and spread out over the surface.

jurisdictions do not control the capture of such waters under their appropriation and permit systems, meaning that individuals are allowed to capture such waters that cross their land and put them to beneficial use. Rules do generally apply when one attempts to avoid such waters.*

18.7   SPRINGS†

A spring is a "place where water wells up from the earth."‡ It has been described as any discharge of water from rock or soil to the surface.

18.8   OCEANS

Oceans are large bodies of saline water not surrounded by land. The oceans touching the United States are the Pacific, along the western coast, and Atlantic, along the eastern coast. Other oceans include the Indian, Arctic, and Antarctic Oceans. Typically, a nation has jurisdiction, under international law, over sea waters three nautical miles from shore (known as its territorial sea). The United States and some other nations also exercise limited jurisdiction over high seas, known as the contiguous zone, which for the United States is nine miles out.§ The United States also exercises exclusive fishing authority for up to 200 nautical miles from its territorial sea.¶

18.9   EPHEMERAL STREAMS

These are streams that flow only part of the year, often during the rainy season and remain dry for much, if not most, of the year. Though they do not flow constantly, they are still streams because they flow in a defined bed with banks, as opposed to surface runoff, which has no bed or banks.

18.10   SWAMPS, MARSHES, BOGS, AND WETLANDS (SEE ALSO 10.13 AND 10.15 ON WETLANDS)

Marshes are low lands that are generally wet and have few trees, if any. They are characterized by tall grasses, sedges, cattails, and rushes.** They may be a transition zone between land and water. Marsh water tends to be neutral or alkaline as opposed to bog water, which is often acidic. Marshes provide a rich, oxygenated environment with fertile soils. Unlike the woody plants in swamps, marsh plants tend to be largely herbaceous†† or nonwoody.

---

* See §§ 6.8, 11.8 on drainage.
† See 78 Am Jur 2d, Waters §§ 176–81 with 1999 update.
‡ *Oxford Pocket Dictionary and Thesaurus,* American Edition.
§ 13 Fed Proc Forms, Navigable Waters § 51:3.
¶ 13 Fed Proc Forms, Navigable Waters § 51:3.
**Rushes are marsh plants with slender stems, often used in making baskets and some chair seats.
††Herblike, not woody, green and leaflike.

Bogs are a type of wet, spongy ground, composed largely of decaying vegetation, usually peat moss. Bogs tend to be acidic. They occasionally have trees within them and are home to numerous shrubs and herbs. Drainage in a bog is generally somewhat worse than in swamps. The soil of a bog is not usually good for growing crops. The undecayed peat moss in a bog stores large amounts of carbon, preventing its release into the atmosphere.

Swamps are seasonally flooded lands that tend to have dense woody plants including a variety of trees. Drainage in a swamp is somewhat better than in a bog.

Wetlands generally include marshes, swamps, bogs, and other water-logged lands. These areas provide habitat for a wide variety of plants and animals and often act as buffers against potential disasters, soaking up water that might otherwise cause flooding and filtering out pollutants from the water and atmosphere. As wetlands are drained, we often disrupt the natural drainage system, creating new flood and erosion dangers.

18.11   RAIN

Rain is water that condenses in the air and falls to the earth as liquid drops. Rainfall generally occurs when atmospheric pressure is high or temperature is low. In order to have rain, there must be sufficient moisture in the air for the water to condense at the existing temperature and air pressure.* As the temperature increases, air tends to rise, lowering the air pressure below. As temperature decreases, air sinks raising the air pressure below.

Temperature is often lower at higher elevations. When water moisture is forced over high-altitude landmasses, it often causes temperatures to lower and increases the air pressure at that altitude, while the clouds are pushed over mountains. This causes rainfall on the sea side of the mountains. The opposite side of the mountains sees a decrease in rain as the clouds settle back to lower elevations where temperatures are usually higher and air is less compressed.

18.12   SNOW

Snow occurs when water freezes into crystals at high altitudes, changing directly from gas to solid. Snow then falls to the earth, in the form of white flakes. In extremely polluted air, snow has been known to have some color to it, when it falls.

18.13   HAIL AND SLEET

Unlike snow, hail stones are solid pieces of ice. They develop when rain freezes as it falls toward the earth. Hail stones can be quite large in parts of the country, sometimes forming pellets greater than the size of baseballs. These pellets can cause injury and property damage when they fall. Sleet is a combination of snow and rain. It may sometimes be referred to as wet snow and can be hazardous to traffic conditions.

---

* Water typically condenses around tiny dust particles floating in the air.

18.14   ARTIFICIAL WATER BODIES

There are numerous types of man-made water bodies. The most common include reservoirs, ditches, canals, and ponds. These bodies typically hold limited public rights, if any, and create greater liability in the owners or developers for injury to others.* There is also artificial groundwater.

---

* See § 11.10.

19 Prominent Federal Agencies Dealing with United States Waters

19.1 UNITED STATES GEOLOGICAL SURVEY (USGS)*

Created in 1879, the United States Geological Survey is charged with mapping the land and studying its composition. The USGS issues regular reports on water uses and supply, both nationally and broken down by states.

19.2 UNITED STATES ARMY CORPS OF ENGINEERS (USACE)†

Created in 1775, today the U.S. Army Corps of Engineers is responsible for the design, construction, and operation of numerous civil works projects, including dams, locks, and other flood-control measures. The Corps has great discretion in approving (or not approving) permits for construction and use of fill materials and permanent structures in navigable waters.

19.3 ENVIRONMENTAL PROTECTION AGENCY (EPA)‡

The EPA was officially established by executive order in 1970. It is responsible for monitoring numerous pollutants, making certain that pollution levels remain safe. The EPA regulates the environment, preventing and repairing harm. The EPA has a strong impact on national policy by establishing acceptable levels for toxins and pollutants and monitoring discharge levels. The EPA has authority under the Clean Water Act§ and Safe Drinking Water Act¶ to regulate pollution levels in water.

19.4 THE BUREAU OF LAND MANAGEMENT (BLM)**

The BLM, which is under the Department of the Interior, traces its origin back to 1785. Its primary purpose is to manage federal lands. Management of federal lands

---

* http://www.usgs.gov/.
† http://www.usace.army.mil/missions/index.html.
‡ http://www.epa.gov/epahome/aboutepa.htm.
§ 33 U.S.C.A. §§ 1251.
¶ 42 U.S.C.A. 300f-300j.
** http://www.blm.gov/wo/st/en.html.

also includes necessary water rights to support those lands, largely in the form of federal reserved rights.

19.5 THE BUREAU OF RECLAMATION (USBR)*

The United States Bureau of Reclamation is responsible for water-resource management in the western United States to encourage settlement of the arid lands. The bureau has constructed dams and diversions to deliver water and power throughout the West. Today, the agency is primarily focused on power generation rather than delivery of water, but with recent water shortages and droughts USBR remains prominent in planning for water resources. It is under the Department of Interior.

19.6 UNITED STATES FISH AND WILDLIFE SERVICE (USFWS)†

The function of the FWS is to provide safe, healthy habitats for fish and other animals and plants. The agency operates numerous fisheries and wildlife refuges. FWS plays a prominent role under the Endangered Species Act.‡ The FWS is under the Department of the Interior.

19.7 OTHER PROMINENT FEDERAL AGENCIES

Other agencies with roles regulating water include the United States Forest Service (USFS),§ National Oceanic and Atmospheric Administration (NOAA),¶ the Bureau of Indian of Affairs,** the National Parks Service,†† the Federal Emergency Management Agency (FEMA),‡‡ and the United States Coast Guard.§§

---

* http://www.usbr.gov/; See Chapter 16 for reclamation matters.
† http://www.fws.gov/index.html.
‡ 16 U.S.C.A. 1531.
§ http://www.fs.fed.us/. The Forestry Service is in charge of national forests and controls waters necessary for the preservation of those forests.
¶ http://www.noaa.gov/. The National Oceanic and Atmospheric Administration conducts most of its operations at sea and is not primarily concerned with inland water bodies or the rights that follow them.
** http://www.doi.gov/bureau-indian-affairs.html. See Chapter 9 for issues related to Indian Law.
†† http://www.nps.gov/. The National Parks Service retains sufficient waters to operate the parks and preserve them for the parks' intended purpose.
‡‡ http://www.fema.gov/
§§ http://www.uscg.mil/. The Coast Guard is primarily concerned with shipping and issues of maritime law, which are not addressed in this volume.

Terms

The terms listed here are often defined by statute in different jurisdictions, and definitions may vary from state to state. The definitions listed below may vary from those used in certain jurisdictions, and readers should consult the statutes of their own jurisdiction to find out if there are special meanings applied to terms in that jurisdiction. Some terms may have slightly different meanings within different sections in a single jurisdiction, and the reader should be cautious of the context in which the term is used.

abandoned well: A well with no present or future use or purpose, when use of the well is permanently discontinued.

abandonment of a water right: A termination of a water right by permanent discontinuance of its use. Conditional rights may terminate by failure to develop them. Such rights usually require reasonable diligence (also called due diligence).

absolute water right: A perfected water right, meaning it has been put to beneficial use.

absorption: The process of taking a liquid substance into a solid, such as when a sponge absorbs water. Often used to refer to the amount of water that can be held by rock and soil.

abstract of title: A summary of facts on the record that can be relied on as evidence of title and conveyance of real property.

abutment: 1. Side of a dam resting against a valley. 2. A supporting structure of a bridge, such as an arch. 3. Point where support and the thing being supported meet.

accretion: Gradual buildup or loss of earthen matter due to natural water flows of a stream.

acequia: Form of irrigation that originated in the American Southwest. These are a type of community ditch, largely unique to New Mexico.

acid: A substance with a pH of less than 7, this being neutral. Acids have more (H^+) ions than (OH) ions.

acidic: A condition in water or soil resulting in a pH less than 7.

acre: An area the size of 43,560 square feet. There are 640 acres in a section.

acre-foot: The amount of water needed to cover one acre of land at a depth of one foot (325,851 gallons).

activated sludge: A cleaning method for removing organic matter from water by using microscopic plants and animals.

active storage capacity: The amount of water in a reservoir that, at capacity, can be released by gravity flow without spilling.

act of God: An event that is so rare as to be unforeseen. Such events often cause damage to property or loss of life. An act of God is considered a defense in most jurisdictions against liability for the damages the event causes.

adjudication: The judicial process by which water rights are confirmed in Colorado. In other states this is the process by which rights are litigated when in dispute, usually when there are shortages or when other interested parties stand to be harmed by the continued exercise of the right.

admiralty: 1. Law of the sea, or maritime law. 2. Court holding jurisdiction over maritime law.

advanced treatment: Additional treatment for cleaning water following primary and secondary methods.

adverse use: A use against the interest of one or more rights in a drainage basin. Many jurisdictions, particularly in the West, no longer recognize such uses.

advertise: The means by which notice is provided to the public of a proposed use of water or extension, in one or more newspapers.

aeration tank: A chamber where air is injected into wastewater to assist microorganisms in cleaning it.

affluent stream: 1. A stream that feeds into another stream or lake. 2. A tributary stream.

aggradation: The raising of streambeds and other land beneath a water body by deposit of material, eroded and transported from elsewhere.

air injection: Movement of water from unsaturated zone to saturated zone by use of compressed air pumped into the soil.

algae: Simple form of plant life that often grows in wet, damp places.

alkali: 1. A substance with a pH over 7, capable of neutralizing acid by liberating hydroxide ions in water. These substances often form corrosive solutions, such as rust. 2. The soluble salts found in soil or water.

alkalinity: Water's ability to neutralize an acid.

alluvial: Soil or material deposited by flowing water.

alluvial fan: A fan-shaped formation created by deposits from water onto a gentle plain.

alluvial stream: 1. The changing *S*-shaped pattern of a channel subject to meandering alterations by constant erosion and deposition of materials from the banks and bed by flowing water. 2. Material deposited from upstream, such as silt, clay, sand, and gravel.

alluvium: 1. Soil deposits, usually fertile, left behind by a flood. 2. Deposits of materials such as clay, silt, sand, and gravel, left by a body of water on a floodplain, a delta, or a streambed, or at the base of a mountain.

American rule: A rule that requires reasonable use of groundwater so as to protect other owners.

anisotropy: Having differing physical properties based on direction or axis.

annual flood: The highest water level of a stream during a year.

annulus: The space in a well between the casing and outer wall of the borehole.

application: A written request to state governing official, usually the state engineer, for a right to use, store, or alter water.

apportionment: Division of water resource by a court or other authorized individual.

appropriate (*verb*)**:** Taking the necessary legal action to make a valid claim to use water from a natural body.

appropriation: The amount of water one has the legal right to use for beneficial use. This is often measured by the rate of flow. An appropriation may refer to any taking of water under either the riparian doctrine or through other means, such as eminent domain. It is most commonly associated with a vested right to use some amount of water for some beneficial purpose under the western United States appropriation doctrine.

appropriation doctrine: The system of water allocation used in the western United States and a few central states. The doctrine involves allocation of water rights based on the principle of first in time as opposed to riparian rights, which use land ownership along the body of water as the basis for allocating rights. Water rights in western states are nearly always based on a permit system, allowing the states to keep track of priorities of use. In times of shortage, the earlier users will have their rights satisfied first. Some variations exist among the several states using this doctrine, preferring certain uses to others.

approval: Granting a water right by the state regulatory agency or official, usually the state engineer. Some water rights may be appurtenant to their place of use.

appurtenance: In water law, when a water right attaches to the land it is used on. This usually means that, absent some agreement to the contrary, a right follows the land when the land is transferred.

appurtenant: A water right is appurtenant when it is attached to something, usually land. The right is said to be for the benefit of the land it is used on. Example: Water is appurtenant to the land.

appurtenant structure: 1. A structure that is said to belong to something else. 2. structures related to or attached to a dam, such as spillways, outlets, bridges, and access ways.

appurtenant to place of use: In many jurisdictions, when the land benefited is sold, the water right follows the land it is used on, unless specifically exempted.

aquaculture: Farming of aquatic life, such as kelp, fish, shellfish, or algae.

aqueduct: A conduit or channel designed for the transport of water from a distant source, often involving gravity.

aquiclude: 1. A confining bed that holds water, preventing it from reaching springs and wells. Aquicludes constitute the boundaries of an aquifer. 2. An impermeable rock formation that acts as a barrier to the flow of water.

aquifer: 1. A porous rock-and-soil formation holding water in substantial quantities. 2. An underground area used to supply water for use on the surface.

aquifer (confined): 1. An aquifer that is under pressure, such that water will often rise naturally, when accessed. 2. Impermeable soil and rock that surrounds water both above and below, such that it will rise to the surface on its own when penetrated.

aquifer (unconfined): An aquifer at atmospheric pressure at its surface.

aquifer system: A regional system of interbedded permeable and less permeable rock-and-soil formations that may comprise one or more aquifers with interspersed aquitards, which are separated by less permeable rock formations, limiting flow between aquifers.

aquitard: A less permeable bed of saturated rock and soil that impedes the flow of groundwater to wells and springs. Aquitards may transmit water to or from adjacent aquifers.

arable land: Land that is or reasonably may become suitable for agriculture and crop production.

area-capacity curve: A graph that shows the interrelation between surface area, volume, and elevation of a body of water in a reservoir.

arid: Area where irrigation is generally necessary to make land useful for agriculture, due to lack of precipitation.

arroyo: Type of drainage ditch common in New Mexico that carries off excess rainwater. Usually made of concrete, with flat bottoms and steep slanted sides.

artesian: An aquifer that is under pressure, because part of its water is at a higher altitude than the well, such that water will rise under pressure, without pumping.

artesian well: A well tapping into a confined aquifer. Such wells often create natural flows of water to the surface.

artificial recharge: The process whereby water is artificially inserted into an aquifer from some other source.

artificial waterway: A waterway that is created by human activity and does not exist naturally.

A.S.C.E.: American Society of Civil Engineers.

assessment: An amount of money paid by shareholders in a mutual irrigation company for costs of maintenance and operations.

assign: Transfer of entire interest in personal property.

assignment: Transfer of a water right that has not yet been perfected from one entity to another.

assimilation: The self-cleaning ability of a body of water. Its ability to remove pollutants and impurities.

attenuation: The slowing diversion or narrowing of the flow of water in a river or stream.

augmentation plan: A plan by which junior appropriators can obtain water under the approval of a water court in Colorado without harming senior appropriators.

augmentation water: Water that is in a stream to offset diversion by a junior that might otherwise harm seniors.

automatic controller: A timer capable of operating valve stations and controlling the time of water application.

average annual yield (water): The average supply of water produced in a year by a given stream or development.

average winter consumption (AWC): The average amount of water used by a consumer during the winter period. This is an indicator of indoor use.

avulsion: Similar to accretion, avulsion is the rapid alteration of land by a sudden event, causing soil to either build or recede.

A.W.R.A.: American Water Resources Association.

A.W.W.A.: American Water Works Association.

axis of dam: An imaginary symmetrical division across the center of the crest of a dam.

backflow: When water backs up in the opposite direction than the normal flow of a stream.

backflow prevention: Prevention of foreign substances from entering the water-distribution pipelines, usually accomplished with a mechanical obstacle or air gap.

backwater: Stream water that is prevented from flowing forward by an obstacle, such as a dam or current. This water sometimes flows backward.

baffle: A device inserted into flowing water to reduce turbulence.

bank: A land slope leading up from a body of water, usually formed by the natural flow of the water over time.

bank storage: Water that is absorbed in the banks of a body of water and that may be returned in time, as seepage, to a lower stage, when water levels at the bank stage drop below the water table.

bar: A long ridge of material, often sand or gravel located at or near the mouth of a river, sometimes rising above the surface of the water.

base flow: The sustained stream flow without direct runoff, including both human-caused and natural flows. The natural base flow is largely the result of groundwater discharge.

base: 1. A base is an elemental substance with a pH higher than 7, meaning it has fewer (H^+) ions than (OH^-) ions. 2. A foundation. 3. Line used as a starting point (see *baseline data*).

baseline data: A starting point for data and statistics to which other points can be compared and referenced.

basin: 1. A dip or hollow point in the earth's surface or beneath it, containing water. 2. An enclosed or partially enclosed body of water. 3. An artificial enclosure for water preventing impacts from tidal activity.

bathing: Activity in water where one washes oneself. At common law, this was a right of riparian landowners. The right still exists to some extent among riparian owners and may also exist in certain public waters. Today it is more common for people to wade in the water for recreational purposes, such as to cool off.

batture: An elevated stream or seabed.

bay: An indentation in land, smaller than gulf but larger than a cove and filled with water, which often serves as safe haven for ships.

beach: Sandy area along the shoreline of a body of water.

bed: The earth's surface beneath a body of water, commonly referred to as the bottom.

bedload: The sediment and particles transported in or near the bed of a body of water.

bedrock: Solid layer of rock beneath the soil and other loose material.

benchmark: 1. Data used for comparative purposes (see *base*). 2. A permanent mark of known coordinates, used as a point of reference.

beneficial use: The nonwasteful use of water for its appropriative purpose. In some jurisdictions, beneficial use is specifically defined. It usually includes irrigation, municipal uses, and industrial uses, among others.

berm: 1. A horizontal deposition of sediments on a beach formed by storm waves, which may be sloping toward land. 2. A narrow shelf extending into a stream or channel.

B.I.A.: U.S. Bureau of Indian Affairs.

biochemical oxygen demand (BOD): The amount of oxygen utilized in organic matter over a certain time at a given temperature.

biosolids: Solid matter that remains in water after waste water treatment.

B.L.M.: U.S. Bureau of Land Management.

blue hole: A void or sinkhole that forms in carbonate rocks, open to the earth's surface, containing tidal waters. This environment does not favor most sea life but often contains bacteria.

bog: Waterlogged, spongy ground, with soil composed of largely old, decomposed vegetation.

bolson: A valley raised by aggradation or alluvial flows, often surrounded by mountains, flowing into a central lake, with no surface outlet.

braided stream: A grouping of interweaving stream channels, converging and diverging, often separated by sandbars and islands.

breach: An opening in a dam or dike, which may be a fissure or a crack.

bridge: A structure used to span waterways or ravines. There are various types of bridges, the longest of which today are usually suspension or floating bridges. Some bridges can be raised to aid navigability of the waterway.

brine: Water containing amounts of salt in excess of 35,000 mg/L.

brook: A natural stream of water, somewhat smaller than a river or creek.

bubblers: A type of irrigation head delivering water to adjacent soil.

Bureau of Land Management: The United States Bureau of Land Management was created to manage federal land holdings. It falls under the Department of the Interior.

Bureau of Reclamation: The United States Bureau of Reclamation was created to provide water to lands in the arid West in order to encourage settlement. In recent years the bureau has seen a shift in purpose. It falls under the Department of the Interior.

bypass flow: Flowing water that passes a diversion structure or storage facility.

calibrated watershed: A watershed with enough devices and records to compute relationships between precipitation and stream flow.

calling the river: An action taken by a senior right holder to shut down at least part of the junior appropriations on a stream in order to secure his own appropriation.

camber: An arched surface material on the crest of a dam to protect the freeboard from settlement.

canal: An open channel constructed for the transport of water.

capacity: 1. The maximum amount of water that can be stored in a reservoir, without spilling over. 2. The flow of water that can be carried in a channel without flooding. 3. Water flow produced by a well per foot of drop in the well's water level.

capillary action: The means of water transport through solid matter such as soil, rock, plant roots, and blood vessels due to surface tensions, adhesion, and cohesion. This action is essential to transport of nutrients in plants and animals.

capture: Increased water in an aquifer from recharge or decreased discharge.

casing: A tube structure inserted into a borehole to maintain a well opening.

catchment area: 1. The intake area of an aquifer. 2. Flood-control reservoir or basin, also used for livestock and environmental water management.

cathodic protection: A means of protecting well casings from corrosion.

cave: A natural open space into the earth. These spaces may be created by splitting of rocks due to movement of the earth, corrosion due to water and ice, or other geological activities. Caves are generally considered large enough for people to enter and have openings to the earth's surface. Many caves are formed from limestone solutions.

cavern: A large, usually dark cave.

cavitation: 1. Erosion process in a streambed or channel that occurs when vapor bubbles collapse in the channel wall. 2. The wear on a hydraulic structure when high gradient is present.

cease and desist order: A formal written order from the state engineer to refrain from the continued performance of an illegal act.

cenote: 1. A deep natural well in Central America, formed by the collapse of limestone on the surface. 2. A water-filled sinkhole in the Yucatan Peninsula.

certificate: A document signifying that a water right has been perfected. The certificate serves as an official record and is usually recorded with the local county recorder's office or its equivalent.

cfs (cubic foot per second, or second-foot): A measurement of flow equal to one cubic foot per second passing a certain point. (448.8 gallons per minute).

chain: 1. A device used by surveyors to measure distance, consisting of 100 equal links in a chain, measuring 66 feet. (Note that an engineer's chain is equal to 100 feet.) 2. A measure of distance equal to 66 or 100 feet.

chain of title: A chronological history of record title for a parcel of real estate.

change: 1. An alteration of a water right. 2. The process of altering a water right. Changes usually affect the point of diversion, period of time, place of use, or type of use.

channel: An open conduit for flowing water.

channelization: Alteration of a channel to improve the flow of water, usually by deepening or straightening.

check dam: A small dam used to divert water or slow the flow of the stream.

chemigation: Process by which irrigation system carries pesticides and fertilizers to farmlands.

Cipoletti weir: A type of weir used for measuring the flow in an open channel.

civil law rule: A rule in the law of drainage that entitles owners of land to the natural drainage of the land. Most jurisdictions that follow this rule allow some alterations if they are reasonable in light of consequences to other lands.

Clean Water Act: 33 U.S.C.A. §§ 1251–1376, federal law stating United States policy and means for restoring and maintaining the quality of the nation's waters, including oceans, lakes, streams, rivers, wetlands, and groundwater. This is largely a pollution control and cleanup law.

closed basin: 1. A water basin where the topography of the land prevents any visible outflow. 2. A basin where neither surface nor groundwater escapes the basin (hydrologically closed).

cloud: A collection of water or ice particles in the air well above the earth's surface.

coagulation: Clumping solids together in order to more easily filter them out of water.

coast water: 1. Tidewater capable of being navigated by an ocean vessel. 2. Water open to the ocean and influenced by the tides, which can be navigated by ships or boats leaving the ocean.

C.O.E.: U.S. Army Corps of Engineers.

cofferdam: A temporary enclosure used to pump water out of an area or part of a body of water and expose the ground beneath the body.

coliform: A bacteria that originates in the digestive system of mammals. It is an indicator of disease when found in water.

collection pipe: A pipe or tube used to collect seepage and runoff from drains and other devices and direct it further downstream of a dam.

Colorado River Compact: An agreement between the states of Arizona, California, Colorado, Nevada, New Mexico, Wyoming, and Utah regarding allocation of the waters of the Colorado River.

Commerce Clause: Provision of U.S. Const. Art. I, § 8, cl. 3, giving Congress the exclusive authority to regulate commerce among the states, with foreign nations and with Indian tribes.

commercial water use: Water used by commercial enterprises such as motels, restaurants, office buildings, industry, etc., which comes from both the public and private sources.

Common Enemy Doctrine: Doctrine in drainage law that allows owners to divert unwanted waters from their lands in a manner that is strongly reminiscent of all owners for themselves. Owners may take what actions they wish on their own land to get rid of surface waters. This doctrine has been tempered in most jurisdictions by a rule of reasonableness.

common law: Body of law that derives from judicial decisions and ancient custom, but not from statutes (or constitutions). The common law often refers to the English legal system and its traditions. It is also known as the body of law to which no statute applies.

compact: An agreement or contract between two or more parties. *Compact* is the terminology usually applied to agreements between the states.

interstate compact: An agreement between two or more states. Interstate compacts must be approved by the United States Congress and therefore have the force of law behind them.

compact call: A call by a lower state to compel an upstream state to cease diversions of water from a river system that is necessary to satisfy the compact rights of a downstream state.

Compact Clause: U.S. Const. Art. I, § 10, cl. 3, forbidding a state from entering into a contractual agreement with other states or foreign powers without the approval of the United States Congress.

compaction: The compression of sediments due to pressure above, *compaction* often refers to the collapse of aquifers as they are drained. This usually refers to the nonrecoverable space for water storage, lost as the land compresses due to overpumping.

compaction residual: The difference between compaction that will occur with a given increase in stress and that which has occurred as of a specified time.

compression: The decrease in thickness of sediments over time, resulting from an increase in pressure above. If the compression is recoverable, it is said to be elastic; if not, it is inelastic.

condensation: Droplets of water that form on a surface when vapor in the air returns to liquid form. This is the opposite of evaporation.

conditional water right: A water right granted by the state, or in Colorado, the water court, to make beneficial use of water that has not yet been put to such use. The right is conditional upon completion of a project and actual application to some beneficial use. Ordinarily, the conditional right may continue so long as the holder of the right is diligent in pursuing the project's completion.

conduit: 1. A pipe or other means of carrying water from one place to another, such as to or from a source, place of use, or treatment facility. 2. A natural or artificial channel for conveying water. 3. A closed channel, such as a pipe or tube, for conveying water under or around a dam.

cone of depression: A depression formed in the water table, when water is withdrawn from a well. It gets its name from the conelike shape of the depression.

confined aquifer: Groundwater below a layer of solid or impermeable rock.

confining bed: A layer of bedrock or other formation that does not transmit water well. The rock is said to confine the water below or within it, such that it cannot readily escape. Also called aquicludes, aquitards, or semiconfining beds.

confining unit: A unit with relatively low permeability and high saturation, that confines water largely to a specific geological area. Confining units tend to create pressure differences between the water and the atmosphere and are associated with artesian aquifers. See also *aquitard*.

confluence: A place where two or more streams meet or join.

conjunctive use: The combined use and management of both ground- and surface waters, which recognizes the interrelation of these uses to one another, in order to optimize use and limit adverse effects.

conservation: The efficient and responsible use of water, resulting in less use and preserving greater resources for other present and future needs. Sometimes called end-use efficiency or demand management.

consolidation: A response in saturated soil to an increased load, which squeezes water from pores in the soil and decreases soil's porosity (storage capacity). See also *compaction*.

consolidated aquifer: An aquifer formed in hard bedrock, such as limestone, dolomite, or sandstone.

consumption pattern: The differences in a customer's use of water over time.

consumptive irrigation requirement (CIR): The amount of water actually required for consumption, by crops during irrigation. This amount does not include sources other than irrigation, such as precipitation, stored soil moisture, groundwater, etc.

consumptive use: 1. The amount of water consumed by a use and no longer available for further use. 2. A use of water that makes it unavailable for further use. Irrigation is generally considered a consumptive use.

contaminate\: To pollute or infect.

Continental Divide: An imaginary line separating North America at the Rocky Mountains, by the directional flow of the rivers either west, to the Pacific, or south and east to the Gulf of Mexico or Atlantic Ocean.

contour line: Line on a map representing elevation constants.

control section: The section of a measuring device where water flows through critical depth (tranquil to turbulent).

conveyance loss: The loss of water that results from conveying water from one location to another, in a ditch or other conduit, through evaporation, seepage, leaking, or transpiration.

Corps of Engineers: Refers to the United States Army Corps of Engineers, the agency charged with regulating many obstructions and construction operations on navigable waters. The Corps authority includes issuing permits for adding fill material to these waters.

cost-benefit analysis: A means of weighing the liabilities of a proposed action against the advantages.

course: Path taken by water as it flows.

cove: A small recess in a shoreline, often used as safe harbor for boats and ships.

creek: A stream of water, generally smaller than a river.

crest: 1. The top of a structure such as a dam, dike, spillway, weir, or other structure over which water must rise before passing the structure. 2. The point of the highest floodwaters on a river. 3. The foamy top of a wave.

critical depth: The depth at which water-flows change from tranquil to turbulent or turbulent to tranquil.

critical management area: An area designated for special management rules because water is being withdrawn in excess of the safe yield or sustainable supply for groundwater.

cubic foot per second (cfs): The amount of water that flows at a velocity of one cubic foot (one foot wide by one foot deep) per second. Often termed "cfs," it is 448.8 gallons per minute or nearly 646,000 gallons per day.

cumulative impact analysis: An analysis of overall environmental, social, and economic impacts of a proposed project and associated activities within a drainage area.

current: Force created by the flow of a river, stream, or other moving body of water.

current meter: An instrument that measures the velocity of flowing water in a stream, channel, or conduit.

cutoff collar: Collar constructed or placed around the outside of a pipe in order to lengthen the path of seepage over the outer surface of the pipe.

dam: 1. An artificial structure intended to control the flow or level of water in a stream. 2. A barrier obstructing the flow of water in some way.

dam inspection: A safety inspection of a dam.

dead storage: Water that cannot be withdrawn from a reservoir without pumping because it is below the lowest outlet (which can be serviced by gravity).

debris: Rubble and remnants of broken-down objects in the water. May include wreckage.

declared undergroundwater basin: Part of the state containing an ascertainable body of water beneath the surface of the earth, with relatively clear boundaries and announced by the state, usually through the state engineer, in assuming jurisdiction over the source of the waters.

decompose: To break down into smaller components.

decree: A court document bearing legal judgment.

decreed rights: Court-determined water rights.

deed: A written document of transfer of ownership of real property.

deformation analysis: An analysis of likely deformation of a dam due to strain resulting from buildup of water, sediments, and movements of the earth.

degradation: 1. Process by which rock and other formations of the earth's surface break down and disintegrate. 2. The removal of bed material from a channel or deepening of the channel.

demand forecast: Prediction of future demand for water and its use, usually based on projected growth and changes in an area.

demand management: Long- or short-term management of water usage by restrictions or conservation programs such as replacing old fixtures and inefficient equipment (e.g., toilets, sinks, washers, etc.), altering landscapes to decrease needs and reduce waste.

dental concrete: A type of concrete used to smooth out rough surfaces on a dam's foundations and abutments.

dependable supply: Water that can be relied on to be available in a particular amount and quality, at a particular time and place.

depletion: 1. The result of greater withdrawal than recharge to a water basin, when overall water supply declines. 2. A system's inflows less outflows.

desalination: The process of removing salts from saline water in order to create freshwater supplies.

desiccation: The loss of stored water from the pores (spaces) in sediments and rocks, caused by compaction, evaporation from exposure to air, cracking, or drying of the soil.

desiccation cracks: Fractures in the surface of the earth from drying soil or sedimentary rock.

designated groundwater: Groundwater that is not available for appropriation or fulfillment of surface rights.

developed water: Water made available by the party claiming right to it.

development: The advancement and improvement of land through construction of buildings and facilities.

dew: Moisture that condenses on cold, smooth surfaces out of the atmosphere.

dewdrop: A single drop of dew.

diffused surface water: Water that collects on the ground and often flows but does not form any type of a watercourse, has no permanent channel or bed, and usually results from one or more of the following: rainfall and other precipitation, snowmelt, or dew and other forms of condensation. Such waters are usually subject to different rules than other, more permanent waters.

dike: An embankment to control flooding.

diligence claim: In some jurisdictions, claims for the use of surface water, prior to a particular date when the law changed may be distinguished. In Utah, a diligence claim is one initiated prior to 1903. The terms used for similar claims may vary from state to state.

dilute: To decrease the amount of a substance as a percentage of a mixture by adding water.

direct flow (or direct right): The uninterupted flow of diverted water to its point of use.

disallowed: Use of the adjudication process to terminate a water right.

discharge: The amount of water passing a location over a period of time. This is most often expressed in terms of cubic feet per second.

disinfect: To kill or destroy disease-causing organisms in water. Purification.

dispersive clays: Clays with detachable particles that can be carried and transported in water.

dissolved oxygen: The amount of oxygen in water that is not attached to another chemical, such as hydrogen. This is usually expressed in milligrams per liter.

dissolved solids: Chemicals that dissolve in water and form a solution (e.g., sugar).

distribution system: A system for distributing water, overseen by the state (usually the state engineer or an appointed water commissioner).

disturbed slopes: Slopes that are altered from their natural state, including vegetation planted or otherwise caused by human activity.

ditch: An artificial channel dug in the earth or rock in order to transport water. Ditches are smaller than canals. They can be lined with impervious materials to prevent seepage, though they are open and subject to evaporation.

ditch rider: An Individual who operates an irrigation system for delivering water to farms within the irrigation project. Operation may include removal of debris, calculations of water needed, operation of gates, etc.

diversion: A change from the natural course of water, often using a canal or ditch. Most diversions are for use on land or for domestic or industrial purposes.

diversion dam: Barrier constructed to divert the water of a stream, either in part or in whole from its natural course.

divide: A land ridge separating two river systems.

divining rod: A forked stick or branch used to indicate the presence of water under the surface of the land, by bending downward when over the source.

division box: A structure that separates the water between two or more irrigation ditches.

dock: 1. Pier or wharf. 2. Area of water next to a pier or wharf, where ships dock.

domestic water use: Water that is used for household purposes, including tap water, bath and shower water, toilet water, etc. Only about 15 percent of the population in the United States supplies its own water. Most is supplied by public organizations, including county water departments.

downpour: Very heavy rainfall.

downstream: In the direction of the stream's current or flow.

dowser: 1. A forked stick used to find underground water. 2. A person who uses the stick or rod.

draft: Diverted water.

drain: 1. A means of withdrawing liquid from a basin or area, often using gravity. 2. To make empty by removal of liquid.

drainway: Conduit by which a liquid, such as water, drains and is carried away.

drainage area: A naturally draining area that flows to a certain point on a river, stream, or creek.

drainage basin: An area drained by a river and its tributaries, also a catchment area, watershed, or river basin.

drainage blanket: A layer of permeable material placed over a foundation.

drawbridge: A bridge that can be raised and lowered in whole or in part to prevent access or enable passage of ships.

drawdown: The lowering of the water table or surface of groundwater due to pumping.

dredging: The deepening of stream and river beds by removing solid materials such as sediment and rock.

driller: One who commences the drilling of wells. Usually requires a permit from the state.

drip irrigation: An irrigation technique involving pipes that fill with water and slowly drip onto crops. This method conserves water by limiting runoff and evaporation losses.

drizzle: Gentle constant rain of fine water droplets.

drop structure: An obstruction within a stream, placed there to stabilize the channel, it facilitates the downward movement of water while decreasing erosion of the banks.

drought: A lengthy period of low precipitation, well below the average for the affected area.

dry: Lacking moisture.

dryland farming: Practice in arid regions intended to produce crops without irrigation, often using moisture-conserving techniques.

due diligence: 1. Sufficient effort to bring about fruition of an intent to appropriate water. 2. Actions showing good faith effort or intention to complete diversion.

duty of water: The reasonable unwasted amount of water that, if applied over time, is necessary to produce the maximum crop harvest on a plot of land.

dynamic analysis: A study that predicts stability and deformation in dams due to seismic loads.

ebb and flow: The movement of the tides.

eddy: A whirlpool-like current at odds with normal currents in an area.

efficiency: The percentage of water diverted for a use that is actually consumed by that use, as opposed to seepage, evaporation, runoff, and other loses. The more efficient a system is, the less waste will exist.

effluent: Discharged polluted water from sewers, industry, or other sources that enters a stream.

effluent exchange: The use of wastewater to replace water diverted upstream.

EIS (Environmental Impact Statement): A detailed statement, evaluating the environmental costs and benefits of a proposed action as well as feasible alternatives.

election to file a water user's claim: An outdated form that was used in some states (e.g., Utah) to allow the state engineer or other officials to file a proof of water right in a court adjudication.

ELU (equivalent livestock unit): A standard of measurement attempting to measure water consumption of livestock (one ELU = one large animal or five medium animals or thirty-three and a third small animals).

embankment: An artificial bank of soil and rock, above the normal surface of the land used to divert, contain, or store water.

emergency action plan: A plan for dealing with emergency situations (e.g., dam breaks, flooding, sudden soil erosion or collapse, equipment failure, etc.), in order to prevent property losses and deaths.

emergency spillway: A secondary means of diverting water around or over a dam, in the event of unusual circumstances, such as failure of the spillway or a sudden increase in the amount of water needing diversion.

eminent domain: The power of the state as a sovereign to take private property for public use. The owner of a seized property must be reasonably compensated for it.

enclave clause: U.S. Const. Art. 1 § 8, cl. 17, clause granting to federal government jurisdiction over certain federal lands as approved by the state in which the lands reside. It has been interpreted with Art. IV § 3, cl. 2 to grant authority over virtually all federal lands.

endangered species: A species of plant or animal threatened with extinction by changes in its range.

Endangered Species Act: 16 U.S.C.A. §§ 1531 et seq. (1973), federal law establishing policy in the United States and legal protection for protect plant and animal species and their habitats from extinction or serious decline.

Energy Policy Act (EPACT): 1992 federal law that regulates water use in toilets (1.6 gallons per flush), showers, and faucets (2.5 gallons per minute) manufactured after January 1994.

English Rule: A rule regarding use of groundwater that states that the owner of land has absolute ownership of all that is in the air above and the ground below. The owner has an absolute right to trap water and use as much as he wishes. Most jurisdictions using this doctrine have tempered it to prevent actual malice. This doctrine has proved impractical in most jurisdictions today and now reasonable use is the predominant doctrine in the United States.

Environmental Assessment: See *Environmental Impact Statement.*

Environmental Impact Statement (EIS): Document required by National Environmental Policy Act (NEPA, 42 USCA § 4332(2)(c)) for all major federal projects, stating environmental costs and benefits of the project and viable alternatives. A lesser document titled *Environmental Assessment* may be completed to determine if an EIS is needed. Its purpose is to assure that government makes informed decisions and its function is largely procedural.

E.P.A.: U.S. Environmental Protection Agency.

ephemeral stream: A short-lived stream or portion of a stream that flows in response to precipitation but remains dry during most other times and usually lasts only a few hours in response to individual storms as distinguished from intermittent streams that may flow for months.

equitable apportionment: The equitable division of water among the states touching on it, by the United States Supreme Court. This doctrine only applies to states and not private individuals and may be applied to other shared resources, such as fish.

erosion: The wearing away of soil and rock from wind and water forces.

escaped waters: Water that is not consumed during application to some beneficial use and can potentially be recaptured for further use.

escarpment: A steep slope separating two largely level areas of land (at different elevations), caused by erosion or faulting.

estuary: The mouth of a river or stream, where freshwater and saltwater mix as the river enters the ocean.

evaporation: The process by which liquid water becomes a vapor from off water and land surfaces.

evaporation net reservoir: The loss of water from a reservoir due to evaporation, after allowing for precipitation. Equals total evaporation minus precipitation on reservoir surface.

evapotranspiration: The total amount of evaporation plus that of transpiration.

excess water: Water that flows in a stream beyond that which is necessary for reasonably beneficial uses. Unappropriated waters not being used for any beneficial purpose by the state.

exchange: 1. The release of water in one place in exchange for water from another. 2. An application to the state to change the water used for a project from the U.S. Bureau of Reclamation or some other like entity.

exfoliation: Process of flaking as applied to either organic material or soil and rock. This process in rock is usually caused by weathering. See *spall.*

exit channel: A downstream channel from a conduit that directs flows to a point where they can be safely released without risk to the dam.

expansion: Undeveloped areas at the time distributor begins operations that eventually are covered by the service based on proximity and reasonable assumptions.

extension: A request for more time to complete a project, resume a use, or complete some other required task related to a water right.

face: Exterior surface of a structure.

fallow (fallow land): Plowed cropland that remains unplanted.

farm efficiency: The required water for crops divided by the amount delivered for irrigation.

faucet: A device for controlling the flow of water from a pipe, such as in a sink.

fecal coliform bacteria: Bacteria present in stomachs and feces of warm-blooded animals. These bacteria indicate possible sewage pollution of the area.

federalism: The distribution of power between national and state governments in a federal system.

FERC: Federal Energy Regulatory Commission; formerly the Federal Power Commission.

Federal Land Policy Management Act (FLPMA): A federal land-use management law, covering national forests and Departments of Interior and

Agriculture lands. The act protects scenic, historical, ecological, environmental, archaeological, and other values.

FEMA: Federal Emergency Management Agency.

fill (*also* fill material): Usually rock, dirt, sand, or gravel placed into a water bed to raise or fill it.

filtration: Process of cleaning water by passing it through porous material, such as sand, in order to trap finished particles.

finished water: Treated (potable) water considered safe and ready for delivery for consumption.

firm annual yield: The yearly dependable supply of water that can be relied on from a natural source.

fish: 1. Animals with vertebrae that live and breathe underwater. 2. The act of catching fish and other sea creatures, often deemed a public right in rivers, lakes, and other public waters.

flashboards: Easily removable timber, concrete, or steel atop a spillway to raise the water level.

flat rate: A rate paid by customers where the amount charged remains constant from bill to bill for all customers, based on a predetermined formula rather than actual consumption.

flats: Flat or level ground, such as tidelands.

floatable stream: A stream that, while not necessarily suitable for boats and other watercraft, may still be used to float items, such as logs, to their destination.

flood: Water overflow on land outside the banks and not usually covered by water.

flood, hundred-year: (See one hundred-year flood).

floodplain: The part of a river valley that is covered with floodwaters when the river overflows.

flood routing: The calculation of stream flow or reservoir change (rise and fall) as floodwaters move downstream.

flood stage: Elevation where overflow of the natural banks reaches an area of measured elevation.

floodwater: Waters in excess of the bank's ability to hold them that flow onto adjoining land outside the channel.

flow: The rate of discharge expressed in terms of volume per time.

flow model: See *groundwater flow model.*

flowing well/spring: A well or spring under sufficient pressure that water rises without pumping.

flowline: The lowest elevation point where water can flow from without pumping, in a conduit, such as a ditch, pipe, or other structure for the conveyance of water.

flow restriction device: A device attached to a meter that regulates flow for the utility, to the consumer.

flume: A sloped channel used to convey water, often constructed of wood or concrete.

fluvial: Of or pertaining to rivers and streams.

fog: Low-forming cloud of water vapor close to the earth's surface.

forebay: 1. A storage basin for regulating water, often near a large reservoir. 2. A basin for transitioning surface water into groundwater.

foreign water: Water from another basin or jurisdiction.

forfeiture: Loss of water right due to statutory period of nonuse. In some jurisdictions there must be an action by the state, requiring notice, before water is forfeit.

foundation of dam: Material on which a dam is built or placed.

fountain: 1. The head of a stream or source of a spring. 2. An upward-shooting stream, projected by artificial means, that may be for drinking, washing, or pure aesthetics.

freeboard: The distance vertically between the top of a structure and the maximum water level for its design.

free-flowing well: An artesian well in which water rises above the surface level of the land. See also *potentiometric surface.*

freshwater: Water with less than 1,000 milligrams per liter of dissolved solids.

friction loss: Energy lost in the flow of water because of friction the water and the walls of the conduit or channel.

frost: Frozen dew or water from the air that condenses on solid surfaces and freezes.

fugitive water: The flow of water from any pipe, faucet, or other conduit for supplying, storing, transporting, or disposing of water or public use.

furrow: A partial surface flooding method of irrigation normally used with clean-tilled crops where water is applied in furrows or rows of sufficient capacity to contain the design irrigation stream.

futile call: The situation when stopping a junior use would not produce any more water for a senior, and so the junior use is allowed to continue.

gabion: A wire cage used in water-control structures, usually filled with stones, to protect banks and channels.

gage (gauge): Instrument used to measure water in some manner (e.g., surface elevation, velocity, or pressure).

gage height: The highest water level where data is collected, also termed stage.

gaging station: Site on a body of water where data is obtained.

gaining stream: A stream or river that gains water from seeping groundwater or springs.

gallery: A structure within a dam's interior that can be accessed for data collection and other work.

gate: A device for controlling the rate of flow in a conduit for water, such as a ditch, canal, or pipe.

geodetic datum: Set of data constants related to the study of the earth.

geohydrology: Study of hydrology as related to groundwaters.

geoid, earth: An imaginary surface coinciding with the earth's mean sea level or the shape of that surface with an ellipsoid flattened at the poles, caused by the rotation and gravitational forces.

Geological Survey: Usually refers to the USGS (United States Geological Survey).

geomembrane: A fabric designed as an impermeable barrier.

geotextile: A fabric created for erosion control, drainage, and soil reinforcement.

geothermal resource: Natural heat energy from beneath the earth's surface.

geyser: An opening in the earth's surface containing naturally heated water that occasionally erupts into a shower of water and or steam.

giardiasis: Disease caused by drinking polluted water with parasite *Giardia*.

GIS: Geographic Information System.

glacier: A large ice formation formed by compaction of snow over long periods of time. Glaciers move slowly down the mountain slope because of their weight and the pull of gravity.

GPS: Global Positioning System.

grade: Slope of a streambed.

gradient: The degree of incline or slope of a streambed.

gravity: Irrigation in which the water is not pumped but flows in ditches or pipes and is distributed by gravity.

gravity flow: A flow of water generated by gravity, ordinarily downhill.

Great Lakes: A series of large lakes in North America, including Lakes Erie, Huron, Superior, Michigan, and Ontario. These are among the largest freshwater bodies in the world.

greenbelt: An area of trees, or natural, undeveloped land within a city, such as a park.

greywater: Wastewater that drains from domestic uses including showers, tubs, sinks, toilets, etc.

groin: Area where the face of the dam meets with the abutments or supports.

groundwater: Water beneath the surface of the earth, sometimes applied only to water beneath the water table.

groundwater, confined: Groundwater that is under pressure, confined beneath a layer of impermeable material.

groundwater, unconfined: Water that is part of an aquifer and has an exposed water table at atmospheric pressure.

groundwater basin: A geographic area with at least one large aquifer.

groundwater flow model: A model created on a computer to calculate the field of hydraulic heads and groundwater flow.

groundwater management plan: A state plan for overseeing and regulating the withdrawal of groundwater, often developed by the state engineer or equivalent state officer.

groundwater mining: Condition when withdrawal of groundwater exceeds recharge, often causing subsidence and permanent decline in the water table.

groundwater recharge: The inflow of water to an aquifer or other underground reservoir from the surface.

groundwater reservoir storage: The stored water within the defined bounds of an aquifer.

grout: Thin mortar, such as concrete, that can be poured into cavities and small places, for example, cracks and crevices.

grout curtain: A barrier used to reduce seepage beneath a dam.

gulch: Small ravine cut by a torrent.

gully: Small ravine, carved by running water and functioning as a drainageway for water after heavy rains.

guzzler: A tool for storing and collecting rainwater to water livestock and other animals.

hail: Pellets of frozen rain that fall to the ground. Also refers to the process of these pellets falling to earth.

hailstone: A single pellet of frozen rain.

hand watering: Use of a handheld or mobile hose to irrigate land.

harbor: 1. A protected area of the water, near the land and deep enough to anchor ships. 2. Protected area of water with port facilities and docks.

hardness: Indication of mineral concentration in the water, often calcium and magnesium. Hard water requires more soap for cleaning.

harvested water: Collected runoff, stored for reuse for additional uses.

haze: Dust, moisture, smoke, and vapor in the atmosphere, tending to obscure visibility.

head: 1. Elevation difference between pipe intake and discharge points (indicates water pressure). 2. Difference in elevation between water at some point upstream and water at some point down stream. 3. Measure of energy in water resulting from elevation differences (gravity), pressure, and velocity.

head hydraulic: Measure of potential flow for water.

headgate: Device for controlling the flow of water into a conduit. Also the diversion point of water from a river for irrigation.

head loss: Decrease in head caused by efriction, a lowering of elevation, a decrease in pressure, or a drop in velocity.

headwaters: 1. The source of a steam. 2. A group of small streams that join together into a river.

heat exchange: Process for transferring heat in fluids from low-temperature reservoir to higher temperature reservoir. An example of this is the radiator of a car where water or antifreeze is used to cool the car by transferring heat to the air.

high water use turf: A layer, of high water use grass, including the roots, that is regularly mowed. Such grasses include varieties of bluegrass, ryegrass, fescue, bentgrass, etc.

historic flow: Past flows that have already occurred and were or could have been recorded at a gaging station if it were operational.

hose: A flexible tube for carrying water. It usually may be moved from place to place and rolled up for storage. Hoses may take numerous attachments, including sprinkler heads, nozzles, and other tools.

hundred-year flood: See *one-hundred-year flood*.

hydraulic conductivity: A measure of a medium's (material or object's) ability to transmit water, measured as the volume of water that will move through the pores in the object or material in a specified period of time, with a specific gradient (slope) through a measured area. This is both a measure of the property of the liquid and the medium through which it travels.

hydraulic fracturing: Fractures created in stone and soil material caused by water pressure.

hydraulic gradient (groundwater): The slope of a water table in a specific direction.

hydraulic head: See *head*.

hydraulic height: The vertical distance on a dam from the streambed on the downstream side to the surface of the water at the crest of the primary spillway.

hydraulics: Science of the use of water in pipes often for power generation, using the pressure created by the water to regulate tools.

hydrocompaction: Process by which dry, buried deposits compact when they become wet for the first time. The process involves a decrease in volume and increase in density. The drop in land surface is often termed shallow subsidence or near surface subsidence.

hydroelectric power water use: Use of falling water for generating electric power by turning large turbines.

hydrogen: Flammable gas, one of the two chemical components of water.

hydrograph: Graph showing properties of water (e.g., stage, flow, velocity, etc.) over time.

hydrographic survey: 1. Survey to determine characteristics of the bodies of water in an area. 2. The elevation and position of high-water marks. 3. Characteristics of a well. 4. Map of water-use locations in an adjudication proceeding.

hydrologic cycle: The cycle of water from liquid or solid on the earth's surface into the air as vapor and back to earth as precipitation.

hydrology: Study of water properties, distribution, and motion on and in the earth, including its atmosphere.

hydrometer: Instrument for measuring liquid density.

hydropower: Electric power produced by turbines moved by falling water.

ice: Ice is the solid form of water. Title to ice generally vests in the owner of the soil below it. When ice is harvested on navigable or public waters, however, title to the soil below is usually irrelevant and the harvester must ordinarily obtain a permit from the relevant state or federal agency. If the harvester is a riparian with riparian rights to the water, he has only an equal right to harvest ice with all other riparian owners.

impermeable: Does not permit fluids to pass through.

impervious: Cannot be penetrated by water.

impoundment: Water confined by physical structure such as a dam, dike, or other barrier.

incidental recharge: Groundwater recharge incidental to human activity not intended as recharge.

incremental damage assessment (IDA): An analysis of likely impact of dam failure during an extreme hydrologic event.

industrial water use: A use of water for industrial purposes such as washing, cooling, mixing, or transporting goods, including metals, oil, chemicals, paper, etc.

infiltration: The flow of water from the earth's surface to the pores and crevices of the soil beneath.

infiltration rate: The rate of water absorption into the soil per unit of time, often expressed in inches per hour.

inflow: The flow into something.

inflow design flood (IDF): Hydrograph of floods, used to determine the size of a dam's spillway.

influent: Water entering a plant for treatment.

injection: Artificial recharge of groundwater, using surface waters.

injection well: A well constructed to inject treated wastewater into the ground for storage or recharge. These wells are usually drilled below freshwater in an aquifer or inserted into aquifers that are not used for drinking water.

inland waters: See *international waters.*

inlet: 1. An indentation of a shoreline such as a small bay. 2. Narrow passage between islands. 3. A stream or bay that leads inland. 4. A drainage passage.

inlet channel: An upstream channel from a conduit or spillway.

inspection: Entry and examination for purpose of ascertaining violations of the water code.

in-stream flow: A nonconsumptive use of water that requires no diversion.

in-stream use: Water uses that require no diversion, such as fishing, swimming, transportation, etc.

integrated resource planning (IRP): The use of environmental, engineering, social, financial, economic, and other considerations, involving stakeholders and users, to plan for future uses and needs.

interference: Condition when a new well adversely affects an existing well, by dropping the water table in the earlier well or by coming into contact with the area of influence of the earlier well, in any way disrupting operations of the prior well. Interference also applies to any activity that disrupts, blocks, stops, or otherwise impedes any legal use of water for any reason. Interference may constitute a tort or even a crime where it is unauthorized and against a vested right, and may result in an award of damages.

intermittent stream: A stream that flows only part of the year, often in response to the rain season when precipitation is higher and groundwater is discharged. Distinguished from ephemeral streams that exist only a short period following a single storm.

interstate compact: See *compact.*

interstices: Cracks and openings in rock and soil where water can be stored.

internal waters: See *international waters.*

inundation map: Map illustrating the area that would be at risk from floodwaters, in severe storms or from dam failure.

irrecoverable losses: Water lost to various processes such as evapotranspiration, evaporation, or salt sink.

irrigated area: The area of land that receives irrigation water from the diversion conduit.

irrigation: The application of flowing water to arable lands for crops, often by ditch or pipes. Some of the different types of irrigation are listed below:*

> **center-pivot:** Automated sprinkler irrigation achieved by rotating the sprinkler pipe or boom, supplying water to the sprinkler heads or nozzles, as a radius from the center of the circular field to be irrigated. The pipe is supported above the crop by towers at fixed spacings and propelled by pneumatic, mechanical, hydraulic, or electric power on wheels or skids in fixed circular paths at uniform angular speeds.

---

* Descriptions of different irrigation methods taken from http://www.seo.state.nm.us/water_info_glossary.html, New Mexico Office of the State Engineer.

Water, which is delivered to the center or pivot point of the system, is applied at a uniform rate by progressive increase of nozzle size from the pivot point of the system to the end of the line. The depth of water applied is determined by the rate of travel of the system. Single units are ordinarily about 1,250 to 1,300 feet long and irrigate about a 130-acre circular area.

drip: An irrigation system in which water is applied directly to the root zone of plants by means of applicators (orifices, emitters, porous tubing, perforated pipe, and so forth) operated under low pressure. The applicators can be placed on or below the surface of the ground or can be suspended from supports.

flood: The application of irrigation water where the entire surface of the soil is covered by ponded water.

furrow: A partial surface flooding method of irrigation normally used with clean-tilled crops where water is applied in furrows or rows of sufficient capacity to contain the design irrigation stream.

gravity: Irrigation in which the water is not pumped but flows in ditches or pipes and is distributed by gravity.

sprinkler: A planned irrigation system in which water is applied by means of perforated pipes or nozzles operated under pressure so as to form a spray pattern.

subirrigation: A system in which water is applied below the ground surface either by raising the water table within or near the root zone or by using a buried perforated or porous pipe system that discharged directly into the root zone.

traveling gun: Sprinkler irrigation system consisting of a single large nozzle that rotates and is self-propelled. The name refers to the fact that the base is on wheels and can be moved by the irrigation or affixed to a guide wire.

irrigation conveyance loss: The water loss in transit from the source to the place of use.

irrigation efficiency: The amount of diverted water for irrigation that is actually consumed by the intended vegetation. The lower runoff and wasted water, the higher efficiency.

irrigation leaching requirement: The quantity of water needed to transport excess salt from soil in order to create a healthier soil for crops.

irrigation requirement: The amount of water needed to grow a specific crop (excludes precipitation).

irrigation return flow: Unconsumed water from an irrigation project that remains as runoff and eventually returns to a ground or surface body of water.

irrigation water use: The use of water by artificial conduit to water crops or other vegetation. See also *irrigation.*

island: A landmass surrounded entirely by water.

jetty: A structure, extending toward the sea, built for the protection of a pier, shore, or bank.

joint tenancy: Co-ownership by multiple individuals with a right of survivorship, for the last remaining owner, to the whole property or that portion included in the joint tenancy (if less than the whole).

junior rights: More recent water rights that are entitled to take only after older rights are satisfied. (See *priority*.)

KAF: One thousand acre-feet.

karst: A limestone formation characterized by sinkholes, ravines, and underground streams, formed by erosion.

karstification: Largely chemical action caused by water that produces karst topography.

karst, mantled: Terrain of limestone (karst) features covered by soil and thin mineral deposits.

kilogram: One thousand grams.

kilowatt-hour (KWH): 1,000 watts an hour. This is often used by utilities for defining rates (e.g., cents per kilowatt hour).

landscape area: A parcel of land less the buildings, driveways, and other paved land.

lake: A large, landlocked body of water.

lapse: The termination of an application for a water-right application because conditions were not met.

lateral: A lesser ditch used for transporting water that splits from a major ditch.

leaching: The removal of salts and other dissolvable material by irrigation and drainage of the land.

left (or right) bank: Facing downstream, the left or right hand side of a stream.

lentic waters: Nonflowing waters such as ponds or lakes (standing water).

levee: A barrier along the banks of a stream or other body of water for containing floodwaters.

license: Formal permission to do something, issued from a government or other authorized authority.

liquefaction: 1. The process of making something into a liquid. 2. Loss of strength in saturated soil from a major force (e.g., an earthquake).

littoral: Region on the shore of a nonflowing water body, such as a lake.

littoral rights: Rights to the water, of those whose land touches a lake or other nonflowing body of water, based on land ownership. Similar to riparian rights for those near the stream.

livestock water use: The water used to water livestock, feedlots, dairy cows, fish farms, etc.

load: The amount of solid material carried in a stream at a particular time.

lock: Enclosed chamber of canal, dam, or other structure, with gates for raising and lowering water levels to aid passing ships

log boom: Device that protects a spillway from large floating debris.

losing stream: A stream that is losing water due to seepage.

losses incidental to irrigation: The amount of water in excess of that beneficially consumed in an irrigation project that is depleted by irrigation.

lotic waters: Flowing waters.

low-level outlet: Conduit used to lower the water level in a reservoir or to release water downstream.

lower basin states (Colorado River Compact): Arizona, Nevada, and California.

low water use plants: Plants that can survive without artificial water once established.

lysimeter: A tool for measuring the movement of water downward (quantity) through a block of soil.

mainstem: The main stretch of a river or stream formed by smaller tributaries.

manometer: Instrument for measuring pressure.

maritime law: Law related to shipping and the open sea.

marsh: Low land that floods in wet weather and usually remains wet even during dry weather.

master: A judicially appointed officer who acts as a referee or serves some other function in order to assist the court. The master reports to the court, in writing, but may perform many functions such as taking testimony, investigating, making computations, and holding hearings on pretrial matters.

master's report: The formal report to a court, provided by an appointed officer, usually presenting findings of fact and making recommendations and conclusions of law.

maximum capacity: The maximum amount of water a reservoir can store when filled.

maximum contaminant level (MCL): Standard set by the Environmental Protection Agency (EPA) for water quality under the Safe Drinking Water Act. MCL represents the highest tolerable amount of a substance in drinking water, under the law, which is intended to avert excess risk to human health.

mayordomo: The chief official in charge of a community ditch or acequia.

M.C.E.: Maximum credible earthquake.

mean depth: The average depth of water in a channel or conduit. Mean depth is equal to the cross sectional area divided by width.

meander: The winding turn or bend of a stream.

meander line: A survey line used to measure property that follows the shape of a body of water, such as a stream. The meander line is not a property line.

measuring weir: A shaped notch for measuring water as it flows through.

medium and low water use turf: A layer of earth's surface covered in grasses that require mid- to low levels of water. These grasses include Bermuda, Zoysia, and buffalo grass.

medium water use plants: Plants requiring some artificial watering for a healthy life.

meteoric water: Groundwater that originated largely from precipitation.

meteorology: Study of atmosphere and weather.

mg/L: Milligrams per liter. This refers to the amount of a substance contained in a liter of water. One milligram per liter is equal to one part per million.

microbes: Microscopic plants and animals. Also *microorganisms*.

milligram (mg): One-thousandth of a gram.

milligrams per liter (mg/L): A measurement of the concentration of solid mater in a liter of water. It represents 0.001 gram of a constituent in one liter of water. It is approximately equal to one part per million (PPM).

million gallons per day (mgd): A rate of flow of water equal to 133,680.56 cubic feet per day, or 1.5472 cubic feet per second, or 3.0689 acre-feet per day. A flow of one million gallons per day for one year equals 1,120 acre-feet (365 million gallons).

millrace: A powerful stream that moves a mill wheel.

miner's inch: A term seldom used any more, referring to the rate of flow of water. The measurement varies among jurisdictions, between 0.02 and 0.029 cfs depending on the state.

mining (groundwater): Withdrawal of water from the ground at a faster rate than the source's recharge.

mining water use: Water used in mining operations such as quarrying rock or mineral extraction.

mist: A cloud of minute water particles suspended in the air near the earth's surface.

mitigation: Action taken to limit or minimize harms.

model: A simulated process consisting of descriptions, concepts, statistics, etc. Models are used when actual observations are difficult to make. They approximate reality as a sort of artificial observation.

moisture equivalent: Ratio of water retained by saturated soil under pressure of one thousand times gravity to the weight of the dry soil, as a percentage.

monitor well: A well used to test groundwater for quantity and quality.

moraine: Glacial deposits, consisting of stones and other debris.

mouth of a river: The end of a river where it empties into either another river or the sea.

movable bed: A streambed that consists of material capable of being moved by a stream.

mulch: Any material, including leaves, bark, straw, grass clippings, etc., that is applied to protect the roots of plants from heat or cold or to reduce evaporation of water.

multifamily residential: Building in which two or more families reside in separate dwellings.

multiple use: The coordinated management of surface and subsurface resources likely to best meet the present and future needs of the people.

municipal and industrial water use: Water used for industry or city water supply provided by the municipal distribution system.

municipal discharge: Treated wastewater released by a publicly owned treatment facility.

municipal water: Water used by a municipality for use within its service area.

municipal water supplier: A supplier of water for municipal uses.

municipal water system: A public water system serving an area of the general public by providing water for domestic uses.

nappe: Stream free-falling from a dam.

narrows: 1. Narrow channel of water connecting two larger bodies. 2. Narrow part of a river.

National Pollution Discharge Elimination System (NPDES) permit: Permit regulating discharge into the nation's waterways under § 401 of the Clean Water Act.

National Environmental Policy Act (NEPA): 42 U.S.C.A. § 4321 et seq., federal law establishing national policy to promote healthy environment, requires integrated consideration of alternatives and cost-benefit analysis in all federal projects by drafting an Environmental Impact Statement (EIS).

natural flow: The rate of water movement past a point in a natural stream with no artificial effects (diversion, storage, import, export, or return flow).

natural recharge: The natural replacement of groundwater in an aquifer from the surface supply, including precipitation and stream flows.

natural replacement: The gradual replacement and upgrade of old fixtures with new, more efficient fixtures as the old ones wear out. These would include toilets, faucets, etc.

natural resource: Natural material or element that can be marketed in some way and sold for value, such as timber, water, or minerals.

Nature Conservancy, The: International organization funded privately, which purchases lands and retains them for conservation purposes.

native waters: The naturally occurring waters within a watershed.

navigable: 1. Capable of supporting vessels and watercraft for travel purposes, including commerce. 2. Capable of being directed by human control.

navigable in fact: Can be actually used in its natural state for travel or commerce.

navigable water: 1. At common law includes any water body influenced by the ebb and flow of the tide. 2. Body of water capable of use for commerce and travel. The United States Constitution gives Congress broad jurisdiction over such waters.

Navigable Water of the United States: Waters in the United States that form a continuous highway for commerce between the states or with foreign nations, either alone or in combination with other waterways.

navigation servitude: 1. Easement benefiting the federal government for purposes of regulation without the need for compensation for interference with private ownership rights. By act of Congress many such takings are now compensated. 2. Easement based on state police power or Public Trust Doctrine allowing regulation with minimal compensation. State servitude is less than federal servitude.

negligence: An unintentionally harmful act, done without due care.

Nephelometric Turbidity Unit (NTU): Unit for measuring cloudiness (turbidity) of water, based on the amount of light reflected off particles.

net demand: Future expected demand after conservation needs and other reductions. Usually expressed in thousand acre-feet (KAF).

nonconsumptive use: Use that does not consume the water, such as in-stream uses for recreation, fishing, swimming or boating, fish promulgation, hydroelectric power, etc.

non-native waters: Imported waters that originated in another watershed or drainage basin.

non-point-source (NPS) pollution: Pollution lacking a generalized source. Such pollution is often carried to water bodies by surface runoff, following precipitation.

nonpotable: Water that is not suitable for drinking due to pollutants, including minerals and infective agents.

nonuse: Passing of a statutory period when water has not been used and is subject to forfeiture. Some states require notice and an action before the right is lost.

nonuse application: Application to the state seeking protection of a right that is not likely to be used within the statutory period necessary to prevent forfeiture.

nonporous: No pores or crevices where water can pass.

nonproduction well: A well used for nonconsumptive purposes, such as recharge, testing, or heat exchange.

nontributary groundwater: Groundwater that is not in any groundwater basin and will not deplete the flow of any natural stream. Some jurisdictions define it within reasonable terms, setting minute limits such as $1/10$ of a percent of the annual withdrawal rate (Colorado).

noria: A waterwheel with attached buckets used to remove water from a stream.

normal freeboard: The distance between the primary spillway and overflow crest above a dam.

N.R.C.S.: See *Natural Resources Conservation Service.*

notice of violation: Formal written notice from the state to the alleged violator, that the law is being or has been violated.

nozzle: Tool on the end of a hose for projecting water.

nuisance: 1. An interference with the use or enjoyment of another's property. A nuisance may result in tort liability and or other criminal or civil penalties. 2. Either an act or failure to act (when there was a duty) interfering with the use or enjoyment of property. 3. Specific tort arising from interference with use or enjoyment of property when interference is unreasonable (e.g., excess noise, pollution, obstruction, etc.).

N.W.S.: National Weather Service.

observation well: Well used for monitoring water changes in aquifers by taking samples for analyses.

obstruction: Anything blocking either in part or in whole, the water in its free flow, effecting that flow or activity on it in any way. Commonly includes bridges, dams, dikes, fences, rocks, logs, vegetation, etc.

offstream: Water diverted from source to another location for use. Water use that requires diversion.

one-hundred-year flood: Flood having a 1 percent probability of being equaled or exceeded in any given year.

one-hundred-year precipitation: Precipitation having a 1 percent probability of being equaled or exceeded in any given year.

organic matter: Any substance from a living organism based on carbon compounds.

orifice: Opening used to control or measure of water.

osmosis: Movement of water through a thin membrane for desalinizing water.

outfall: Location of discharge of a sewer, stream, or drain. Also refers to the outlet for treated municipal water.

outlet: Point where water flows from a larger body by natural forces such as gravity (though conduit may be artificially created), to some other location (usually at a lower elevation).

overappropriated: When a drainage system has more rights or claims than supply.

overdraft: Groundwater withdrawal that exceeds recharge. See *groundwater mining*.

overflow: The amount of water rising above the banks and flowing onto the land, often referred to as floodwaters.

> **seasonal overflows:** Floodwaters that occur regularly within certain times of the year.

> **periodic overflows:** Floodwaters that occur occasionally within no easily defined season or time of year.

oxbow: An old meander in a riverbed that has been cut off from the stream. Also refers to the U-bend in a river.

oxygen: Colorless, odorless gas, one of the two chemical components of water.

oxygen demand: The required oxygen to meet the needs of biological and chemical processes in a body of water.

pH: A measure of the relative acidity or alkalinity of water. 7 is neutral, lower than 7 is acidic, and higher than 7 indicates a base.

paleokarst: Karstified area buried by sediment deposits.

Pareto Optimality: In economics, the situation where no one can be made better without harming someone else. The term comes from the Italian economist Vilfredo Pareto (1848–1923).

Pareto Superiority: In economics, an exchange that is beneficial to some but harms none. The exchange shifts to Pareto optimality when no such exchange is possible.

Parshall flume: Tool for measuring the flow of water in an open channel.

partial penetration: Well type that draws water directly from an upper part of an aquifer, without accessing the entire aquifer.

participation agreement: Agreement where developer of land pays cost of distributing water in the area.

particle size: The diameter, in millimeters of suspended sediment. Classifications: clay—0.00024–0.004 millimeters (mm); silt—004–0.062 mm; sand—0.062–2.0 mm; and gravel—2.0–64.0 mm.

particulates: Minute pieces of solid matter (particles) suspended in water, not easily seen by the naked eye.

parts per billion: The number of "parts" by weight of a substance per billion parts of water.

parts per million: The number of parts (by weight) of a substance per million parts of water. This is the unit of measurement preferred for measuring pollutant concentrations in water.

pathogen: Disease-causing agent, usually a tiny organism such as a virus, bacteria, or fungi.

peak flow: The maximum volume of water flowing across a point in a river or stream at one time. See also *maximum stage*.

penstock: Gate used to control the flow of water; also refers to a tube or trough used in conjunction with a water wheel, for carrying water.

per capita use: The average per person use of water over a period of time such as a day, week, month, or year.

perched groundwater: Groundwater in saturated soil, beneath unsaturated soil, and overlying an impervious layer of material that acts to prevent downward flow.

perched water table: The water table of a smaller body of groundwater above a larger more general body of groundwater.

percolating water: Water that seeps through soil and rock crevices outside a defined channel.

percolation: 1. The slow movement of water within the pores and spaces of rock and soil, absent a defined channel. 2. Slow seepage of water through a filter. 3. The flow of water from a stream toward an aquifer as groundwater recharge.

perennial stream: A stream that flows all year.

perfected water right: A water right that has been approved by the state and put to beneficial use, and in many jurisdictions is then certified, usually by the state engineer or a court.

perforation of wells: Holes in the well casing that allow water into the well.

permanent change application: Application to permanently change the point of diversion, place of use or nature of use, filed with the state.

permeability: The ability of water to travel through a material, such as rocks or soil. The more permeable a material, the faster water will move through it.

pervious zone: A part of an embankment or dam that is made of permeable material.

phreatic: Of or relating to groundwater.

phreatic surface: Surface of groundwater; also the water table.

phreatophyte: Deep-rooted plant that obtains water from a permanent water supply or the water table.

pier: Structure built from the land out over water, often supported by pillars or pontoons, that supports buildings, businesses, and moored ships.

piezometer: Tool for measuring water pressure or compressibility, often used to measure pressure in the soil.

piezometric surface: See *potentiometric surface.*

pipe: Hollow cylinder used to convey water or steam from one location to another. Pipes are distinguished from hoses in that they are far less flexible.

piping: Progress of internal erosion of a material caused by seepage of water into the material.

pitot tube: Tool used to measure velocity of flowing water.

place of use: Specific location where water is to be used. It is cited in an application and often again in documents to perfect the right. The water right specifies the place of use when granted as well.

playa: The often sandy, muddy, or salty flat-floored (level) area at the bottom of a desert basin, usually filled with water.

plumb: A weight on the end of a line, which is used to determine the depth of water.

plutonic: Igneous rock that crystallizes at a great depth in the earth, under high pressure. Sometimes used to describe any igneous rock.

pluvial: Of or related to rain.

point discharge: The rate of discharge at a point, as opposed to the mean rate over a period of time.

point of diversion: The location listed in a water right where water is diverted for use.

point source: Nonmobile source of large-scale pollution to either air or water from emissions.

point-source pollution: Pollution in air or water that comes from a single source, such as an industrial or sewage outflow pipe.

pollution: Introduction of impurities to water. Pollutants are often unhealthy and in large amounts can create serious problems for plants, animals, and people.

polychlorinated biphenyls (PCBs): Group of chemical toxins used in industrial compounds and often found in industrial wastes. PCBs continue to turn up in fish and other animals, though largely banned in the 1979 Toxic Substances Control Act (TSCA).

pond: A relatively small body of nonflowing water that may be naturally or artificially created.

pool: A small body of still water.

pore pressure: The pressure of water in the spaces within rock and soil.

porosity: Ratio of the volume of pores in a rock or soil in relation to the entire mass of the rock or soil.

porous interval: Part of a piezometer (water-pressure tool for pores in stone) that interacts with infiltrating water.

posted water: Water that is reserved for the use of the owner of the land it rests on. This right is secured by posting a notice to others prohibiting them from using it.

potable water: Water that is suitable for drinking, uncontaminated water.

potential yield: The highest rate that a well can yield water under set conditions.

potentiometric surface: Level that water will rise to in an encased well.

power consumption: Water consumed in the cooling of power generators or the production of electricity. Water used as a cooling agent is often lost as steam, vented into the air.

power filing: Filing for a right to generate electric power.

P.O.T.W.: Publicly owned treatment works; domestic sewage treatment facilities.

practicably irrigable acreage: A standard used by courts to determine water rights on Indian reservations. The amount of land that can be practically irrigated in a manner that makes sense.

preatophyte: Water-loving plant, vegetation that consumes high volumes of water (e.g., cottonwood trees). (Also *phreatophyte*.)

precipitation: Moisture falling to the ground in the form of rain, snow, hail, sleet, dew, or frost.

precipitation rate: The amount of water falling to the earth over a period of time (e.g., inches per hour).

prescription: 1. An effected change in rights in real property based on hostile occupation over time. 2. Process of acquiring title to something by open and continuous use for a statutory period. 3. Negative prescription. Extinction of a right for failure to use it for a long period of time.

prescriptive right: A right obtained by prescription. Such rights usually require statutory periods of occupation that remain unbroken, are hostile to the true owner's rights, without permission, and operate to the exclusion of others including the true owner.

pressure gage: A device for measuring the pressure in solids, liquids, or gases.

pressure head: Energy within a fluid caused by pressure.

pressure pipe: Pipe used for the distribution of water for domestic purposes.

pressure regulating valve (PRV): Where water pressure is too high and might cause appliances to malfunction, a PRV reduces the pressure, allowing individuals to regulate it to a lower pressure.

pressure zone: Geographic area of water distribution where pressure builds up sufficiently to cause water to flow through the system.

primacy: In rank of importance, the first or most important.

primary wastewater treatment: The first treatment process for removal of solids and contaminants from wastewater, often involving filtration, settling, and screening.

principal spillway: The primary spillway for normal operations.

priority: The order of right based on first in time of all water users from a particular source.

priority date: The date of origin for a water right.

prior appropriation doctrine: The rule that water rights are determined based on first in time to make beneficial use, independent of ownership of land touching the water body.

private water: Nonnavigable water owned and controlled by private individuals for private use and enjoyment. The public has no right to access private waters.

probable maximum flood (PMF): The flooding that might be expected from the most severe flood conditions possible in a region.

probable maximum precipitation (PMP): The amount of precipitation possible under the most severe conditions in a drainage area.

projected savings: The estimated amount of water saved when suppliers and customers implement more efficient practices.

proof due date: The date by which proof of beneficial use must be made; also, the date by which water must be put to beneficial use.

proof of appropriation: Document attesting to actual use of water for beneficial use under the permit system.

proof of beneficial use: See *proof of appropriation.*

proof of change: Same as proof of appropriation except applied to a changed use instead of a new use.

protest: A document stating opposition to a water right application. Most formal protests must be filed in accordance with state laws and rules in order to be valid.

protestant: An individual filing a protest against the granting of a water-right application.

provisional well: Authority to drill a well for test purposes, under a pending, as yet unapproved application for water.

pseudostatic analysis: Method that approximates the dynamic stability of a structure, using static loads.

public domain: Lands acquired by the federal government from the original thirteen states, the Louisiana Purchase, cession from Spain, the Oregon Territory, cession from Mexico, purchase from Texas, Alaska, or the Gadsden Purchase.

public interest: Water rights generally must not be contrary to the public interest. Public Interest is often a vague term related to the Public Trust Doctrine, which states that government owns the lands and water in trust for the public. Strictly speaking, Public Interest would seemingly involve a cost-benefit analysis of the use but it is not always weighed in such a manner. In some jurisdictions it is a very loose term, meaning little more than beneficial use. Many jurisdictions define beneficial use in statutes or case law.

public right-of-way: An area of land acquired by a governing body within the state for use by members of the public, including use by people (sidewalks, paths, curbs), vehicles (streets), stormwater (drains), or goods.

public supply: Water that is withdrawn by government or other public corporations, utilities, or agencies to supply the needs of the public, including domestic uses, some commercial and power uses, certain industrial uses, and other public purposes.

Public Trust Doctrine: Doctrine under which states acquire title to waters and the land beneath them in trust for the public and cannot act contrary to that trust.

public water use: Water from the public supply used for public purposes such as firefighting, street cleaning, parks, and public pools.

public water: Water preserved for public uses, including navigation.

puddle: Small pool of water, often rainwater, collected on the ground, that tends to exist for only very short periods of time, often following rainfall.

pump lift: The distance water must travel from the water table belowground to the surface of the overlying land.

pump test: Test for determining the characteristics of a well or aquifer.

quantification: The procedure for determining the quantity and limits of a water right, based on beneficial use, usually measured in acre-feet per year.

quicksand: Loose, wet sand that exerts upward pressure on the water, causing objects that fall into it to be sucked down.

quitclaim deed: A deed with no warranties or guarantees, conveying only the current rights or interests the grantor possesses to the property.

race: 1. A channel built to transport water. 2. Strong, swift currents.

race notice state: The doctrine that in the event of conflicting deeds, the first to properly record will be given effect. This does not ordinarily include known fraud or wrongdoing on the part of the buyer. It is intended to protect reliance on the record.

radial flow: Flow of water to a vertical well, in an aquifer.

rain: Atmospheric moisture that condenses and falls to the earth in drops. The term refers to both the drops and the process.

raindrop: A single drop of rain.

rainwater: Water collected from rain.

ramping: The gradual shifting of flow changes in streams for safety downstream.

range lines: The grid lines running north and south in the U.S. Public Land Survey System.

rated capacity: The amount of treated water a treatment plant is able to produce under normal conditions.

rating curve: A curve showing relation between gage height and stream discharge at a particular gaging station.

raw water: Untreated water.

reach (of river): The length of a river or channel, with approximately uniform depth and discharge. This length can generally be serviced by a single gage to obtain satisfactory measurements for the whole. Also the length of the river between gaging stations.

reasonable use: Doctrine that weighs the cost and value of consumptive use of water by riparian owners. This doctrine allows some consumptive uses of water if they are reasonable under the circumstances as opposed to preserving the river or stream in its natural condition for nonconsumptive uses alone.

recapture: The capture of water that has escaped from some prior use so that it might be applied to further use by the original owner or his or her appointed successor. Ordinarily, water must be recaptured, if at all, before it leaves the owner's land or reaches a public body of water, such as a natural stream.

recharge: Water added to an aquifer from either natural or artificial sources.

recharge area: See *recharge zone.*

recharge basin: A surface basin used to recharge groundwater by increasing infiltration of water into the groundwater basin.

recharge, groundwater: Water flowing into an aquifer.

recharge zone: Area of land that can enable water to reasonably infiltrate an aquifer.

reclaimed wastewater: Wastewater that once treated is used for additional beneficial purposes, such as irrigation.

reclamation: Reclaiming of wasteland and desert for cultivation and settlement.

record: The process of recording means to record with the county recorder or like agency. Records provide constructive notice of title, interests, and claims, regarding real property, to others.

recorder: Tool that records continuous water levels or other water-related data.

recoverable groundwater: The amount of water than may be practically withdrawn for use, from the ground.

recycled water: Water used more than once before returning to the natural water system.

red tag: A written order to cease drilling operations of a well, until the violation ceases or the matter is resolved.

refuse: Trash.

regimen: Orderly system of stream flow. Often refers to the equilibrium between erosion and deposit.

regional engineer: In some states, an employee of the state engineer, responsible for administering water rights within a particular area of the state.

reinstate: Where a right that has lapsed for any reason is reactivated by the state governing agency. Priority date may be changed to the date of reinstatement.

regulating agency: Agency in charge of a state's water rights or any agency regulating fields that affect such rights, e.g., environmental regulations. Usually refers to executive agency as opposed to public utilities that often fall under the jurisdiction of such agencies.

regulation: The artificial management of stream flows.

reliction: Drying of the land due to removal of the water and involving no change in elevation.

renovate/replace: To replace or update an existing well. Such actions often require a permit from the state.

repair of a well structure: The replacement and repair of worn parts of an existing well, such as casings, screens, seals, etc.

replacement of a well structure: To replace an existing well with a new well, abandoning the old well.

reserved rights: Water rights retained by the federal government for benefit of federal lands or for the satisfaction of treaty obligations, including implied rights for Native Americans. See also *Winter's Doctrine*.

Reserved Rights Doctrine: Rule that federal lands set aside for a particular purpose include a reservation of sufficient water to accomplish that purpose.

reservoir: A nonflowing body of water, such as a pond, lake, or basin (including artificial bodies), for storing and controlling water supplies.

reservoir area: The surface area of a reservoir at a given elevation of water.

reservoir stage: Depth or elevation of water in a reservoir as related to known information.

reservoir surcharge: Water above the spillway of a reservoir.

residual freeboard: Distance between the top of a dam and the maximum height of the water surface during a hydrologic event.

res ipsa loquitur: Doctrine in tort law that presumes negligence by the fact that an event occurred. Often applied when artificially stored waters are released to the detriment of neighboring properties.

response spectrum: A graphical representation of motions (i.e., displacement, velocity, acceleration) in a body of water, caused by seismic events.

responsible party: The individual or individuals who receive and are liable for the water bill at the time of a violation. May be the owner, manager, supervisor, or person in charge of property.

restricted plants: Plants that are regulated and restricted in some areas because of their high water use and consumption.

Resume: A monthly publication in Colorado by the water court summarizing water rights and applications filed in that court during that month.

return flow: Flow that, after a diversion, is not consumed and returns to the stream channel.

return seepage: Percolating water that moves from irrigation canals and ditches into the groundwater basin and back to a natural channel.

retrofit: To modify in order to improve efficiency or performance. Usually used to refer to replacement of plumbing fixtures such as faucets, pipes, valves, toilets, etc.

reuse: To use more than once, also recycle. The interception of water that would otherwise return to a natural stream or body for further beneficial use.

reverse osmosis: 1. A method for removing salt from water by running the water through a membrane that the salt cannot pass through. (Related to *electrodialysis,* which differs in that it uses an electrically charged membranes to separate the ions). 2. A method of advanced water treatment that uses semipermeable membranes to remove pollutants from solutions of higher concentration to those of lesser concentration.

revetment: A retaining wall of stone or concrete used to protect soil from erosion.

riffle: 1. Shallow rapids in an open stream. 2. Area of water with ripples or waves.

right of survivorship: Right of successive ownership.

riparian: Located on the bank of a body of water, usually a stream or river.

Riparian Doctrine: System of water rights allocation in the eastern United States, basing rights on ownership of the land bordering the water.

riparian proprietor: A landowner whose land touches a natural body of water.

riparian right: The right of a property owner to make reasonable use of water as it crosses his land based on ownership of that land.

Riparian Rights Doctrine: Rule stating that property owners whose land borders a natural body of water have certain equal rights to use the water as it crosses their land.

riprap: A barrier of large rock intended as a protection against waves, ice, and scour.

risk assessment: An effort to define risks of a situation or decision to public health and the environment.

river: A large natural body of flowing water in well-defined channel with banks, cut naturally and flowing generally in a single direction.

river basin: The drainage area of a river and its tributaries.

river commissioner: See *water commissioner.*

rod: Surveyor's measurement equal to 16.5 feet.

routing: Determination of the outflow of a stream by looking at known values of the flow upstream and plugging them into appropriate formulas. 2. Formula used to determine the impact of stored water in a channel and its movement on the shape and motion of a wave of floodwater along a river's channel.

runoff: Precipitation in any of its forms, including snowmelt, when it settles on the surface in flowing form.

Safe Drinking Water Act (SDWA): Federal law passed in 1974 to regulate treatment of drinking water (water subject to use as), including testing for contaminants.

safety factor: A quantity of water added to projections as a safety margin to protect against unforeseen shifts in supply and demand.

safe yield: Although definitions vary among jurisdictions, ordinarily this refers to the amount of water that can be replenished each year after extraction and use.

saline: Water containing salt (sodium chloride).

saline water: Water containing significant amounts of dissolved solids. The following are the parameters used by USGS: Freshwater—less than 1,000 parts per million (ppm); slightly saline water—from 1,000 ppm to 3,000 ppm; moderately saline water—from 3,000 ppm to 10,000 ppm; highly saline water—from 10,000 ppm to 35,000 ppm; brine—more than 35,000

salts: Dissolved solids in water.

saltwater intrusion: Saltwater that takes the place of freshwater underground when wells withdraw too much of the freshwater too close to a saline body.

salvaged water: Water that would otherwise be lost, which is preserved for use by some activity.

sandbag: A sack filled with sand and used to contain floodwaters, for ballast or other purposes.

saturated: Filled with water to maximum storage capacity, beyond which substance will hold no more.

saturated flow: Flow of water in soil caused by saturation of some part of the soil and flow toward an area of lower pressure.

saturated zone: Area belowground where the pores and crevices of rock and soil are completely filled with water under higher than atmospheric pressure.

saturation point: The point at which no more water can be absorbed by the soil or an aquifer. Maximum storage capacity of an aquifer.

scour: Washing or flushing out by swift waters; eroding of banks by moving water.

seal: Impermeable material placed around the rim or space between well casing and hole to prevent surface or shallow groundwater from passing.

secondary wastewater treatment: The second phase of wastewater treatment involving biological processes for reducing solids in the water. This process can remove up to 90 percent of oxygen demanding substances.

section: An area of land usually equal to one square mile or 640 acres. In Utah, townships are comprised of thirty-six sections.

Section 404: Part of the Clean Water Act (33 U.S.C.A. § 1251 et seq.), which restricts dredging and filing of wetlands or the disruption of beds and banks of streams.

sediment: Material that at some point is carried in water, and that may settle and be deposited in the bed or on a floodplain or other area of land over which the water travels.

sediment load: The amount of sediment being carried and moved by flowing water in a stream or river.

sediment pool: Part of a reservoir where sediments accumulate.

sedimentary rock: Rock formed from compressed sediment, deposited by water over time; includes sandstone and shale. Many such rocks have distinct layers from successive deposits.

sedimentation: The use of gravity to settle solids near the bottom of a body of water.

sedimentation tanks: Tanks in which floating waste is skimmed from the top of wastewater and other solids are permitted to settle and then removed.

seepage: 1. Water that seeps into the soil from canals, ditches, reservoirs, etc. 2. The movement of water through the soil and rock, outside a defined channel.

seeps: Springs with small discharge rates.

segregation: To divide a water right into multiple rights.

seismic: Related to movement in the earth, such as an earthquake.

self-supplied water: Water supplied by the user for his own purposes and not by a municipality (e.g., well water for a single-home use).

septic tank: Tank for storing domestic waste, used primarily where sewers are not present. The tank includes a settling tank for early removal of solids.

settle: A means of cleaning water by causing impurities to settle near the bottom.

settling basin: The enlargement of part of a stream so that materials might more easily settle, instead of remaining suspended in the water.

settling pond (water quality): An open pond where wastewater is placed and allowed to stand while suspended solids sink to the bottom. The water is then allowed to overflow its enclosure to the next treatment site.

severance: The separation of land from its riparian water rights, usually by contract or other agreement. May also be applied to appropriation rights for specific nonriparian lands when the water right is severed from the designated land.

sewage: Wastewater from drains and sewers.

sewage reuse: Capture and reuse of treated wastewater from a treatment plant. Some municipalities may resale this water for irrigation and other uses.

sewage treatment plant: A facility for treating domestic wastewater to remove solid and dissolved materials in the interest of public and environmental health. The four basic classifications of removed materials are: greases and fats; solid waste; dissolved substances; and microorganisms. The means of removal often include mechanical methods, bacterial decomposition and chlorine treatment.

sewage treatment return flow: Treated water that reenters the hydrologic system.

sewer: An underground system of pipes for the transport of wastewater for treatment and disposal

shallow well: A well with a pumping head less than thirty feet.

share: Stockholding in a mutual ditch or irrigation company, the total number of shares varies from company to company. The more shares a holder owns, the higher his water entitlement will be. The shares themselves are not water rights. The rights are held by the company for its members.

sheet erosion: The removal of a thin layer of surface soil by runoff and rainfall.

sheriff's deed: Title conveyance of debtor property following a foreclosure sale.

shoal: A shallow area of water.

shore: 1. Part of the land that touches on a large body of water. 2. The area between the ordinary high and low water marks.

shower: Brief fall of rain or other precipitation.

shutoff nozzle: Tool on the end of a hose for shutting off the flow of water.

Sierra Club: National environmental organization founded in 1892.

silt: A smaller sedimentary partical than sand and larger than clay.

siltation: Deposition of small fine soil material and rock particles in the bed of a stream or lake.

sink: A basin or tub used for washing things and storing water.

sinkhole: A geological phenomenon created when limestone and other minerals beneath the surface dissolve in water during drainage, weakening support for the surface and eventually collapsing.

siphon: To draw liquid from one container to another by means of a tube or hose. Also refers to the tube or hose.

site identification (USGS): A grid system used by USGS, showing latitude and longitude, using fifteen digits, the first six denote degrees, minutes, and seconds of latitude, while the next seven represent the same for longitude and the last two are for site identification.

skimming: A shallow overflow diversion of water from a conduit or stream.

sleet: Snow and rain falling together.

slope protection: Protection of bank slopes from erosion.

slough: 1. Location of deep mud or a wet, marshlike area. 2. Inlet from a river.

sludge: The solids that remain after treatment of wastewater, also termed biosolids. These solids are removed by gravity and can be used in fertilizers.

sluice: A channel built to conduct water, with a valve for regulating water flow. Also refers to the body of water behind a floodgate.

slump: Movement of overlying soil, usually due to saturation, often resulting from resting on a smooth impermeable surface.

slurry wall: A structure used to prevent the lateral movement of groundwater.

small dam: A dam used to impound water, generally at levels of 20 acre-feet or less or 18 feet in height or less. Often used for overnight storage.

smog: Polluted fog creating numerous irritations and problems.

snow: Frozen water vapor that falls to the earth in light flakes.

snow survey: Process of forecasting eventual runoff by calculating the water in snowpacks at various locations.

snow water equivalent: The amount of liquid water that would exist if snowpack were melted.

snowmelt: The amount of solid snow that melts and flows away or pools, in the form of liquid water.

soak: 1. To sit in water for a prolonged period of time. 2. To absorb water or be extremely wet.

soggy: Containing a high amount of moisture.

solute: A substance dissolved in another substance (e.g., water) to form a solution.

solution: Mixture of two substances, a solvent (e.g., water) and a solute (e.g., sugar) where in one substance disolves into the other.

solvent: A substance (e.g., water, also known as the universal solvent) that dissolves a solute within it, forming a solution.

sounding: To measure the depth of water.

source: 1. The place of origin of a river or body of water, such as glacial melt, springs, a large natural reservoir or other modes of collection. The source of the Mississippi River is the Boundary Waters in Northern Minnesota and Canada. 2. Place or body of origin of water designated for beneficial use.

spall: A rock chip that flakes away from the rock's surface, the result of weathering, usually by exfoliation. See *exfoliation*.

special master: A water master appointed by the court to assist in a particular dispute.

specific capacity: The amount of water produced by a well in gallons per minute per foot of drawdown following a period of continuous pumping.

specific conductance: Measure of ability of water to conduct electricity. Can be used to determine the amount of dissolved solids in water. Used to indicate chemical presence in groundwater and potential leakage from waste and disposal facilities.

specific storage: The amount of water released or taken in by an aquifer system per volume per change in head.

spill water: Water released from a reservoir for lack of space to store it.

spillway: A channel through which excess water escapes from a reservoir or other body of water.

spit: 1. A light rain or snowfall. 2. A narrow point of land that extends into water.

spray: Minute water droplets blown through the air.

spray irrigation: Common method of irrigation using high-pressure sprayers to apply water to crops. Water is often lost to evaporation due to the height of the spray and exposure to the air.

spring: Water flowing from the earth and appearing on the surface as a small stream.

sprinkler head: Tool for watering vegetation by projecting droplets of water through the air.

sprinkler irrigation: Type of irrigation system using pressurized pipes, nozzles, and sprinkler heads to apply water to crops and vegetation.

squall: Sudden, violent winds, usually accompanied by precipitation of some kind. Windstorm of 25-plus miles per hour lasting at least one minute. Term usually applies to storms at sea. Also *white squall*.

staff gage: A graduated scale used to determine the elevation of water surfaces in a stream, lake, or reservoir.

stage: Elevation of a water surface above an established reference point, such as a water mark on the banks of a stream or lake, showing frequent shifts in water levels at different times of the year.

standard operating plan: Written plan for operation and maintenance of a dam and attached structures.

standing master: An individual, usually an expert, appointed to assist a court in an ongoing matter.

standpipe: Large pipe in a pipe system, into which water is added to create pressure or from which water is released to relieve pressure.

start card: A card stating an intention to drill used in some jurisdictions.

state engineer: State officer (in many jurisidictions) in charge of administering water rights and appropriations.

static head: Elevation difference between the centerline of a discharge pipe, where water exits, and the surface of the pumped body.

static level: The stable level in a well that is not pumped. This level is beyond the influence of other pumping operations.

static water pressure: The pressure of nonflowing water in a municipal water supply.

steady state: State where in flow equals the out flow, also known as equilibrium.

steam: Gaseous form of water; also a source of power in early engines.

stilling basin: Structure at the foot of a fall or drop (such as a spillway) to remove excess energy or force in the falling water.

stock certificate: Document issued by an irrigation or ditch company, representing the ownership rights of the holder. Assessments are often required for company expenses based on total shares owned.

stock pond tank: Natural or artificial pond used to water cattle and other livestock. Sometimes refers to fish operations.

stoplog: Hydraulic engineering tool, used in floodgates to adjust water level and control the flow in a stream, canal, or reservoir. Typically, long, rectangular timbers stacked into premade slots.

storage: 1. Artificial water stored in surface or underground reservoirs for future use. 2. Water captured naturally in a drainage basin.

storage capacity: The amount of water that can be stored in a place, at the elevation of the primary spillway, including both active and dead storage.

storage coefficient: The amount of water released or taken in by an aquifer with a change in the head, measured by surface area.

storm: Heavy fall of rain, sleet, or snow often accompanied by heavy winds, lightning, and thunder.

storm drain: A system of pipes for carrying rainwater away from city streets.

storm sewer: A drainage system for street and surface runoff, including snowmelt.

strain: Change in the area (including volume and length) of a body of water due to stress, often expressed as displaced water divided by original area.

stream: A flowing body of water, often considered somewhat smaller than a river. Streams are natural bodies that flow in defined channels at least part of the year.

stream alteration: To change the stream from its natural flow in some way by altering its channel.

stream channel: The hollow bed of a stream, including the banks.

stream gaging: Means of determining the amount of water flowing in a stream by use of gages, meters, weirs, etc.

stream gaging station: Location on a stream where discharge information is regularly taken.

stream regimen: The stable condition of a stream and its channel as related to erosion.

streambed: The channel through which a natural stream flows.

stream flow: Water flow in a natural channel.

stress: The amount of force per unit of area pressing on a surface or the cause of that force (pressure per unit of area—e.g., pounds per square inch, etc.).

stress applied: The stress or force (weight per unit area) above a specific plane in an aquifer at a given level. The force is caused by the weight of sediment and water above the water table plus saturated land above the plane plus or minus flow force.

stress effective: Pressure on the grain-to-grain contacts of deposits affecting porosity and storage capability. The weight per unit area of sediments and moisture above the water table plus the submerged weight per unit area of sediments between the water table and a specified depth and seepage stress.

stress, geostatic (lithostatic): The weight in a unit area of sediment and water, which acts through gravity. Also the sum of effective stress and water pressure at a depth.

stress, preconsolidation: The amount of stress a geological deposit can withstand without further permanent deformation. See virgin stress.

stress, seepage: The force per unit area transferred by friction from water to the soil or rock as the water flows through it.

structural height: The vertical dimension of a dam from the streambed to the top of the dam.

sublimation: Change from solid to vapor or vapor to solid without taking a liquid form.

submerged lands: Lands covered by water.

subsidence: The lowering of elevation at the surface of land as a result of over-pumping of water from aquifers below. Subsidence is often irreversible.

subsidence, near surface: See *hydrocompaction*.

subsidence, shallow: See *hydrocompaction*.

subterranean water: Water beneath the ground's surface.

supply management: Means whereby a utility maximizes use of untreated water.

surf: Swell of the sea or line of foamy water caused by the water's breaking upon the shore.

surface supply: Water supply from streams, rivers, lakes, and reservoirs.

surface tension: Attraction of molecules on the surface of a liquid, creating a barrier between the air and the liquid.

surface water: Water on the surface of the earth, as opposed to groundwater or atmospheric moisture.

surplus water: Water that is not consumed in the irrigation process; runoff from irrigated land.

survey marker: Permanent mark on a dam or connected structure for measuring changes in vertical or horizontal movement of water.

suspended sediment: Fine particles of soil suspended in water over a lengthy period of time, due to currents and turbulence, without settling to the bottom.

suspended sediment concentration: The mass of dry sediment suspended in water as a percentage of the complete mass of water and sediment (usually milligrams of dry sediment per liter of water-sediment mixture).

suspended sediment discharge: The amount of suspended solids to cross a point in the stream over a specific measure of time (e.g., cubic feet per second, milligrams per liter).

suspended solids: Undissolved solids, suspended in water, which can be removed by filtration.

sustainability: A management program that meets present and future needs.

sustained yield: See *safe yield*.

swamp: Waterlogged ground, often muddy.

swimming: A recreational activity in water where one moves through the water by moving one's limbs. Sometimes includes wading and other activities.

system loss: The amount of water lost to leaks, seepage, theft, etc., as a percentage of the whole.

tailings: The waste that remains after removing metal from ore.

tailwater: 1. Diverted water that remains unused. 2. Water downstream from a structure.

tailwater recovery: The collection process of irrigation runoff for reuse.

tap: A physical connection to a public water-distribution system.

tap allocation: A process of distribution of connections to a public water supply when circumstances requiring rationing.

T.D.S.: Total dissolved solids.

temporary application: An application to appropriate or change a water right that would be of a temporary duration (e.g., one year).

temporary irrigation system: Irrigation system that is installed for temporary use and permanently disabled within specified time (e.g., 36 months in New Mexico).

tenancy in common: A co-ownership of property where two or more persons hold an undivided interest in the whole.

termination: A court-ordered end to a water right.

territorial sea: See *territorial waters.*

territorial waters: The waters under a particular jurisdiction (e.g., state or nation), including those inland and those typically three miles out to sea.

tertiary wastewater treatment: Also known as advanced treatment; it is additional treatment of wastewater after primary and secondary treatment and may include chemical treatment or radiation.

test borings: Holes drilled into the earth to test properties of materials below the surface.

test well: A well drilled for test or exploratory purposes.

thalweg: 1. An underground stream that flows beneath a surface stream, in the same general direction. 2. Line on a map connecting the lowest part of a valley. 3. Middle of channel that forms boundary between states or nations.

thermal pollution: The heat pollution caused by an increase in temperature of the water, often the result of industrial or power generation processes.

thermoelectric power water use: Water used to generate thermoelectric power (power produced by heat generation such as burning of coal or oil).

tidal wave: See *tsunami.*

tide: The rise and fall of the sea in response to gravitational forces of the moon, also termed the ebb and flow of the sea.

tideland: Land between the average high and low tides, covered periodically by the ebb and flow. These lands are usually flat and sometimes called flats or flatlands.

tidewater: Water that is influenced by the tides, such that it rises and falls with the ebb and flow. Usually applied to bays, coves, and rivers.

tideway: The land resting between the high- and low-water marks.

tilting gate: Gate counterbalanced to open and close with changes in water elevation and pressure.

time of concentration: Time of travel for water from its furthest point in the watershed to the gaging station.

toe of dam: Where the face and foundation of a dam are joined (may also be termed upstream or downstream toe based on which face of the dam is being described).

topographic map: Map showing features of the land often using contour lines to show changing elevation.

torrent: Violent, fast-moving stream of water.

total dissolved solids (TDS or salt): The total amount of all salts dissolved within water. These salts exert pressures and can be lethal to marine life.

total head: Energy within water due to pressure, velocity, and elevation (gravity).

total sediment load: The total amount of solid material carried by water in a stream.

total storage: The volume of water stored below the maximum intended height for the water surface.

totalizing meter: Device for measuring water volume in a flow.

township: A county subdivision that exists in some states.

township line: Grid running east and west in Public Land Survey System.

tract: An usually large area of land or water.

trans-basin diversion: Transfer of water from basin of origin to another basin.

transition zone: Zone of filter material between two nonfilter zones.

trans-mountain diversion: Transfer of water from one watershed to another often from the western slope to the front range.

transmissibility (groundwater): The ability of rock to transmit water under pressure, the flow of which is usually measured in gallons per day through a one-foot strip of the aquifer.

transmissivity: The ability of an aquifer to transmit water, usually measured in gallons per minute, crossing a vertical section of the aquifer, a single foot wide and extending down through the entire aquifer.

transpiration: The evaporation of water, absorbed by plants, into the atmosphere by means of surfaces on the plant, such as pores.

transport: The carrying of solids in water flows to a point downstream.

trap efficiency of reservoirs: Percentage of sediments retained to inflow.

trash rack: A screen to prevent entry of floating or submerged debris into an area or piece of equipment.

treated water: Filtered and disinfected water, also potable water.

tremie: Device for placing concrete and like substances under water.

tremie pipe: Device for carrying materials down a drill hole or rounded space.

tributary: A small river or stream flowing into a larger river. Large rivers often consist of many smaller tributaries.

tributary drainage: The drainage area of water pulled by gravity into a stream or river.

tributary groundwater: Groundwater that is connected to a natural stream or river.

tsunami: A wave of unusual size and destructive capacity usually cased by earth-quake or volcanic activity.

tub: A large open container for storing water, including a bathtub.

turbid water: Muddy or dirty water containing solid matter that may prevent light from traveling through it.

turbidity: Amount of solid particles suspended in water, scattering light rays through the water. The greater the turbidity, the cloudier the water. Turbidity is measured in nephelometric turbidity units (NTUs).

turbulence: Fluid in an irregular flow or a state of commotion.

unaccounted-for water: The difference between the amount of water exiting a treatment facility and the amount measured at consumer meters, including system losses, unmetered beneficial uses (e.g., firefighting, etc.), system maintenance, and water theft.

unappropriated water: Water in public bodies within the state, subject to appropriation, exceeding existing rights.

unconfined aquifer: An aquifer that is not confined by a layer of impermeable material above it.

unconstrained demand: What demand would be if supply were unlimited.

underground water claim: A claim for the use of underground water prior to its inclusion in the appropriation system.

ungated outlet: Outlet with uncontrolled flow through a dam.

unit hydrograph: Hydrograph showing the rates of runoff for an inch of storm runoff from the drainage area.

unsaturated zone: The area below the surface and above the water table where land is not fully saturated.

uplift: Upward pressure of water.

Upper Basin States (Colorado River Compact): Colorado, New Mexico, Utah, and Wyoming.

upstream: In the direction from which water is flowing, which is against the current or flow of a stream.

urban runoff: Water from an urban area that runs off the surface into storm or sewer drains or open waterways but does not enter the soil.

U.S.B.R.: U.S. Bureau of Reclamation.

U.S.F.S.: U.S. Forest Service.

U.S.F.W.S.: U.S. Fish and Wildlife Service.

U.S.G.S. (U.S. Geological Survey): Agency established in 1879. The Water Resources Division collects, organizes, interprets, and publishes research on water uses and issues.

use: 1. To bring something into service. 2. A service for which something is employed.

> **highest and best use:** The use that generates the greatest amount of profit, used in determining the fair market value of land seized by eminent domain.

> **public use:** The right of the public to beneficial use of property condemned by eminent domain. Also public purpose.

V-notch: Weir used for gaging discharge in small streams.

vadose zone: An area from the surface to the water table, containing water at below atmospheric pressure.

valve: A device for controlling or stopping water flowing in or from a pipe or hose.

vapor: A semigaseous state of water that includes minute particles of liquid water. Vapors are generally below the critical temperature for gaseous water.

velocity: Speed, usually measured in distance traveled per unit of time (e.g., second).

Venturi meter: Used to measure flow (velocity) of water. Consists of gradually contracting closed conduit that causes a buildup of pressure near the head.

vested water right: A perfected water right, one that is active and valid.

viscosity: The resistance of a liquid to forces, such as flow.

vortex: A mass of revolving water forming a whirlpool.

vug: A small cavity in rock often lined with crystal formations.

wake: Wave pattern left behind by a boat or ship.

warranty deed: A deed that offers guarantees in the form of covenants regarding title and right to possession.

waste: Water that is unnecessarily lost to beneficial use by neglect, inefficiency, or intentional misconduct. Most jurisdictions penalize waste when it becomes significant.

wastewater: 1. Runoff and other appropriated water that is ultimately lost to the appropriator, through seepage, evaporation, or other processes, and leaves the appropriator's property. 2. Water left over after some process, including sewage, chemical treatment, or manufacturing.

wastewater-treatment return flow: The wastewater intentionally returned to the environment after being treated by a treatment facility.

water balance: A water balance equation can be used to manage water supplies. It predicts the amount of water available in a basin with an equation such as $P = Q + E + \text{Delta } S$, where P equals precipitation, Q equals runoff, E equals evapotranspiration, and Delta S is the change in storage capacity of rock and soil.

water bayley: A colonial official who collected dues for fish taken from colony's waters.

water budget: An accounting of water taken in to water going out, considering changes in storage of a water body.

water commissioner (also referred to as river commissioner): Title for state employee in charge of distribution of water to rights holders within a specific distribution system.

water conservation: The wise and efficient use of water. Avoidance of waste.

watercourse: A body of flowing water within a defined channel, having a bed and banks.

> **ancient watercourse:** A watercourse, where the channel has existed for as far back as can be remembered.

> **artificial watercourse:** A man-made watercourse. After a sufficient length of time such a course may be treated as a natural course, meaning riparian rights might attach.

> **natural watercourse:** A naturally created stream, flowing in a defined channel with banks and a bed. This does not include intermittently flowing

surface waters without a defined channel. Natural streams may be subject to riparian rights in jurisdictions that recognize such rights.

water cycle: The cycle of water as it moves from atmosphere to earth and back to the atmosphere in various forms including precipitation, runoff, percolation, evaporation, storage, and consumption by plants and animals.

water district: A subdivision created by state and local governments to provide water to the public.

water, duty of: The amount of water needed to irrigate the land, expressed as a flow rate per area.

water duty: See *duty of water.*

water exports: The transport of water from one watershed to another by means of canals, pipes, ditches, etc.

waterfront: Land, often housing buildings, which touches or fronts the water. It may include piers and docks.

water gage: 1. A wall or bank for the purpose of holding back water. 2. An instrument, such as a reservoir, boiler, or tank, for measuring depth or quantity of water, including its height.

watergavel: A fee paid for a right to use a body of water for some privilege, such as fishing.

water main (or distribution main): A 12-inch diameter pipe or smaller along or under public streets or other means of distribution of water to consumers.

water mark: A mark in the rock or soil, showing the high or low point where water rises or falls.

    high-water mark: 1. Often computed as the mean or average mark, this refers to the part of the shoreline touched by the high tide or the water's furthest reach on shore. 2. In a lake, created by construction and use of a dam, this refers to the highest point on shore that the water can reach under normal circumstances, as a result of the dam. 3. The impression, in a riverbank that shows where the water deprives the land of agricultural value, due to coverage.

    low-water mark: 1. The part of the shoreline, on the sea, that marks the waters furthest retreat toward the sea, or the greatest exposure of the land when the water retreats. 2. The recession level of water at its lowest stage in a river.

water marketing: Trading of water rights in the open market.

watermaster: Individual hired to oversee water distribution in a particular area. A watermaster may be assigned by the state, the court, a private group of rights holders, a corporation, or other interested individuals or bodies.

waterpower: 1. The energy force created by flowing or falling water. 2. The right of a riparian owner to the fall in the stream as it crosses his land, referring to the difference in level between the surface as water enters the land and the surface where it leaves the land.

water purveyor: Someone who sells water to the public for drinking and consumption, usually a public corporation.

water quality: The suitability of water for a particular purpose. This describes the physical attributes of the water, such as the foreign components within the water.

water reclamation: The recycling of used water, including treatment.

water right: The right, granted by the regulating authority (usually the state), to make some use of water from either a natural stream or artificial canal, for some beneficial use.

waterscape: A passage wherein water can flow.

water share: See *share.*

watershed: A feature in land, determined naturally by the topography of the land and its elevation. A watershed is the natural drainage area that feeds into a particular body of water, such as a stream or lake. Some large watersheds contain many smaller watersheds within them.

waterstops: Material, usually strips, used to prevent leakage in the joints of concrete.

water supplier: The owner or operator of a water system for providing water to others for some legal purpose.

water table: The highest elevation of saturated rock and soil on or below the earth's surface. Wells must reach beneath the water table in order to be productive.

water use: A purpose for which water is used. May include domestic, irrigation, industrial, in-stream uses, or other purposes. A water use is a human influence or interaction on or with the water cycle.

water user's claim: A claim of right submitted to the court in connection with an adjudicative procedure (administrative or judicial).

water waste: The use of water for a nonbeneficial purpose.

waterway: See *watercourse.*

water witch: A person who uses a rod or dowser to find water beneath the ground.

waterworks: Complete system for purification and dispersal of water for domestic uses, including reservoirs and pipes.

water year: A 12-month period from October until September. The year is designated by the year that includes nine of the twelve months.

watt hour (wh): A unit of measurement equal to one watt of power steadily supplied over an electrical circuit for one hour.

wave: A moving ridge or swell in a sea or other large body of water.

weir: A device that sits vertically in an open-channel, for measuring the rate of flow in a body of water. (See also *flume.*)

weir box: A box housing, usually made of wood or concrete, designed to house a weir. It is rectangular in shape, open at both ends, and set lengthwise in a canal.

well (water): An excavation for accessing groundwater from aquifers. There are numerous methods of access, including drilling, boring, digging, etc., all of which create wells, which are essentially deep holes in the ground for accessing water.

well capacity: The maximum rate under present conditions that a well can produce water.

well development: Improvement of a well by drawing out fine material in order to increase the discharge capacity of the aquifer.

well driller: A duly licensed individual who constructs wells.

well drilling: 1. Act of drilling a well. 2. Action regarding a well including, construction, repair, renovation, deepening, cleaning, or developing it. Sometimes also refers to act of abandoning it.

well drilling rig operator: The person who operates well drilling equipment under the direction of licensed driller.

well field: An area of land containing numerous wells for municipal supply of water, irrigation, or a heat-exchange system.

well interference: Negative impact of neighboring junior wells due to drawdown of the water table, often requiring deepening of wells.

well log: An official well construction report submitted by drillers, including description or rock formations and depths of these formations.

wellhead: 1. Source of a well. 2. A structure or device at the surface, from which water is pumped or flows.

Western States Water Council: Created in 1965 by resolution of the Western Governors Conference, the organization is intended to foster effective cooperation among western states, to maintain state prerogatives, to identify legitimate federal interests, to foster communication of ideas, and to provide analysis of policy decisions. The Council is made up of appointed members from the eighteen western states. It is divided into three committees: the Water Resource Committee, the Water Quality Committee, and the Legal Committee.

wet: Containing significant moisture.

wet line: The length of a sounding line beneath the water surface, used to measure the depth of water in a stream or lake.

wetlands: Lands inundated with water, including swamps, marshes, bogs, etc., at such frequency to support vegetation adapted to saturated conditions of the soil.

white squall: A whirlwind or turbulence at sea, not accompanied by cloud cover. It takes its name from the whitecap of the disturbance.

wilderness: Land that is without permanent human improvements or habitation (uncultivated and uninhabited), which is often protected in such a way as to preserve the natural condition of the land.

wing wall: The side walls of a structure designed for prevention of sloughing of banks and channels by confining and directing materials from the bank.

Winters **Doctrine:** Doctrine created by the United States Supreme Court in the landmark case of 1908, finding federal reserved water rights not specifically stated in contracts creating the Indian reservations are implied.

withdrawal: Water removed from its source, whether ground or surface, to be put to use elsewhere.

xeriscaping: A landscape method using drought-resistant plants adopted to arid areas. This is a popular way of conserving water in many areas.

yield: Mass per unit time per unit area.

yield, operational: See *yield, optimal.*

yield, optimal: An amount of groundwater that is determined by policy and viable alternatives to be the ideal amount for withdrawal from a groundwater basin each year. This amount is based on goals, objectives, and other management decisions as opposed to annual recharge, which is the basis for perennial pield.

yield, perennial: The sustainable amount of renewable water that can be beneficially consumed and used from an aquifer each year. This amount is generally equal to the average annual recharge entering the aquifer.

yield, safe: See *safe yield.*

zone of influence: The area affected or capable of being affected by pumping operations in a well, lowering the water table. (See also *cone of depression*).

zone of saturation: Underground area where all openings in rock and soil are filled with water. The top of this zone is usually referred to as the water table.

μg/L (micrograms per liter): A measurement of the amount of substance in a liter of water, in terms of weight per volume. One μg/L is equal to one part per billion.

Table of Cases

Table of Statutes*

---

* See list of state codes above for further information.

Acknowledgments*

6 POF2d pp. 301–44, *Unreasonable Alteration of Surface Drainage* (1975), with 1989 update (American Jurisprudence Proof of Facts 2d).

6 POF2d pp. 595–645 *Contamination of Subterranean Water Supply by Sewage* (1975) w/1989 update (American Jurisprudence Proof of Facts 2d).

6 Am Jur Proof of Facts 3d *Act of God,* §§ 1–57, pp. 319–456 (1989) (American Jurisprudence Proof of Facts 2d).

8 POF2d *Municipality's Failure to Maintain Sewers Properly,* pp. 101–44 (1976), with 1989 update (American Jurisprudence Proof of Facts 2d).

9 Fed Proc Forms 29 (NEPA & CWA).

10 POF *Rain and Other Weather Phenomena,* pp. 49–136 (1961), with 1989 update.

11 Fed Proc, L Ed § 32:16-32:40 (NEPA).

11 Fed Proc, L Ed § 32:258-32:424 (CWA).

13 Fed Proc Forms 50 (ESA).

13 Fed Proc Forms 51 (Navigable Waters).

17 POF2d *Failure to Manage Dam or Reservoir to Prevent Flooding,* pp. 133–90 (1978), with 1989 update.

19 POF2d *Dam Failure as Result of Negligent Design or Maintenance,* pp. 75–161 (1979), with 1989 update (American Jurisprudence Proof of Facts 2d).

21 POF 2d, *Change in Shoreline* §§ 1–42 (1980), with 1989 update (American Jurisprudence Proof of Facts 2d).

25 Fed Proc, L Ed § 56:2017-56:2057 (ESA).

25 Fed Proc, L Ed § 57:229–57:247 (Navigable Waters Permit Procedures).

25 POF, *Water Pollution—Sewage and Industrial Wastes* §§ 1–51 at 233–332 (1970) with 1989 update (American Jurisprudence Proof of Facts).

78 Am Jur 2d Waters (2006); (1975); (1988); (1999).

Barker, Barbara, Irene Scharf, and Kelly Kunsch, *Washington Practice, Methods of Practice,* vol. 1B, § 58 Water Law (3rd ed. 1989 with 1993 update). (1B Wash Prac § 58).

Barron, Jerome A., and C. Thomas Dienes, *Constitutional Law in a Nutshell,* Chapter 3 (State Power) (5th ed. 2003).

Campbell-Mohn, Breen, and Futrell, *Environmental Law From Resources to Recovery,* §§ 8.1 (3) (b) (iii), and 8.2 (1993).

Canby, William C., Jr., *American Indian Law in a Nutshell* (3rd ed., 1998).

Corbin, Arthur L., *Corbin on Contracts: One Volume Edition* (1951).

Dobbs, Dan B., *Remedies: Damages, Equity, Restitution* (1973).

Engdahl, David E., *Constitutional Federalism in a Nutshell* (2nd ed., 1987).

Findley, Roger W. & Farber, Daniel A., *Environmental Law in a Nutshell* (5th ed., 2000).

Garner, Bryan A., ed., *Black's Law Dictionary* (7th ed., 1999).

Getches, David H., *Water Law in a Nutshell* (3rd ed. 1997).

Glicksman, Robert L., *Modern Public Land Law in a Nutshell,* 8, 128–48 (3rd ed., 2006).

Gould, George A., and Grant, Douglas L., *Cases and Materials on Water Law* (5th ed., 1995).

Juergensmeyer, Julian Conrad, Donald G. Hagman, *Urban Planning and Land Development Control Law Second Edition,* §§ 13.11–13.21 (1986).

---

* Note: This list includes print materials only. Web sites are listed in footnotes and following figures and tables. All Web pages listed in this book are current as of March 2007 or later.

Killian, Johnny H., ed., *The Constitution of the United States of America: Analysis and Interpretation 1982* (1987) with 1988 supplement, The Congressional Research Service (1989); Senate Document No. 99-16, 99th Congress 1st Session.

Kurian, George T., *Datapedia of the United States 1790–2005* (2nd ed., 2001).

Laitos, Jan G. & Tomain, Joseph P., *Energy and Natural Resources Law in a Nutshell,* 356–97 (1992).

Laitos, Jan G., *Natural Resources Law,* Ch. 11, §§ 11.1–11.2, 378–421 (2002).

Little, Daniel, *Varieties of Social Explanation: An Introduction to the Philosophy of Social Science,* Westview Press, Boulder, Co. (1991).

Nowak, John E., Ronald D. Rotunda, *Constitutional Law* §§ 8.1, 8.7 (Commerce Clause), 2.12 (Federal Jurisdiction) (6th ed., 2000).

Plater, Zygmunt, Robert H. Abrams, William Goldfarb, Robert L. Graham, Lisa Heinzerling, David A. Wirth, *Environmental Law and Policy: Nature, Law, and Society* (3rd ed., 2004).

Prosser, William L., *Law of Torts* 4th ed., § 93 (1971).

Prosser, William L., and Page Keaton, *Prosser and Keaton on Torts* 5th ed. (1984, 1988).

Reynolds, Osborne M., *Local Government Law,* § 193 (1982, 1996 update).

Rodgers, William H., *Environmental Law* §§ 4.1–4.9 (Water Pollution) (2nd ed., 1994).

Rotunda, Ronald D., *Modern Constitutional Law: Cases and Materials,* The American Casebook Series (6th ed., 2000).

Sax, Joseph L., Robert H. Abrams, Barton H. Thompson Jr., John D. Leshy, *Legal Control of Water Resources: Cases and Materials* (3rd ed., 2000).

Statistical Abstract of the United States, 234 No. 390 (Dep't Commerce ed. Hoover Business Press 120th ed., 2000).

Sullivan, Kathleen M., and Gerald Gunther, *Constitutional Law,* University Casebook Series (15th ed., 2004).

Tarlock, Dan, Charles J. Meyers, *Water Resource Management: A Casebook in Law and Public Policy* (2nd ed., 1980).

Tarlock, Dan, David H. Getches, James N. Corbridge, *Water Resource Management: A Casebook in Law and Public Policy* (4th ed., 1993).

Thompson West, publisher, *Selected Environmental Law Statutes: 2005–2006 Edition.*

Torcia, Charles E., *Wharton's Criminal Law* 15th Edition, vol. 4 §§ 530, 533, and 534 (1996), 4 Wharton's Crim Law §§ 530, 533, ad 534.

USCS Constitution, Article 1, § 8, cl 3 (2004).

Williams, Jerre S., *Constitutional Analysis in a Nutshell,* Chapter 6 (State Powers) (1979).

Wright, Charles Alan, Law of Federal Courts § 22 p. 126 (5th ed., 1994).

FURTHER RESEARCH AVAILABLE

12 Fed Proc, L Ed § 34 (Farms, Ranches, and Agricultural Products).

19 Fed Proc, L Ed § 46 (Indians and Indian Affairs).

29 Fed Proc, L Ed § 66 (Public Lands and Property).

Corpus Juris Secundum, vols. 93–94 (Waters).

Federal Practice Digest 4th vol. 77C (Navigable Waters), 95A (Waters).

20 *American Jurisprudence Legal Forms* 2d, 260–61 (Waters and Waterworks and Water Companies).

24B *American Jurisprudence Pleading and Practice Forms Annotated,* Waters.

25 *American Jurisprudence Pleading and Practice Forms Annotated,* Waterworks and Water Companies.

ONLINE RESOURCES*

Findlaw—http://library.findlaw.com/1999/Jan/1/241492.html
Western Water Law—http://www.westernwaterlaw.com/index.htm
Megalaw—http://www.megalaw.com/top/water.php
California Coastal Commission—http://www.coastal.ca.gov/desalrpt/dchap1.html
http://groundwater.sdsu.edu/
Southern Nevada Water Authority—http://www.snwa.com/html/cons_wsl.html
Western States Water Council—http://www.westgov.org/wswc/

FEDERAL AGENCIES

Environmental Protection Agency—http://www.epa.gov/water/index.html
Bureau of Indian Affairs—http://www.doi.gov/bureau-indian-affairs.html
Bureau of Land Management—http://www.blm.gov/nstc/WaterLaws/abstract1.html
FEMA—http://www.fema.gov/
National Oceanic and Atmospheric Administration—http://www.noaa.gov/
National Parks Service—http://www.nps.gov/
United States Geological Survey—http://www.usgs.gov/
United States Bureau of Reclamation—http://www.usbr.gov/main/water/
United States Army Corps of Engineers—http://www.usace.army.mil/
United States Fish and Wildlife Service—http://www.fws.gov/index.html
United States Forestry Service—http://www.fs.fed.us/
See earlier listing for state agencies and their contact information.

---

* The online listings here were not all used in compiling this book but are identified for the benefit of the reader. Web pages used in writing this book are cited in footnotes or below figures and charts obtained therefrom.

Water Codes and Water Departments

STATE WATER CODES*

Alabama: Titles 9 (Title 33 navigation)
http://alisondb.legislature.state.al.us/acas/ACASLogin.asp

Alaska: Title 46
http://www.legis.state.ak.us/cgi-bin/folioisa.dll/stattx06/query=*/doc/{t19801}

Arizona: Title 45
http://www.azleg.state.az.us/ArizonaRevisedStatutes.asp?Title=45

Arkansas: Title 15 Subtitle 2
http://www.arkleg.state.ar.us/NXT/gateway.dll?f=templates&fn=default.
htm&vid=blr:code

California: Water Code
http://www.leginfo.ca.gov/cgi-bin/calawquery?codesection=wat&codebody=
&hits=20

Colorado: Water Code
http://www.michie.com/colorado/lpext.dll?f=templates&fn=main-h.htm&cp=
Title 37
http://www.michie.com/colorado/lpExt.dll?f=templates&eMail=Y&fn=main-h.htm
&cp=cocode/59836

Connecticut: Title 25
http://www.cga.ct.gov/2005/pub/Title25.htm

Delaware: Title 23
http://www.delcode.state.de.us/title23/index.htm#TopOfPage

Florida: Title XXVIII & Title XXI
http://www.leg.state.fl.us/statutes/index.cfm?App_mode=Display_Index&Title_
Request=XXVIII#TitleXXVIII
http://www.leg.state.fl.us/statutes/index.cfm?App_mode=Display_Statute&URL=
Ch0298/titl0298.htm&StatuteYear=2006&Title=%2D%3E2006%2D%3E
Chapter%20298

---

* Please note that while most of these sites are relatively easy to figure out, not all links are complete and some sites may be difficult to navigate.

Georgia: Title 52
http://w3.lexis-nexis.com/hottopics/gacode/default.asp

Hawaii: 0174C
http://www.capitol.hawaii.gov/hrscurrent/Vol03_Ch0121-200D/HRS0174C/
HRS_0174C-0031.HTM

Idaho: Titles 42, 43, 70
http://www3.state.id.us/idstat/TOC/42FTOC.html
http://www3.state.id.us/idstat/TOC/43FTOC.html
http://www3.state.id.us/idstat/TOC/70FTOC.html

Illinois: Chapter 615
http://www.ilga.gov/legislation/ilcs/ilcs2.asp?ChapterID=47

Indiana: Title 14, Art. 15
http://www.in.gov/legislative/ic/code/title14/ar15/

Iowa: Chapters 161, 357, 466
http://nxtsearch.legis.state.ia.us/NXT/gateway.dll/2007%20Iowa%20Code/
2007code/1/27534?f=templates&fn=defaultURLquerylink.htm

Kansas: Chapters 42, 82a
http://www.kslegislature.org/legsrv-statutes/articlesList.do

Kentucky: Chapters 104
http://www.lrc.ky.gov/KRS/104-00/CHAPTER.HTM

Louisiana: Civil Code 655 to 658
http://www.legis.state.la.us/

Maine: Title 38
http://janus.state.me.us/legis/statutes/38/title38ch0sec0.html

Maryland: State Code
http://www.michie.com/maryland/lpext.dll?f=templates&fn=main-h.htm&2.0
Natural Resources Title 8
http://www.michie.com/maryland/lpExt.dll?f=templates&eMail=Y&fn=main-h.htm
&cp=mdcode/198fd/1a5f8

Massachusetts: Chapter 21G
http://www.mass.gov/legis/laws/mgl/gl-21g-toc.htm

Michigan: Chapter 323
http://www.legislature.mi.gov/(S(bzup02555qtlfn3g3gqefcuq))/mileg.aspx?page=
getobject&objectname=mcl-chap323

Minnesota: Chapters 103A to 114B
http://ros.leg.mn/revisor/pages/statute/statute_chapter.php?year=2006&start=
103A&close=114B&history=&border=0

Mississippi: State Code
  http://michie.com/mississippi/lpext.dll?f=templates&fn=main-h.htm&cp=
  Title 51 Waters, Water Resources, Water Districts, Drainage, And Flood Control
  http://michie.com/mississippi/lpExt.dll?f=templates&eMail=Y&fn=main-h.htm
    &cp=mscode/fce2

Missouri: Title XV
  http://www.moga.state.mo.us/STATUTES/STATUTES.HTM#T15

Montana: Title 85
  http://data.opi.state.mt.us/bills/mca_toc/85.htm

Nebraska: Chapter 46
  http://law.justia.com/nebraska/codes/s46index/s46index.html

Nevada: Title 48
  http://www.leg.state.nv.us/NRS/Index.cfm

New Hampshire: Title L
  http://www.gencourt.state.nh.us/rsa/html/NHTOC/NHTOC-L.htm

New Jersey: Title 58
  http://lis.njleg.state.nj.us/cgi-bin/om_isapi.dll?clientID=53910968&Depth=
    2&depth=2&expandheadings=on&headingswithhits=on&hitsperheading=
    on&infobase=statutes.nfo&record={17660}&softpage=Doc_Frame_PG42

New Mexico: Chapter 72
  http://nxt.ella.net/NXT/gateway.dll?f=templates$fn=default.htm$vid=nm:all

New York: Article 15
  http://public.leginfo.state.ny.us/menugetf.cgi?COMMONQUERY=LAWS

North Carolina: Chapter 159G
  http://www.ncga.state.nc.us/gascripts/Statutes/StatutesTOC.pl?Chapter=0159G

North Dakota: Chapter 61
  http://www.legis.nd.gov/cencode/t61.html

Ohio: State Code
  http://codes.ohio.gov/orc
  Title XV Conservation of Natural Resources
  http://codes.ohio.gov/orc/15

Oklahoma: Title 82
  http://www.lsb.state.ok.us/

Oregon: Title 45, Chapters 536–558
  http://www.leg.state.or.us/ors/vol13.html

Pennsylvania: Title 32 Part II
  http://members.aol.com/StatutesP2/32.html

Rhode Island: Title 46
  http://www.rilin.state.ri.us/Statutes/TITLE46/INDEX.HTM

South Carolina: Title 49
http://www.scstatehouse.net/code/titl49.htm

South Dakota: Title 46 & 46A
http://legis.state.sd.us/statutes/DisplayStatute.aspx?Type=Statute&Statute=46
http://legis.state.sd.us/statutes/DisplayStatute.aspx?Type=Statute&Statute= 46A

Tennessee: State Code
http://www.michie.com/tennessee/lpext.dll?f=templates&fn=main-h.htm&cp=
Title 69 Waters, Waterways, Drains And Levees
http://www.michie.com/tennessee/lpExt.dll?f=templates&eMail=Y&fn=main-h.htm
&cp=tncode/2c893

Texas: Chapter 3 Article 7808-7880, Water Code
http://tlo2.tlc.state.tx.us/statutes/wl.toc.htm
http://tlo2.tlc.state.tx.us/statutes/wa.toc.htm

Utah: Title 73
http://le.utah.gov/~code/TITLE73/TITLE73.htm

Vermont: Title 25
http://www.leg.state.vt.us/statutes/chapters.cfm?Title=25

Virginia: Title 62.1
http://leg1.state.va.us/cgi-bin/legp504.exe?000+cod+TOC6201000

Washington: Title 90
http://apps.leg.wa.gov/rcw/default.aspx?Cite=90

West Virginia: State Code
http://www.legis.state.wv.us/WVCODE/Code.cfm
Chapter 20 Natural Resources
http://www.legis.state.wv.us/WVCODE/code.cfm?chap=20&art=1

Wisconsin: Chapter 30
http://nxt.legis.state.wi.us/nxt/gateway.dll?f=templates&fn=default.htm&vid=WI
:Default&d=index&jd=top

Wyoming: Title 41
http://legisweb.state.wy.us/statutes/statutes.aspx?file=titles/Title41/Title41.htm

WESTERN STATE WATER DEPARTMENTS AND PERMITTING AGENCIES

Alaska Department of Natural Resources: Division of Mining, Land & Water
http://www.dnr.state.ak.us/mlw/water/wrfact.htm
Water Resources Anchorage Office
550 W. 7th Ave., Suite 1020
Anchorage, AK 99501-3577
Phone (907) 269-8503 / Fax (907) 269-8947

Arizona Department of Water Resources
http://www.water.az.gov/WaterManagement/
Arizona Department of Water Resources
500 N. Third Street
Phoenix, AZ 85004
Phone: (602) 417-2400
Long Distance within Arizona: (800) 352-8488 / Fax: (602) 417-2401

California Environmental Protection Agency: Division of Water Rights
http://www.waterrights.ca.gov/
State Water Resources Control Board
Division of Water Rights
1001 I Street Sacramento, CA 95814
P.O. Box 2000 Sacramento, CA 95812
P.O. Box 100 Sacramento, CA 95812
Phone: (916) 341-5300 / (916) 341-5250 / Fax: (916) 341-5252

Colorado Division of Water Resources
http://water.state.co.us/wateradmin/prior.asp
Colorado Division of Water Resources
1313 Sherman Street, Rm 818
Denver, CO 80203
Phone: 303-866-3581 / Fax: 303-866-3589

Idaho Department of Water Resources
http://www.idwr.idaho.gov/water/rights/
The Idaho Water Center
322. E. Front Street
Boise, ID 83720-0098
Phone: (208) 287-4800 / Fax: (208) 287-6700

Kansas Department of Agriculture: Division of Water Resources
http://www.ksda.gov/appropriation/
Topeka Field Office
Division of Water Resources
109 S.W. 9th Street, First Floor
Topeka, KS 66612-2216
Phone: (785) 368-8251 / PH: (785) 296-3556 / Fax: (785) 296-4619

Mississippi Department of Environmental Quality: Office of Land & Water Resources
http://www.deq.state.ms.us/MDEQ.nsf/page/L&W_Home?OpenDocument
Southport Center
2380 Highway 80 West
Jackson, MS 39204
Office of Land and Water Resources
P.O. Box 10631

Jackson, MS 39289-0631
Phone: (601) 961-5202 / Fax: (601) 354-6938

Montana Department of Natural Resources and Conservation
http://dnrc.mt.gov/
DNRC Water Resources Division
Water Rights Bureau
1424 9th Ave
P.O. Box 201601
Helena, MT 59620
Phone: 406-444-6610 / Fax: 406-444-0533

Nebraska Department of Natural Resources
http://www.dnr.ne.gov/
The Nebraska Department of Natural Resources
301 Centennial Mall South
Lincoln, NB 68509-4676
Phone: 402-471-2363 / Fax: 402-471-2900

New Mexico Office of the State Engineer
http://www.seo.state.nm.us/index.html

Office of the State Engineer
130 South Capitol Street
Concha Ortiz y Pino Building
P.O. Box 25102
Santa Fe, NM 87504-5102
Phone: (505) 827-6166 / Fax: (505) 827-3806

Water Resource Allocation Program
407 Galisteo Street
Bataan Memorial Building Rm.102
P.O. Box 25102
Santa Fe, NM 87504-5102
Phone: (505) 827-6120 / Fax: (505) 827-6682

Nevada Division of Water Resources
http://water.nv.gov/
Division of Water Resources
901 South Stewart St., Suite 2002
Carson City, NV 89701
Phone: 775-684-2800 / Fax: 775-684-2811

North Dakota State Water Commission
http://www.swc.state.nd.us/4dlink9/4dcgi/redirect/index.html
North Dakota State Water Commission
900 East Boulevard Avenue, Dept 770

Bismarck, ND 58505-0850
Phone: (701) 328-2750 Fax: (701) 328-3696

Oklahoma Water Resources Board
http://www.owrb.ok.gov/index.php
Oklahoma Water Resources Board
3800 North Classen Blvd.
Oklahoma City, OK 73118
Phone: 405-530-8800 / Fax: 405-530-8900

Oregon Water Resources Department
http://www.oregon.gov/OWRD/index.shtml
Water Resources Department
725 Summer Street NE, Suite A
Salem, OR 97301
Phone: 503-986-0900 / Fax: 503-986-0904

South Dakota Department of Environment & Natural Resources
http://www.state.sd.us/DENR/des/waterrights/wr_permit.htm
Joe Foss Building
523 E Capitol
Pierre SD 57501
Phone: (605) 773-3151 / Fax: (605) 773-6035

Texas Commission on Environmental Quality
http://www.tceq.state.tx.us/
Texas Commission on Environmental Quality
PO Box 13087, MC-160
Austin, TX 78711-3087
Phone: (512) 239-4691 / Fax: (512) 239-4770

Utah Division of Water Rights
http://www.waterrights.utah.gov/wrinfo/default.htm
Utah Division of Water Rights
1594 W. North Temple Suite 220
P.O. Box 146300
Salt Lake City, UT 84114-6300
Phone: (801) 538-7240 / Fax: (801) 538-7467

Washington Department of Ecology
http://www.ecy.wa.gov/programs/wr/rights/water-right-home.html
Washington State Department of Ecology
P.O. Box 47600
Olympia, WA 98504-7600
Phone: 360-407-6000

Wyoming State Engineers Office
http://seo.state.wy.us/
122 West 25th Street
4th Floor East
Cheyenne, WY 82002
Phone: (307) 777-6150 / Fax: (307) 777-5451

Appendix 1

Brief Discussion of Some Important Cases in Federal Water Law

For more case law, see footnotes to main text.

GIBBONS V. OGDEN, 22 U.S. (9 WHEAT.) 1, 6 L.ED. 23 (1824)*

This was the first big case dealing with Congress' authority over navigable waters. The case defined federal power under the Commerce Clause to include any activity that affects matters beyond state lines. While the court in this case found reason to think federal authority over commerce to be exclusive under the Constitution, it was unnecessary to decide that matter. Subsequent courts have found concurrent authority, with state laws trumped by conflicting federal law according to the Supremacy Clause of the U.S. Constitution. This case dealt with competing licenses to traverse the same waters handed out by the federal and state governments. The court included navigation within the term commerce and a federal license ultimately trumped the license given by the State of New York. Traditionally, Congress could regulate activities that directly affected interstate commerce. The courts until the New Deal legislation of the 1930s held this view. The New Deal† signified a shift away from the notion that only direct impacts on interstate commerce were implicated, toward one that any impact is included in the commerce power. While somewhat limited by recent court decisions,‡ the commerce power of Congress remains the primary basis for the courts finding federal jurisdiction under the Constitution and is the primary basis for federal jurisdiction over the waters of the United States, even where interstate commerce is not directly implicated.

---

* See § 5.9.

† *Stafford v. Wallace*, 258 U.S. 495, 42 S.Ct. 397, 66 L.Ed. 735 (1922) (using stream of commerce); *NLRB v. Jones & Laughlin Steel Corp.*, 301 U.S. 1, 57 S.Ct. 615, 81 L.Ed. 893 (1937) (beginning to inquire into extent of burden on interstate commerce); *Wickard v. Fliburn*, 317 U.S. 111, 63 S.Ct. 82, 87 L.Ed. 122 (1942) (upholding regulation of wheat grown at home for personal consumption because of aggregate impact on interstate commerce, even though none of the wheat entered the stream of commerce at all).

‡ *United States v. Lopez*, 514 U.S. 549, 115 S.Ct. 1624, 131 L.Ed.2d 626 (1995); *United States v. Morrison*, 529 U.S. 598, 120 S.Ct. 1740, 146 L.Ed.2d 658 (2000).

NEW ORLEANS V. UNITED STATES, 35
U.S. 662, 10 PETERS 662, (1836)

New Orleans attempted to sale certain lands claimed by the United States by virtue of cessation from France (Louisiana Purchase). The land was part of a "quay" (embankment) along the Mississippi. A quay is a vacant space between the first row of buildings and the water's edge used for receiving or shipping goods. The quay in question was no longer adjacent to the Mississippi and was considered prime real estate. Based on French and Spanish law, which existed in the territory prior to the United States acquisition, the court found that when Louisiana became a state the United States could no longer exercise authority over the public lands. This case appears to be a misreading of the Constitution and the Equal Footing Clause but is the basis for state ownership of the public lands in the Louisiana Territories. This case was limited to quays and does not prevent other federal land holdings.

POLLARD V. HAGAN, 44 U.S. 212, 3 HOW. 212, 11 L.ED. 565 (1845)*

An Alabama action for ejectment by plaintiff against defendant, this case stands for the idea that states acquire title to noncoastal tidelands upon admission to the Union. The dispute arose when the United States granted title to the land and later the state also granted competing title. The issue was who owned the land, or, in other words, which government (federal or state) had the right to dispose of it. This case was limited to noncoastal tidelands. These lands apparently were considered part of the water for navigable purposes. The majority of the court found that the United States no longer owned the land after Alabama became a state based on the Equal Footing Clause. These lands pass to the states on admission to the Union.

THE DANIEL BALL, 77 U.S. (10 WALL) 557, 19 L.ED. 999 (1871)†

The Supreme Court found jurisdiction for a federal law regarding safety regulations on board ship though the ship traveled only on intrastate routes within Michigan, because it was part of an interstate journey. The ship was an instrumentality of the shipping of goods from one state to another and was therefore using waterways of interstate commerce.

WINTERS V. UNITED STATES, 207 U.S. 564,
28 S.CT. 207, 52 L.ED. 340 (1908)‡

In the treaty with the United States government, the tribe did not reserve water rights to service the arid lands of the reservation. The water was first used by settlers and then by residents of the reservation. The court found that water rights were implied in the contract between the United States and the tribe and ruled that those rights could

---

* See § 1.7.
† See § 1.7.
‡ See §§ 9.1, 9.13, 9.14.

not be destroyed by the Equal Footing Doctrine, upon admission of Montana as a state. The rights of settlers, perfected under state law were invalid, so far as the water was reserved by the federal government for use on the reservation. Even though the settlers began using the water before the Indians, the reserved water rights existed as of the time the reservation was created or earlier and not as of the date water was first put to beneficial use.

UNITED STATES V. RANDS, 389 U.S. 121, 88 S.CT. 265, 19 L.ED.2D 329 (1967)*

Landowner leased land to the State of Oregon, granting to the state,\ the option to purchase the land as a port site on the Columbia River. The United States government took the land by eminent domain before the option to buy was exercised, as part of a comprehensive plan for the river. The compensation was only for about one-fifth the claimed value of the land, if used as a port. The Ninth Circuit held that the land should be compensable for its port value, but the United States Supreme Court reversed, holding that to find otherwise would create a private claim in the public domain. This decision was overruled by legislative act,† so that today land is condemned at its optimal value.

UTAH V. UNITED STATES, 403 U.S. 9, 91 S.CT. 1775, 29 L.ED.2D 279 (1971)‡

The test for navigation is whether the waterway is in fact navigable or used commercially for transportation, even if for only part of the course or part of the year.§ The Great Salt Lake was found to meet this test, because at the time of statehood there was evidence that some small boats did in fact use the lake for navigation. The United States Supreme Court stated that navigability applies to lakes as well as rivers. It found that the Great Salt Lake, which only had a minimal amount of small boat traffic, all over a century ago, and was wholly in Utah, with no connections to other bodies of water, was still navigable in a constitutional sense. These cases were crucial in establishing federal authority over all major bodies of water and many lesser bodies.

TENNESSEE VALLEY AUTH. V. HILL, 437 U.S. 153, 98 S.CT. 2279, 57 L.ED.2D 117 (1978)¶

This case was brought to save the snail darter from extinction. The tiny fish was threatened by the Tellico Dam project. The United States Supreme Court found that

---

* See § 3.10. This decision was effectively overturned by Congress after much criticism. See § 111 of the Rivers and Harbors Act, 33 U.S.C.A. § 595a.
† § 111 of the Rivers and Harbors Act, 33 U.S.C.A. § 595a.
‡ See §§ 1.8, 11.3.
§ See *United States v. Utah*, 283 U.S. 64, 75 L.Ed. 844, 51 S.Ct. 438 (1931).
¶ See § 10.10.

the Endangered Species Act requires federal agencies to consider threats to species and their habitat before implementing further actions. The court upheld an injunction preventing final construction of the largely completed dam. Ultimately, the project was completed based on special congressional authorization after the government lost in the courts, and failed to gain approval from a special committee authorized to approve important projects in spite of the environmental cost. The snail darter did disappear in that area, though other colonies were later found to exist elsewhere. This case illustrates the incredible power of the Endangered Species Act to stop even large, expensive, nearly completed federal projects, in order to protect threatened or endangered species and their designated "critical habitat."

LUCAS V. SOUTH CAROLINA COASTAL COUNCIL, 505 U.S. 1003, 112 S.CT.2886, 120 L.ED.2D 798 (1992)

A property owner with two beachfront lots was prevented from building on those lots by a local ordinance. The United States Supreme Court ruled that a zoning ordinance that prevents the owner of land from any economically viable use is a per se taking. In this case the owner was due compensation for the time in which he was denied economic use of his land. While this was not directly a water law case, water is property and it would constitute a taking where the right is vested and regulations deny the holder all economic value of that right.

RAPANOS V. UNITED STATES, 547 U.S. __, 126 S.CT. 2208 (2006)*

(See memorandum on following page, discussing authority of the Army and EPA under CWA.) This heavily divided opinion is somewhat confusing to define. The CWA makes it illegal to discharge dredge or fill material into navigable waters without a permit from the Army Corps of Engineers.† This opinion is about the definition of "navigable waters" used by the Corps In this decision the court ruled 5 to 4 that two developers may not be required to acquire a permit, where the waters being filled in are not navigable, or permanent in nature. The Court stated that the CWA refers only to permanent water bodies. As a result the cases were remanded back to the lower courts. One case involved lands separated by an impermeable man-made berm. The other involved a connection by man-made ditches. In his concurrence, Justice Kennedy implied that an actual connection of such surface waters might be enough to find jurisdiction. The case's 4-4-1 outcome makes this concurring opinion significant. It is not certain that this case implies an actual constitutional limitation on federal authority. It may mean little more than a limitation on the defined statutory authority granted by the CWA. It is unclear whether the decision would be different should congress modify the CWA to expand the definition of "navigable waters." The likelihood is that the Court would rule differently if the definition were expanded. Nothing in the opinion forbids the states from restricting development, in this traditionally state matter.

---

* See §§ 1.8, 10.7.
† 33 U.S.C.A. §§ 1311 (a), 1342 (a), 1362 (7) (definition of navigable waters)

Post *Rapanos* Memorandum on Clean Water Act Jurisdiction of Crops and EPA

Clean Water Act Jurisdiction
Following the U.S. Supreme Court's Decision
in
Rapanos v. United States & Carabell v. United States

This memorandum provides guidance to EPA regions and U.S. Army Corps of Engineers ["Corps"] districts implementing the Supreme Court's decision in the consolidated cases Rapanos v. United States and Carabell v. United States[1] (herein referred to simply as "Rapanos") which address the jurisdiction over waters of the United States under the Clean Water Act.[2] The chart below summarizes the key points contained in this memorandum. This reference tool is not a substitute for the more complete discussion of issues and guidance furnished throughout the memorandum.

Summary of Key Points

The agencies will assert jurisdiction over the following waters:
- Traditional navigable waters
- Wetlands adjacent to traditional navigable waters
- Non-navigable tributaries of traditional navigable waters that are relatively permanent where the tributaries typically flow year-round or have continuous flow at least seasonally (e.g., typically three months)
- Wetlands that directly abut such tributaries

The agencies will decide jurisdiction over the following waters based on a fact-specific analysis to determine whether they have a significant nexus with a traditional navigable water:
- Non-navigable tributaries that are not relatively permanent
- Wetlands adjacent to non-navigable tributaries that are not relatively permanent
- Wetlands adjacent to but that do not directly abut a relatively permanent non-navigable tributary

The agencies generally will not assert jurisdiction over the following features:
- Swales or erosional features (e.g., gullies, small washes characterized by low volume, infrequent, or short duration flow)
- Ditches (including roadside ditches) excavated wholly in and draining only uplands and that do not carry a relatively permanent flow of water

The agencies will apply the significant nexus standard as follows:
- A significant nexus analysis will assess the flow characteristics and functions of the tributary itself and the functions performed by all wetlands adjacent to the tributary to determine if they significantly affect the chemical, physical and biological integrity of downstream traditional navigable waters
- Significant nexus includes consideration of hydrologic and ecologic factors

---

[1] 126 S. Ct. 2208 (2006).
[2] 33 U.S.C. §1251 et seq.

Post *Rapanos* Memorandum on Clean Water Act Jurisdiction of Corps and EPA. Available at http://www.epa.gov/owow/wetlands/pdf/RapanosGuida nce6507.pdf.

Background

Congress enacted the Clean Water Act ("CWA" or "the Act") "to restore and maintain the chemical, physical, and biological integrity of the Nation's waters."[3]   One of the mechanisms adopted by Congress to achieve that purpose is a prohibition on the discharge of any pollutants, including dredged or fill material, into "navigable waters" except in compliance with other specified sections of the Act.[4] In most cases, this means compliance with a permit issued pursuant to CWA §402 or §404. The Act defines the term "discharge of a pollutant" as "any addition of any pollutant to navigable waters from any point source[,]"[5] and provides that "[t]he term 'navigable waters' means the waters of the United States, including the territorial seas." [6]

In Rapanos, the Supreme Court addressed where the Federal government can apply the Clean Water Act, specifically by determining whether a wetland or tributary is a "water of the United States." The justices issued five separate opinions in Rapanos (one plurality opinion, two concurring opinions, and two dissenting opinions), with no single opinion commanding a majority of the Court.

The Rapanos Decision

Four justices, in a plurality opinion authored by Justice Scalia, rejected the argument that the term "waters of the United States" is limited to only those waters that are navigable in the traditional sense and their abutting wetlands.[7]   However, the plurality concluded that the agencies' regulatory authority should extend only to "relatively permanent, standing or continuously flowing bodies of water" connected to traditional navigable waters, and to "wetlands with a continuous surface connection to" such relatively permanent waters.[8]

Justice Kennedy did not join the plurality's opinion but instead authored an opinion concurring in the judgment vacating and remanding the cases to the Sixth Circuit Court of Appeals.[9] Justice Kennedy agreed with the plurality that the statutory term "waters of the United States" extends beyond water bodies that are traditionally considered navigable.[10] Justice Kennedy, however, found the plurality's interpretation of the scope of the CWA to be "inconsistent with the Act's text, structure, and purpose[,]" and he instead presented a different standard for evaluating CWA jurisdiction over

---

[3] 33 U.S.C. § 1251(a).
[4] 33 U.S.C. § 1311(a), §1362(12)(A).
[5] 33 U.S.C. § 1362(12)(A).
[6] 33 U.S.C. § 1362(7). See also 33 C.F.R. § 328.3(a) and 40 C.F.R. § 230.3(s).
[7] Id. at 2220.
[8] Id. at 2225-27.
[9] Id. at 2236-52. While Justice Kennedy concurred in the Court's decision to vacate and remand the cases to the Sixth Circuit, his basis for remand was limited to the question of "whether the specific wetlands at issue possess a significant nexus with navigable waters." 126 S. Ct. at 2252. In contrast, the plurality remanded the cases to determine both "whether the ditches and drains near each wetland are 'waters,'" and "whether the wetlands in question are 'adjacent' to these 'waters' in the sense of possessing a continuous surface connection...." Id. at 2235.
[10] Id. at 2241.

wetlands and other water bodies.[11] Justice Kennedy concluded that wetlands are "waters of the United States" "if the wetlands, either alone or in combination with similarly situated lands in the region, significantly affect the chemical, physical, and biological integrity of other covered waters more readily understood as 'navigable.' When, in contrast, wetlands' effects on water quality are speculative or insubstantial, they fall outside the zone fairly encompassed by the statutory term 'navigable waters.'"[12]

Four justices, in a dissenting opinion authored by Justice Stevens, concluded that EPA's and the Corps' interpretation of "waters of the United States" was a reasonable interpretation of the Clean Water Act.[13]

When there is no majority opinion in a Supreme Court case, controlling legal principles may be derived from those principles espoused by five or more justices.[14] Thus, regulatory jurisdiction under the CWA exists over a water body if either the plurality's or Justice Kennedy's standard is satisfied.[15] Since Rapanos, the United States has filed pleadings in a number of cases interpreting the decision in this manner.

The agencies are issuing this memorandum in recognition of the fact that EPA regions and Corps districts need guidance to ensure that jurisdictional determinations, permitting actions, and other relevant actions are consistent with the decision and supported by the administrative record. Therefore, the agencies have evaluated the Rapanos opinions to identify those waters that are subject to CWA jurisdiction under the reasoning of a majority of the justices. This approach is appropriate for a guidance document. The agencies intend to more broadly consider jurisdictional issues, including clarification and definition of key terminology, through rulemaking or other appropriate policy process.

---

[11] Id. at 2246.

[12] Id. at 2248. Chief Justice Roberts wrote a separate concurring opinion explaining his agreement with the plurality. See 126 S. Ct. at 2235-36.

[13] Id. at 2252-65. Justice Breyer wrote a separate dissenting opinion explaining his agreement with Justice Stevens' dissent. See 126 S. Ct. at 2266.

[14] See Marks v. United States, 430 U.S. 188, 193-94 (1977); Waters v. Churchill, 511 U.S. 661, 685 (1994) (Souter, J., concurring) (analyzing the points of agreement between plurality, concurring, and dissenting opinions to identify the legal "test ... that lower courts should apply," under Marks, as the holding of the Court); cf. League of United Latin American Citizens v. Perry, 126 S. Ct. 2594, 2607 (2006) (analyzing concurring and dissenting opinions in a prior case to identify a legal conclusion of a majority of the Court); Alexander v. Sandoval, 532 U.S. 275, 281-282 (2001) (same).

[15] 126 S. Ct. at 2265 (Stevens, J., dissenting) ("Given that all four justices who have joined this opinion would uphold the Corps' jurisdiction in both of these cases – and in all other cases in which either the plurality's or Justice Kennedy's test is satisfied – on remand each of the judgments should be reinstated if *either* of those tests is met.") (emphasis in original).

Agency Guidance[16]

To ensure that jurisdictional determinations, administrative enforcement actions, and other relevant agency actions are consistent with the Rapanos decision, the agencies in this guidance address which waters are subject to CWA § 404 jurisdiction.[17] Specifically, this guidance identifies those waters over which the agencies will assert jurisdiction categorically and on a case-by-case basis, based on the reasoning of the Rapanos opinions.[18]   EPA and the Corps will continually assess and review the application of this guidance to ensure nationwide consistency, reliability, and predictability in our administration of the statute.

1. Traditional Navigable Waters (i.e., "(a)(1) Waters") and Their Adjacent Wetlands

---

Key Points

- The agencies will assert jurisdiction over traditional navigable waters, which includes all the waters described in 33 C.F.R. § 328.3(a)(1), and 40 C.F.R. § 230.3 (s)(1).
- The agencies will assert jurisdiction over wetlands adjacent to traditional navigable waters, including over adjacent wetlands that do not have a continuous surface connection to traditional navigable waters.

EPA and the Corps will continue to assert jurisdiction over "[a]ll waters which are currently used, or were used in the past, or may be susceptible to use in interstate or

---

[16] The CWA provisions and regulations described in this document contain legally binding requirements. This guidance does not substitute for those provisions or regulations, nor is it a regulation itself. It does not impose legally binding requirements on EPA, the Corps, or the regulated community, and may not apply to a particular situation depending on the circumstances. Any decisions regarding a particular water will be based on the applicable statutes, regulations, and case law. Therefore, interested persons are free to raise questions about the appropriateness of the application of this guidance to a particular situation, and EPA and/or the Corps will consider whether or not the recommendations or interpretations of this guidance are appropriate in that situation based on the statutes, regulations, and case law.

[17] This guidance focuses only on those provisions of the agencies' regulations at issue in Rapanos -- 33 C.F.R. §§ 328.3(a)(1), (a)(5), and (a)(7); 40 C.F.R. §§ 230.3(s)(1), (s)(5), and (s)(7). This guidance does not address or affect other subparts of the agencies' regulations, or response authorities, relevant to the scope of jurisdiction under the CWA. In addition, because this guidance is issued by both the Corps and EPA, which jointly administer CWA § 404, it does not discuss other provisions of the CWA, including §§ 311 and 402, that differ in certain respects from § 404 but share the definition of "waters of the United States." Indeed, the plurality opinion in Rapanos noted that "… there is no reason to suppose that our construction today significantly affects the enforcement of §1342 … The Act does not forbid the 'addition of any pollutant *directly* to navigable waters from any point source,' but rather the 'addition of any pollutant *to* navigable waters.'" (emphasis in original) 126 S. Ct. 2208, 2227. EPA is considering whether to provide additional guidance on these and other provisions of the CWA that may be affected by the Rapanos decision.

[18] In 2001, the Supreme Court held that use of "isolated" non-navigable intrastate waters by migratory birds was not by itself a sufficient basis for the exercise of federal regulatory jurisdiction under the CWA. See Solid Waste Agency of Northern Cook County (SWANCC) v. U.S. Army Corps of Engineers, 531 U.S. 159 (2001). This guidance does not address SWANCC, nor does it affect the Joint Memorandum regarding that decision issued by the General Counsels of EPA and the Department of the Army on January 10, 2003. See 68 Fed. Reg. 1991, 1995 (Jan. 15, 2003).

foreign commerce, including all waters which are subject to the ebb and flow of the tide."[19] These waters are referred to in this guidance as traditional navigable waters.

The agencies will also continue to assert jurisdiction over wetlands "adjacent" to traditional navigable waters as defined in the agencies' regulations. Under EPA and Corps regulations and as used in this guidance, "adjacent" means "bordering, contiguous, or neighboring." Finding a continuous surface connection is not required to establish adjacency under this definition. The Rapanos decision does not affect the scope of jurisdiction over wetlands that are adjacent to traditional navigable waters because at least five justices agreed that such wetlands are "waters of the United States."[20]

2. Relatively Permanent Non-navigable Tributaries of Traditional Navigable Waters and Wetlands with a Continuous Surface Connection with Such Tributaries

> **Key Points**
>
> - The agencies will assert jurisdiction over non-navigable tributaries of traditional navigable waters that are relatively permanent where the tributaries typically flow year-round or have continuous flow at least seasonally (e.g., typically three months).
> - The agencies will assert jurisdiction over those adjacent wetlands that have a continuous surface connection to such tributaries (e.g., they are not separated by uplands, a berm, dike, or similar feature.)

A non-navigable tributary[21] of a traditional navigable water is a non-navigable water body whose waters flow into a traditional navigable water either directly or indirectly by means of other tributaries. Both the plurality opinion and the dissent would uphold CWA jurisdiction over non-navigable tributaries that are "relatively permanent" – waters that typically (e.g., except due to drought) flow year-round or waters that have a

---

[19] 33 C.F.R. § 328.3(a)(1); 40 C.F.R. § 230.3(s)(1). The "(a)(1)" waters include all of the "navigable waters of the United States," defined in 33 C.F.R. Part 329 and by numerous decisions of the federal courts, plus all other waters that are navigable-in-fact (e.g., the Great Salt Lake, UT and Lake Minnetonka, MN).

[20] Id. at 2248 (Justice Kennedy, concurring) ("As applied to wetlands adjacent to navigable-in-fact waters, the Corps' conclusive standard for jurisdiction rests upon a reasonable inference of ecologic interconnection, and the assertion of jurisdiction for those wetlands is sustainable under the Act by showing adjacency alone.").

[21] A tributary includes natural, man-altered, or man-made water bodies that carry flow directly or indirectly into a traditional navigable water. Furthermore, a tributary, for the purposes of this guidance, is the entire reach of the stream that is of the same order (i.e., from the point of confluence, where two lower order streams meet to form the tributary, downstream to the point such tributary enters a higher order stream). The flow characteristics of a particular tributary will be evaluated at the farthest downstream limit of such tributary (i.e., the point the tributary enters a higher order stream). It is reasonable for the agencies to treat the stream reach as a whole in light of the Supreme Court's observation that the phrase "navigable waters" generally refers to "rivers, streams, and other hydrographic features." 126 S. Ct. at 2222 (Justice Scalia, quoting Riverside Bayview, 474 U.S. at 131). The entire reach of a stream is a reasonably identifiable hydrographic feature. The agencies will also use this characterization of tributary when applying the significant nexus standard under Section 3 of this guidance.

continuous flow at least seasonally (e.g., typically three months).[22]   Justice Scalia emphasizes that relatively permanent waters do not include tributaries "whose flow is 'coming and going at intervals ... broken, fitful.'"[23]   Therefore, "relatively permanent" waters do not include ephemeral tributaries which flow only in response to precipitation and intermittent streams which do not typically flow year-round or have continuous flow at least seasonally. However, CWA jurisdiction over these waters will be evaluated under the significant nexus standard described below. The agencies will assert jurisdiction over relatively permanent non-navigable tributaries of traditional navigable waters without a legal obligation to make a significant nexus finding.

In addition, the agencies will assert jurisdiction over those adjacent wetlands that have a continuous surface connection with a relatively permanent, non-navigable tributary, without the legal obligation to make a significant nexus finding. As explained above, the plurality opinion and the dissent agree that such wetlands are jurisdictional.[24] The plurality opinion indicates that "continuous surface connection" is a "physical connection requirement."[25]   Therefore, a continuous surface connection exists between a wetland and a relatively permanent tributary where the wetland directly abuts the tributary (e.g., they are not separated by uplands, a berm, dike, or similar feature).[26]

---

[22]   See 126 S. Ct. at 2221 n. 5 (Justice Scalia, plurality opinion) (explaining that "relatively permanent" does not necessarily exclude waters "that might dry up in extraordinary circumstances such as drought" or "seasonal rivers, which contain continuous flow during some months of the year but no flow during dry months").

[23]   Id. (internal citations omitted).

[24]   Id. at 2226-27 (Justice Scalia, plurality opinion).

[25]   Id. at 2232 n.13 (referring to "our physical-connection requirement" and later stating that Riverside Bayview does not reject "the physical-connection requirement") and 2234 ("Wetlands are 'waters of the United States' if they bear the 'significant nexus' of physical connection, which makes them as a practical matter *indistinguishable* from waters of the United States.") (emphasis in original). See also 126 S. Ct. at 2230 ("adjacent" means "physically abutting") and 2229 (citing to Riverside Bayview as "confirm[ing] that the scope of ambiguity of 'the waters of the United States' is determined by a wetland's *physical connection* to covered waters...") (emphasis in original). A continuous surface connection does not require surface water to be continuously present between the wetland and the tributary. 33 C.F.R. § 328.3(b) and 40 C.F.R. § 232.2 (defining wetlands as "those areas that are inundated or saturated by surface or ground water at a frequency and duration sufficient to support ... a prevalence of vegetation typically adapted for life in saturated soil conditions").

[26]   While all wetlands that meet the agencies' definitions are considered adjacent wetlands, only those adjacent wetlands that have a continuous surface connection because they directly abut the tributary (e.g., they are not separated by uplands, a berm, dike, or similar feature) are considered jurisdictional under the plurality standard.

3.   Certain Adjacent Wetlands and Non-navigable Tributaries That Are Not Relatively Permanent

---

Key Points

- The agencies will assert jurisdiction over non-navigable, not relatively permanent tributaries and their adjacent wetlands where such tributaries and wetlands have a significant nexus to a traditional navigable water.
- A significant nexus analysis will assess the flow characteristics and functions of the tributary itself and the functions performed by any wetlands adjacent to the tributary to determine if they significantly affect the chemical, physical and biological integrity of downstream traditional navigable waters.
- "Similarly situated" wetlands include all wetlands adjacent to the same tributary.
- Significant nexus includes consideration of hydrologic factors including the following:
    - volume, duration, and frequency of flow, including consideration of certain physical characteristics of the tributary
    - proximity to the traditional navigable water
    - size of the watershed
    - average annual rainfall
    - average annual winter snow pack
- Significant nexus also includes consideration of ecologic factors including the following:
    - potential of tributaries to carry pollutants and flood waters to traditional navigable waters
    - provision of aquatic habitat that supports a traditional navigable water
    - potential of wetlands to trap and filter pollutants or store flood waters
    - maintenance of water quality in traditional navigable waters
- The following geographic features generally are not jurisdictional waters:
    - swales or erosional features (e.g. gullies, small washes characterized by low volume, infrequent, or short duration flow)
    - ditches (including roadside ditches) excavated wholly in and draining only uplands and that do not carry a relatively permanent flow of water

---

The agencies will assert jurisdiction over the following types of waters when they have a significant nexus with a traditional navigable water: (1) non-navigable tributaries that are not relatively permanent,[27] (2) wetlands adjacent to non-navigable tributaries that are not relatively permanent, and (3) wetlands adjacent to, but not directly abutting, a relatively permanent tributary (e.g., separated from it by uplands, a berm, dike or similar feature).[28] As described below, the agencies will assess the flow characteristics and functions of the tributary itself, together with the functions performed by any wetlands adjacent to that tributary, to determine whether collectively they have a significant nexus with traditional navigable waters.

The agencies' assertion of jurisdiction over non-navigable tributaries and adjacent wetlands that have a significant nexus to traditional navigable waters is supported by five

---

[27] For simplicity, the term "tributary" when used alone in this section refers to non-navigable tributaries that are not relatively permanent.

[28] As described in Section 2 of this guidance, the agencies will assert jurisdiction, without the need for a significant nexus finding, over all wetlands that are both adjacent and have a continuous surface connection to relatively permanent tributaries. See pp. 6-7, supra.

justices. Justice Kennedy applied the significant nexus standard to the wetlands at issue in Rapanos and Carabell: "[W]etlands possess the requisite nexus, and thus come within the statutory phrase 'navigable waters,' if the wetlands, either alone or in combination with similarly situated lands in the region, significantly affect the chemical, physical, and biological integrity of other covered waters more readily understood as 'navigable.'"[29] While Justice Kennedy's opinion discusses the significant nexus standard primarily in the context of wetlands adjacent to non-navigable tributaries,[30] his opinion also addresses Clean Water Act jurisdiction over tributaries themselves. Justice Kennedy states that, based on the Supreme Court's decisions in Riverside Bayview and SWANCC, "the connection between a non-navigable water or wetland may be so close, or potentially so close, that the Corps may deem the water or wetland a 'navigable water' under the Act. … Absent a significant nexus, jurisdiction under the Act is lacking."[31] Thus, Justice Kennedy would limit jurisdiction to those waters that have a significant nexus with traditional navigable waters, although his opinion focuses on the specific factors and functions the agencies should consider in evaluating significant nexus for adjacent wetlands, rather than for tributaries.

In considering how to apply the significant nexus standard, the agencies have focused on the integral relationship between the ecological characteristics of tributaries and those of their adjacent wetlands, which determines in part their contribution to restoring and maintaining the chemical, physical and biological integrity of the Nation's traditional navigable waters. The ecological relationship between tributaries and their adjacent wetlands is well documented in the scientific literature and reflects their physical proximity as well as shared hydrological and biological characteristics. The flow parameters and ecological functions that Justice Kennedy describes as most relevant to an evaluation of significant nexus result from the ecological inter-relationship between tributaries and their adjacent wetlands. For example, the duration, frequency, and volume of flow in a tributary, and subsequently the flow in downstream navigable waters, is directly affected by the presence of adjacent wetlands that hold floodwaters, intercept sheet flow from uplands, and then release waters to tributaries in a more even and constant manner. Wetlands may also help to maintain more consistent water temperature in tributaries, which is important for some aquatic species. Adjacent wetlands trap and hold pollutants that may otherwise reach tributaries (and downstream navigable waters) including sediments, chemicals, and other pollutants. Tributaries and their adjacent wetlands provide habitat (e.g., feeding, nesting, spawning, or rearing young) for many aquatic species that also live in traditional navigable waters.

---

[29] Id. at 2248. When applying the significant nexus standard to tributaries and wetlands, it is important to apply it within the limits of jurisdiction articulated in SWANCC. Justice Kennedy cites SWANCC with approval and asserts that the significant nexus standard, rather than being articulated for the first time in Rapanos, was established in SWANCC. 126 S. Ct. at 2246 (describing SWANCC as "interpreting the Act to require a significant nexus with navigable waters"). It is clear, therefore, that Justice Kennedy did not intend for the significant nexus standard to be applied in a manner that would result in assertion of jurisdiction over waters that he and the other justices determined were not jurisdictional in SWANCC. Nothing in this guidance should be interpreted as providing authority to assert jurisdiction over waters deemed non-jurisdictional by SWANCC.

[30] 126 S. Ct. at 2247-50.

[31] Id. at 2241 (emphasis added).

When performing a significant nexus analysis,[32] the first step is to determine if the tributary has any adjacent wetlands. Where a tributary has no adjacent wetlands, the agencies will consider the flow characteristics and functions of only the tributary itself in determining whether such tributary has a significant effect on the chemical, physical and biological integrity of downstream traditional navigable waters. A tributary, as characterized in Section 2 above, is the entire reach of the stream that is of the same order (i.e., from the point of confluence, where two lower order streams meet to form the tributary, downstream to the point such tributary enters a higher order stream). For purposes of demonstrating a connection to traditional navigable waters, it is appropriate and reasonable to assess the flow characteristics of the tributary at the point at which water is in fact being contributed to a higher order tributary or to a traditional navigable water. If the tributary has adjacent wetlands, the significant nexus evaluation needs to recognize the ecological relationship between tributaries and their adjacent wetlands, and their closely linked role in protecting the chemical, physical, and biological integrity of downstream traditional navigable waters.

Therefore, the agencies will consider the flow and functions of the tributary together with the functions performed by all the wetlands adjacent to that tributary in evaluating whether a significant nexus is present. Similarly, where evaluating significant nexus for an adjacent wetland, the agencies will consider the flow characteristics and functions performed by the tributary to which the wetland is adjacent along with the functions performed by the wetland and all other wetlands adjacent to that tributary. This approach reflects the agencies' interpretation of Justice Kennedy's term "similarly situated" to include all wetlands adjacent to the same tributary. Where it is determined that a tributary and its adjacent wetlands collectively have a significant nexus with traditional navigable waters, the tributary and all of its adjacent wetlands are jurisdictional. Application of the significant nexus standard in this way is reasonable because of its strong scientific foundation – that is, the integral ecological relationship between a tributary and its adjacent wetlands. Interpreting the phrase "similarly situated" to include all wetlands adjacent to the same tributary is reasonable because such wetlands are physically located in a like manner (i.e., lying adjacent to the same tributary).

Principal considerations when evaluating significant nexus include the volume, duration, and frequency of the flow of water in the tributary and the proximity of the tributary to a traditional navigable water. In addition to any available hydrologic information (e.g., gauge data, flood predictions, historic records of water flow, statistical data, personal observations/records, etc.), the agencies may reasonably consider certain physical characteristics of the tributary to characterize its flow, and thus help to inform the determination of whether or not a significant nexus is present between the tributary and downstream traditional navigable waters. Physical indicators of flow may include the presence and characteristics of a reliable ordinary high water mark (OHWM) with a

---

[32] In discussing the significant nexus standard, Justice Kennedy stated: "The required nexus must be assessed in terms of the statute's goals and purposes. Congress enacted the [CWA] to 'restore and maintain the chemical, physical, and biological integrity of the Nation's waters' ..." 126 S. Ct. at 2248. Consistent with Justice Kennedy's instruction, EPA and the Corps will apply the significant nexus standard in a manner that restores and maintains any of these three attributes of traditional navigable waters.

channel defined by bed and banks.[33] Other physical indicators of flow may include shelving, wracking, water staining, sediment sorting, and scour.[34] Consideration will also be given to certain relevant contextual factors that directly influence the hydrology of tributaries including the size of the tributary's watershed, average annual rainfall, average annual winter snow pack, slope, and channel dimensions.

In addition, the agencies will consider other relevant factors, including the functions performed by the tributary together with the functions performed by any adjacent wetlands. One such factor is the extent to which the tributary and adjacent wetlands have the capacity to carry pollutants (e.g., petroleum wastes, toxic wastes, sediment) or flood waters to traditional navigable waters, or to reduce the amount of pollutants or flood waters that would otherwise enter traditional navigable waters.[35] The agencies will also evaluate ecological functions performed by the tributary and any adjacent wetlands which affect downstream traditional navigable waters, such as the capacity to transfer nutrients and organic carbon vital to support downstream foodwebs (e.g., macroinvertebrates present in headwater streams convert carbon in leaf litter making it available to species downstream), habitat services such as providing spawning areas for recreationally or commercially important species in downstream waters, and the extent to which the tributary and adjacent wetlands perform functions related to maintenance of downstream water quality such as sediment trapping.

After assessing the flow characteristics and functions of the tributary and its adjacent wetlands, the agencies will evaluate whether the tributary and its adjacent wetlands are likely to have an effect that is more than speculative or insubstantial on the chemical, physical, and biological integrity of a traditional navigable water. As the distance from the tributary to the navigable water increases, it will become increasingly important to document whether the tributary and its adjacent wetlands have a significant nexus rather than a speculative or insubstantial nexus with a traditional navigable water.

Accordingly, Corps districts and EPA regions shall document in the administrative record the available information regarding whether a tributary and its adjacent wetlands have a significant nexus with a traditional navigable water, including the physical indicators of flow in a particular case and available information regarding the functions of the tributary and any adjacent wetlands. The agencies will explain their basis for concluding whether or not the tributary and its adjacent wetlands, when considered together, have a more than speculative or insubstantial effect on the chemical, physical, and biological integrity of a traditional navigable water.

---

[33] See 33 C.F.R. § 328.3(e). The OHWM also serves to define the lateral limit of jurisdiction in a non-navigable tributary where there are no adjacent wetlands. See 33 C.F.R. § 328.4(c). While EPA regions and Corps districts must exercise judgment to identify the OHWM on a case-by-case basis, the Corps' regulations identify the factors to be applied. These regulations have recently been further explained in Regulatory Guidance Letter (RGL) 05-05 (Dec. 7, 2005). The agencies will apply the regulations and the RGL and take other steps as needed to ensure that the OHWM identification factors are applied consistently nationwide.

[34] See Justice Kennedy's discussion of "physical characteristics," 126 S. Ct. at 2248-2249.

[35] See, generally, 126 S. Ct. at 2248-53; see also 126 S. Ct. at 2249 ("Just as control over the non-navigable parts of a river may be essential or desirable in the interests of the navigable portions, so may the key to flood control on a navigable stream be found in whole or in part in flood control on its tributaries....") (citing to Oklahoma ex rel. Phillips v. Guy F. Atkinson Co., 313 U.S. 508, 524-25(1941)).

Swales or erosional features (e.g., gullies, small washes characterized by low volume, infrequent, or short duration flow) are generally not waters of the United States because they are not tributaries or they do not have a significant nexus to downstream traditional navigable waters. In addition, ditches (including roadside ditches) excavated wholly in and draining only uplands and that do not carry a relatively permanent flow of water are generally not waters of the United States because they are not tributaries or they do not have a significant nexus to downstream traditional navigable waters.[36]   Even when not jurisdictional waters subject to CWA §404, these geographic features (e.g., swales, ditches) may still contribute to a surface hydrologic connection between an adjacent wetland and a traditional navigable water. In addition, these geographic features may function as point sources (i.e., "discernible, confined, and discrete conveyances"), such that discharges of pollutants to other waters through these features could be subject to other CWA regulations (e.g., CWA §§ 311 and 402).[37]

Certain ephemeral waters in the arid west are distinguishable from the geographic features described above where such ephemeral waters are tributaries and they have a significant nexus to downstream traditional navigable waters. For example, in some cases these ephemeral tributaries may serve as a transitional area between the upland environment and the traditional navigable waters. During and following precipitation events, ephemeral tributaries collect and transport water and sometimes sediment from the upper reaches of the landscape downstream to the traditional navigable waters. These ephemeral tributaries may provide habitat for wildlife and aquatic organisms in downstream traditional navigable waters. These biological and physical processes may further support nutrient cycling, sediment retention and transport, pollutant trapping and filtration, and improvement of water quality, functions that may significantly affect the chemical, physical, and biological integrity of downstream traditional navigable waters.

Documentation

As described above, the agencies will assert CWA jurisdiction over the following waters without the legal obligation to make a significant nexus determination: traditional navigable waters and wetlands adjacent thereto, non-navigable tributaries that are relatively permanent waters, and wetlands with a continuous surface connection with such tributaries. The agencies will also decide CWA jurisdiction over other non-navigable tributaries and over other wetlands adjacent to non-navigable tributaries based on a fact-specific analysis to determine whether they have a significant nexus with traditional navigable waters. For purposes of CWA §404 determinations by the Corps, the Corps and EPA are developing a revised form to be used by field regulators for documenting the assertion or declination of CWA jurisdiction.

Corps districts and EPA regions will ensure that the information in the record adequately supports any jurisdictional determination. The record shall, to the maximum extent practicable, explain the rationale for the determination, disclose the data and information relied upon, and, if applicable, explain what data or information received greater or lesser weight, and what professional judgment or assumptions were used in

[36]   See 51 Fed. Reg. 41206, 41217 (Nov. 13, 1986).
[37]   33 U.S.C. § 1362(14).

reaching the determination. The Corps districts and EPA regions will also demonstrate and document in the record that a particular water either fits within a class identified above as not requiring a significant nexus determination, or that the water has a significant nexus with a traditional navigable water. As a matter of policy, Corps districts and EPA regions will include in the record any available information that documents the existence of a significant nexus between a relatively permanent tributary that is not perennial (and its adjacent wetlands if any) and a traditional navigable water, even though a significant nexus finding is not required as a matter of law.

All pertinent documentation and analyses for a given jurisdictional determination (including the revised form) shall be adequately reflected in the record and clearly demonstrate the basis for asserting or declining CWA jurisdiction.[38] Maps, aerial photography, soil surveys, watershed studies, local development plans, literature citations, and references from studies pertinent to the parameters being reviewed are examples of information that will assist staff in completing accurate jurisdictional determinations. The level of documentation may vary among projects. For example, jurisdictional determinations for complex projects may require additional documentation by the project manager.

_____
Benjamin H. Grumbles
Assistant Administrator for Water
U.S. Environmental Protection Agency

John Paul Woodley, Jr.
Assistant Secretary of the Army
(Civil Works)
Department of the Army

---

[38] For jurisdictional determinations and permitting decisions, such information shall be posted on the appropriate Corps website for public and interagency information.

Appendix 2

Select Federal Statutes*

(For more federal statutes in this text, see footnotes to main chapters and Table of Statutes)

THE FIFTH AMENDMENT

No person shall be . . . deprived of life, liberty, or property, without due process of law; nor shall private property be taken for public use, without just compensation.†

16 U.S.C.A. § 1531 (B) (ESA § 2)

The purposes of this chapter are to provide a means whereby the ecosystems upon which endangered species and threatened species depend may be conserved, to provide a program for conservation of such endangered species and threatened species, and to take such steps as may be appropriate to achieve the purposes of the treatise and conventions set forth in subsection (a) of this section.‡

16 U.S.C.A. § 1536 (E.S.A. § 7 (2))

(2) Each Federal agency shall, in consultation with and with the assistance of the Secretary, insure that any action authorized, funded, or carried out by such agency (hereinafter in this section referred to as an "agency action") is not likely to jeopardize the continued existence of any endangered species or threatened species or result in the destruction or adverse modification of habitat of such species which is determined by the Secretary, after consultation as appropriate with affected States, to be critical, unless such agency has been granted an exemption for such action by the Committee pursuant to subsection (h) of this section. In fulfilling the requirements of this paragraph each agency shall use the best scientific and commercial data available.§

16 U.S.C.A. § 1536 (E.S.A. § 7)

(c) Biological assessment. (1) To facilitate compliance with the requirements of subsection (a) (2) of this section, each Federal agency shall, with respect to any agency

---

* The full U.S. Code can be accessed at http://www.loc.gov/law/guide/uscode.html#usc.
† See § 3.10.
‡ See § 10.10.
§ See § 10.10.

action of such agency for which no contract for construction has been entered into and for which no construction has begun on November 10, 1978, request of the Secretary information whether any species which is listed or proposed to be listed may be present in the area of such proposed action. If the Secretary advises, based on the best scientific and commercial data available, that such species may be present, such agency shall conduct a biological assessment for the purpose of identifying any endangered species or threatened species which is likely to be affected by such action. Such assessment shall be completed within 180 days after the date on which initiated. . . .*

16 U.S.C.A. § 1538 (E.S.A. § 9 (1))

(1) Except as provided in sections 1535(g)(2) and 1539 of this title, with respect to any endangered species of fish or wildlife listed pursuant to section 1533 of this title it is unlawful for any person subject to the jurisdiction of the United States to (A) import any such species into , or export any such species from the United States; (B) take any such species within the United States or the territorial sea of the United States; (C) take any such species upon the high seas; (D) possess, sell, deliver, carry, transport, or ship, by any means whatsoever, any such species taken in violation of subparagraphs (B) and (C); (E) deliver, receive, carry, transport, or ship in interstate or foreign commerce, by any means whatsoever and in the course of a commercial activity, any such species; (F) sell or offer for sale in interstate or foreign commerce any such species; or violate any regulation pertaining to such species or to any threatened species of fish or wildlife listed pursuant to section 1533 of this title and promulgated by the Secretary pursuant to authority provided by this chapter.

28 U.S.C.A. § 1345 (FEDERAL CIVIL JURISDICTION)

Except as otherwise provided by Act of Congress, the district courts shall have original jurisdiction of all civil actions, suits or proceedings commenced by the United States, or by any agency or officer thereof expressly authorized to sue by Act of Congress.†

33 U.S.C.A. § 1251 (G) (CLEAN WATER ACT § 101)

It is the policy of Congress that the authority of each State to allocate quantities of water within its jurisdiction shall not be superseded abrogated or otherwise impaired by this chapter. It is the further policy of Congress that nothing in this chapter shall be construed to supersede or abrogate rights to quantities of water which have been established by any State. Federal agencies shall co-operate with State and local agencies to develop comprehensive solutions to prevent, reduce, and eliminate pollution in concert with programs for managing water resources.‡

---

* See § 10.10.
† See Wright, Charles Alan, *Law of Federal Courts Fifth Ed.* § 22, p. 126 (1994); This jurisdiction statute is relevant to the forum for adjudicating Native American Water Rights, since the federal government is almost always a party. See § 9.16.
‡ See §§ 10.8, 10.18.

33 U.S.C.A. § 403 (THE RIVERS AND HARBORS APPROPRIATION ACT OF 1899 (RHP) § 10)

The creation of any obstruction not affirmatively authorized by Congress to the navigable capacity of any of the waters of the United States is prohibited; and it shall not be lawful to build or commence the building of any wharf, pier, dolphin, boom, weir, breakwater, bulkhead, jetty or other structures in any . . . navigable river, or other water of the United States, outside established harbor lines, or where no harbor lines have been established, except on plans recommended by the Chief of Engineers and authorized by the Secretary of the Army; and it shall not be lawful to excavate or fill, or in any manner to alter or modify the course, location, condition, or capacity . . . of the channel of any navigable water of the United States, unless the work has been recommended by the Chief of Engineers and authorized by the Secretary of the Army prior to beginning the same.*

33 U.S.C.A. § 595A (RIVERS AND HARBORS ACT)†

In all cases where real property shall be taken by the United States for public use in connection with any improvement of rivers, harbors, canals, or waterways of the United States, and in all condemnation proceedings by the United States to acquire lands or easements for such improvements, the compensation to be paid for real property taken by the United States above the normal high water mark of navigable waters of the United States shall be the fair market value of such real property based upon all uses to which such real property may reasonably be put, including its highest and best use, any of which uses may be dependent upon access to or utilization of such navigable waters.

33 U.S.C.A. § 1311 (THE CLEAN WATER ACT § 301 (B) (2) (E))

as expeditiously as practicable but in no case later than three years after the date such limitations are promulgated under section 1314(b) of this title, and in no case later than March 31, 1989, compliance with effluent limitations for categories and classes of point sources, other than publicly owned treatment works, which in the case of pollutants identified pursuant to section 1314 (a)(4) of this title shall require application of the best conventional pollutant control technology as determined in accordance with regulations issued by the Administrator pursuant to section 1314(b)(4) of this title.‡

33 U.S.C.A. § 1344 (F)(2) (C.W.A. § 404)

Any discharge of dredged or fill material into the navigable waters incidental to any activity having as its purpose bringing an area of the navigable waters into a use to which it was not previously subject, where the flow or circulation of navigable waters

---

* See § 10.15.
† See § 3.10.
‡ See § 10.3.

may be impaired or the reach of such waters be reduced, shall be required to have a permit under this section.*

33 U.S.C.A. § 1362 (THE CLEAN WATER ACT § 502 (14))

The term "point source" means any discernible, confined and discrete conveyance, including but not limited to any pipe, ditch, channel, tunnel, conduit, well, discrete fissure, container, rolling stock, concentrated animal feeding operation, or vessel or other floating craft, from which pollutants are or may be discharged. This term does not include agricultural stormwater discharges and return flows from irrigated agriculture.†

42 U.S.C.A. § 4321 (NEPA § 2)

The purposes of this chapter are: To declare a national policy which will encourage productive and enjoyable harmony between man and his environment; to promote efforts which will prevent or eliminate damage to the environment and biosphere and stimulate the health and welfare of man; to enrich the understanding of the ecological systems and natural resources important to the Nation; and to establish a Council on Environmental Quality.‡

42 U.S.C.A. § 4331 (NEPA § 101)

(a) The Congress, recognizing the profound impact of man's activity on the interrelations of all components of the natural environment, particularly the profound influences of population growth, high-density urbanization, industrial expansion, resource exploitation, and new and expanding technological advances and recognizing further the critical importance of restoring and maintaining environmental quality to the overall welfare and development of man, declares that it is the continuing policy of the Federal Government, in cooperation with State and local governments, and other concerned public and private organizations, to use all practicable means and measures including financial and technical assistance, in a manner calculated to foster and promote the general welfare, to create and maintain conditions under which man and nature can exist in productive harmony, and fulfill the social, economic, and other requirements of present and future generations of Americans.§

42 U.S.C.A. § 4332 (NEPA § 102 (C))

All agencies of the Federal Government shall . . . (C) include in every recommendation or report on proposals for legislation and other major Federal actions significantly affecting the quality of the human environment, a detailed statement by the responsible

---

* See § 10.7.
† See §§ 10.3, 10.7.
‡ See § 10.1.
§ See § 10.1.

official on (i) the environmental impact of the proposed action, (ii) any adverse environmental effects which cannot be avoided should the proposal be implemented, (iii) alternatives to the proposed action, (iv) the relationship between local short-term uses of man's environment and the maintenance and enhancement of long-term productivity, and (v) any irreversible and irretrievable commitments of resources which would be involved in the proposed action should it be implemented.*

43 U.S.C.A. §§ 321–25 THE DESERT LAND ENTRIES ACT OF 1877

Addresses the creation of the appropriative system for determining water rights in the west. Declared all non-navigable water to be available for appropriation in the present states of California, Oregon, Washington, Idaho, Montana, Wyoming, Utah, Colorado, Nevada, Arizona, New Mexico, and North and South Dakota. These appropriations are governed by state law.†

43 U.S.C.A. § 666 THE MCCARRAN AMENDMENT

(a) Consent is hereby given to join the United States as a defendant in any suit (1) for the adjudication of rights to the use of water of a river system or other source, or (2) for the administration of such rights, where it appears that the United States is the owner of or is in the process of acquiring water rights by appropriation under State law, by purchase or exchange, or otherwise, and the United States, is a party to any such suit. The United States, when a party to any such suit, shall (1) be deemed to have waived any right to plead that the United States is not amenable thereto by reason of its sovereignty, and (2) shall be subject to the judgments, orders, and decrees of the court having jurisdiction . . .‡ For a discussion of the Clean Water Act 33 U.S.C.A., see § 10.2–10.8.

THE NORTHWEST ORDINANCE ART. 5

. . . And, whenever any of the said States§ shall have sixty thousand free inhabitants therein, such State shall be admitted, by its delegates, into the Congress of the United States, on an equal footing with the original States in all respects whatever, and shall be at liberty to form a permanent constitution and State government: . . .

---

* See § 10.1.
† See § 4.2.
‡ See § 8.16.
§ Refers to states created from territories ceded by the original 13 states to provide the federal government with revenue. This is not a reference to all future territories of the United States, though it has occasionally been mistakenly interpreted that way.

43 U.S.C.A. §§ 324-25: THE DESERT LAND ENTRIES ACT OF 1877

Appendix 3

Charts and Lists

TABLE 1
Total Water Withdrawals by Source and State, 2000*

| State | Population (in thousands) | By Source and Type — Withdrawals (in million gallons per day) | | | | | | | | | Withdrawals (in thousand acre-feet per year) | | |
| | | Groundwater | | | Surface Water | | | Total | | | Total | | |
		Fresh	Saline	Total	Fresh	Saline	Total	Fresh	Saline	Total	Fresh	Saline	Total
Alabama	4,450	440	0	440	9,550	0	9,550	9,990	0	9,990	11,200	0	11,200
Alaska	627	50.2	90.4	141	111	53.4	164	161	144	305	181	161	342
Arizona	5,130	3,420	8.17	3,430	3,300	0	3,300	6,720	8.17	6,730	7,530	9.16	7,540
Arkansas	2,670	6,920	0.08	6,920	3,950	0	3,950	10,900	0.08	10,900	12,200	0.09	12,200
California	33,900	15,200	152	15,400	23,200	12,600	35,800	38,400	12,800	51,200	43,100	14,300	57,400
Colorado	4,300	2,320	0	2,320	10,300	0	10,300	12,600	0.00	12,600	14,200	0	14,200
Connecticut	3,410	143	0	143	565	3,440	4,010	708	3,440	4,150	794	3,860	4,650
Delaware	784	115	0	115	466	741	1,210	582	741	1,320	652	831	1,480
District of Columbia	572	0	0	0	9.87	0	9.87	9.87	0	9.87	11.1	0	11.1
Florida	16,000	5,020	0	5,020	3,110	12,000	15,100	8,140	12,000	20,100	9,120	13,400	22,500
Georgia	8,190	1,450	0	1,450	4,960	91.7	5,060	6,410	91.7	6,500	7,190	103	7,290
Hawaii	1,210	433	0.85	434	208	0	208	640	0.85	641	718	0.95	719
Idaho	1,290	4,140	0	4,140	15,300	0	15,300	19,500	0	19,500	21,800	0	21,800
Illinois	12,400	813	0	813	12,900	0	12,900	13,700	0	13,700	15,400	0	15,400
Indiana	6,080	656	0	656	9,460	0	9,460	10,100	0	10,100	11,300	0	11,300
Iowa	2,930	679	0	679	2,680	0	2,680	3,360	0	3,360	3,770	0	3,770
Kansas	2,690	3,790	0	3,790	2,820	0	2,820	6,610	0	6,610	7,410	0	7,410

continued

Kentucky	4,040	189	0	189	3,970	0	3,970	4,160	0	4,160	4,660	0	4,660
Louisiana	4,470	1,630	0	1,630	8,730	0	8,730	10,400	0	10,400	11,600	0	11,600
Maine	1,270	80.80	0	80.8	423	295	718	504	295	799	565	330	895
Maryland	5,300	225	0	225	1,200	6,490	7,690	1,430	6,490	7,910	1,600	7,270	8,870
Massachusetts	6,350	269	0	269	783	3,610	4,390	1,050	3,610	4,660	1,180	4,050	5,220
Michigan	9,940	734	0	734	9,260	0	9,260	10,000	0	10,000	11,200	0	11,200
Minnesota	4,920	720	0	720	3,150	0	3,150	3,870	0	3,870	4,340	0	4,340
Mississippi	2,840	2,180	0	2,180	632	148	781	2,810	148	2,960	3,150	166	3,320
Missouri	5,600	1,780	0	1,780	6,450	0	6,450	8,230	0	8,230	9,220	0	9,220
Montana	902	188	0	188	8,100	0	8,100	8,290	0	8,290	9,300	0	9,300
Nebraska	1,710	7,860	4.55	7,860	4,390	0	4,390	12,200	4.55	12,300	13,700	5.10	13,700
Nevada	2,000	757	0	757	2,050	0	2,050	2,810	0	2,810	3,140	0	3,140
New Hampshire	1,240	85.20	0	85.2	362	761	1,120	447	761	1,210	501	854	1,350
New Jersey	8,410	584	0	584	1,590	3,390	4,980	2,170	3,390	5,560	2,430	3,800	6,230
New Mexico	1,820	1,540	0	1,540	1,710	0	1,710	3,260	0	3,260	3,650	0	3,650
New York	19,000	893	0	893	6,190	5,010	11,200	7,080	5,010	12,100	7,940	5,610	13,600
North Carolina	8,050	580	0	580	9,150	1,620	10,800	9,730	1,620	11,400	10,900	1,810	12,700
North Dakota	642	123	0	123	1,020	0	1,020	1,140	0	1,140	1,280	0	1,280
Ohio	11,400	878	0	878	10,300	0	10,300	11,100	0	11,100	12,500	0	12,500
Oklahoma	3,450	771	256	1,030	990	0	990	1,760	256	2,020	1,970	287	2,260
Oregon	3,420	993	0	993	5,940	0	5,940	6,930	0	6,930	7,770	0	7,770
Pennsylvania	12,300	666	0	666	9,290	0	9,290	9,950	0	9,950	11,200	0	11,200
Rhode Island	1,050	28.6	0	28.6	110	290	400	138	290	429	155	326	481

TABLE 1 (continued)
Total Water Withdrawals by Source and State, 2000*

State	Population (in thousands)	Withdrawals (in million gallons per day)									Withdrawals (in thousand acre-feet per year)		
		By Source and Type						Total			Total		
		Groundwater			Surface Water								
		Fresh	Saline	Total	Fresh	Saline	Total	Fresh	Saline	Total	Fresh	Saline	Total
South Carolina	4,010	330	0	330	6,840	0	6,840	7,170	0	7,170	8,040	0	8,040
South Dakota	755	222	0	222	306	0	306	528	0	528	592	0	592
Tennessee	5,690	417	0	417	10,400	0	10,400	10,800	0	10,800	12,100	0	12,100
Texas	20,900	8,470	504	8,970	16,300	4,350	20,700	24,800	4,850	29,600	27,800	5,440	33,200
Utah	2,230	1,020	26.5	1,050	3,740	177	3,920	4,760	203	4,970	5,340	228	5,570
Vermont	609	43.2	0	43.2	404	0	404	447	0	447	501	0	501
Virginia	7,080	314	0	314	4,880	3,640	8,520	5,200	3,640	8,830	5,830	4,080	9,900
Washington	5,890	1,470	0	1,470	3,800	39.9	3,840	5,270	39.9	5,310	5,910	44.7	5,960
West Virginia	1,810	90.9	0	90.9	5,060	0	5,060	5,150	0	5,150	5,770	0	5,770
Wisconsin	5,360	813	0	813	6,780	0	6,780	7,590	0	7,590	8,510	0	8,510
Wyoming	494	541	222	763	4,400	0	4,400	4,940	222	5,170	5,540	248	5,790
Puerto Rico	3,810	137	0	137	483	2,190	2,670	620	2,190	2,810	695	2,460	3,150
U.S. Virgin Islands	109	1.03	0	1.03	10.6	136	147	11.6	136	148	13.0	153	166
Total	285,000	83,300	1,260	84,500	262,000	61,000	323,000	345,000	62,300	408,000	387,000	69,800	457,000

Source: Modified from table available at http://pubs.usgs.gov/circ/2004/circ1268/htdocs/table01.html, March 2007.
* See Figure 21
Note: Figures may not sum to totals because of independent rounding.

TABLE 2
Total Water Withdrawals by Water-Use Category, 2000

State	Public Supply Fresh	Domestic Fresh	Irrigation Fresh	Livestock Fresh	Aquaculture Fresh	Industrial Fresh	Industrial Saline	Mining Fresh	Mining Saline	Thermoelectric Power Fresh	Thermoelectric Power Saline	Total Fresh	Total Saline	Total Total
Alabama	834	78.9	43.1	—	10.4	833	0	—	—	8,190	0	9,990	0	9,990
Alaska	80.0	11.2	1.01	—	—	8.12	3.86	27.4	140	33.6	0	161	144	305
Arizona	1,080	28.9	5,400	—	—	19.8	0	85.7	8.17	100	0	6,720	8.17	6,730
Arkansas	421	28.5	7,910	—	198	134	0.08	2.78	0	2,180	0	10,900	0.08	10,900
California	6,120	286	30,500	409	537	188	13.6	23.7	153	352	12,600	38,400	12,800	51,200
Colorado	899	66.8	11,400	—	—	120	0	—	—	138	0	12,600	0	12,600
Connecticut	424	56.2	30.4	—	—	10.7	0	—	—	187	3,440	708	3,440	4,150
Delaware	94.9	13.3	43.5	3.92	0.07	59.4	3.25	—	—	366	738	582	741	1,320
District of Columbia	0	0	0.18	—	—	0	0	—	—	9.69	0	9.87	0	9.87
Florida	2,440	199	4,290	32.5	8.02	291	1.18	217	0	658	12,000	8,140	12,000	20,100
Georgia	1,250	110	1,140	19.4	15.4	622	30.0	9.80	0	3,250	61.7	6,410	91.7	6,500
Hawaii	250	12.0	364	—	—	14.5	0.85	—	—	0	0	640	0.85	641
Idaho	244	85.2	17,100	34.9	1,970	55.5	0	—	—	0	0	19,500	0	19,500
Illinois	1,760	135	154	37.6	—	391	0	—	—	11,300	0	13,700	0	13,700
Indiana	670	122	101	41.9	—	2,400	0	82.5	0	6,700	0	10,100	0	10,100
Iowa	383	33.2	21.5	109	—	237	0	32.8	0	2,540	0	3,360	0	3,360
Kansas	416	21.6	3,710	111	5.60	53.3	0	31.4	0	2,260	0	6,610	0	6,610

continued

TABLE 2 (continued)
Total Water Withdrawals by Water-Use Category, 2000

State	Public Supply Fresh	Domestic Fresh	Irrigation Fresh	Livestock Fresh	Aquaculture Fresh	Industrial Fresh	Industrial Saline	Mining Fresh	Mining Saline	Thermoelectric Power Fresh	Thermoelectric Power Saline	Total Fresh	Total Saline	Total
Kentucky	525	27.5	29.3	—	—	317	0	—	—	3,260	0	4,160	0	4,160
Louisiana	753	41.2	1,020	7.34	243	2,680	0	—	—	5,610	0	10,400	0	10,400
Maine	102	35.7	5.84	—	—	247	0	—	—	113	295	504	295	799
Maryland	824	77.1	42.4	10.4	19.6	65.8	227	8.31	0.02	379	6,260	1,430	6,490	7,910
Massachusetts	739	42.2	126	—	—	36.8	0	—	—	108	3,610	1,050	3,610	4,660
Michigan	1,140	239	201	11.3	—	698	0	—	—	7,710	0	10,000	0	10,000
Minnesota	500	80.8	227	52.8	—	154	0	588	0	2,270	0	3,870	0	3,870
Mississippi	359	69.3	1,410	—	371	242	0	—	—	362	148	2,810	148	2,960
Missouri	872	53.6	1,430	72.4	83.3	62.7	0	16.9	0	5,640	0	8,230	0	8,230
Montana	149	18.6	7,950	—	—	61.3	0	—	—	110	0	8,290	0	8,290
Nebraska	330	48.4	8,790	93.4	—	38.1	0	128	4.55	2,820	0	12,200	4.55	12,300
Nevada	629	22.4	2,110	—	—	10.3	0	—	—	36.7	0	2,810	0	2,810
New Hampshire	97.1	41.0	4.75	—	16.3	44.9	0	6.80	0	236	761	447	761	1,210
New Jersey	1,050	79.7	140	1.68	6.46	132	0	110	0	650	3,390	2,170	3,390	5,560
New Mexico	296	31.4	2,860	—	—	10.5	0	—	—	56.4	0	3,260	0	3,260
New York	2,570	142	35.5	—	—	297	0	—	—	4,040	5,010	7,080	5,010	12,100
North Carolina	945	189	287	121	7.88	293	0	36.4	0	7,850	1,620	9,730	1,620	11,400
North Dakota	63.6	11.9	145	—	—	17.6	0	—	—	902	0	1,140	0	1,140

Ohio	1,470	134	31.7	25.3	1.36	807	0	88.5	0	8,590	0	11,100	0	11,100
Oklahoma	675	25.5	718	151	16.4	25.9	0	2.48	256	146	0	1,760	256	2,020
Oregon	566	76.2	6,080	—	—	195	0	—	—	15.30	0	6,930	0	6,930
Pennsylvania	1,460	132	13.9	—	—	1,190	0	182	0	6,980	0	9,950	0	9,950
Rhode Island	119	8.99	3.45	—	—	4.28	0	—	—	2.40	290	138	290	429
South Carolina	566	63.5	267	—	—	565	0	—	—	5,710	0	7,170	0	7,170
South Dakota	93.3	9.53	373	42.0	—	5.12	0	—	—	5.24	0	528	0	528
Tennessee	890	32.6	22.4	—	—	842	0	—	—	9,040	0	10,800	0	10,800
Texas	4,230	131	8,630	308	—	1,450	907	220	504	9,820	3,440	24,800	4,850	29,600
Utah	638	16.1	3,860	—	116	42.7	5.08	26.3	198	62.20	0	4,760	203	4,970
Vermont	60.1	21.0	3.78	—	—	6.91	0	—	—	355	0	447	0	447
Virginia	720	133	26.4	—	—	470	53.3	—	—	3,850	3,580	5,200	3,640	8,830
Washington	1,020	125	3,040	—	—	577	39.9	—	—	519	0	5,270	39.9	5,310
West Virginia	190	40.4	0.04	—	—	968	0	—	—	3,950	0	5,150	0	5,150
Wisconsin	623	96.3	196	66.3	70.2	447	0	—	—	6,090	0	7,590	0	7,590
Wyoming	107	6.57	4,500	—	—	5.78	0	79.5	222	243	0	4,940	0	5,170
Puerto Rico	513	0.88	94.5	—	—	11.2	0	—	—	0	2,190	620	2,190	2,810
U.S. Virgin Islands	6.09	1.69	0.50	—	—	3.34	0	—	—	0	136	11.6	136	148
Total	43,300	3,590	137,000	1,760	3,700	18,500	1,280	2,010	1,490	136,000	59,500	345,000		62,300

Source: Modified from table available at http://pubs.usgs.gov/circ/2004/circ1268/htdocs/table02.html, March 2007.

Note: Figures may not sum to totals because of independent rounding. All values are in million gallons per day. —, data not collected.

TABLE 3
Surface-Water Withdrawals by Water-Use Category, 2000*

State	Public Supply Fresh	Domestic Fresh	Irrigation Fresh	Live-stock Fresh	Aqua-culture Fresh	Industrial Fresh	Industrial Saline	Mining Fresh	Mining Saline	Thermoelectric Power Fresh	Thermoelectric Power Saline	Total Fresh	Total Saline	Total
Alabama	553	0	28.7	—	1.44	777	0	—	—	8,190	0	9,550	0	9,550
Alaska	50.7	.25	.02	—	—	3.80	3.86	27.4	49.5	28.9	0	111	53.4	164
Arizona	613	0	2,660	—	—	0	0	4.43	0	26.2	0	3,300	0	3,300
Arkansas	289	0	1,410	—	10.4	66.8	0	2.57	0	2,170	0	3,950	0	3,950
California	3,320	28.6	18,900	227	380	5.65	13.6	2.71	.46	349	12,600	23,200	12,600	35,800
Colorado	846	0	9,260	—	—	96.4	0	—	—	122	0	10,300	0	10,300
Connecticut	358	0	13.4	—	—	6.61	0	—	—	186	3,440	565	3,440	4,010
Delaware	49.8	0	7.89	.22	0	42.5	3.25	—	—	366	738	466	741	1,210
District of Columbia	0	0	0.18	—	—	0	0	—	—	9.69	0	9.87	0	9.87
Florida	237	0	2,110	1.51	.21	74.7	1.18	57.8	0	629	12,000	3,110	12,000	15,100
Georgia	968	0	392	17.7	7.72	333	30.0	2.05	0	3,240	61.7	4,960	91.7	5,060
Hawaii	7.60	7.22	193	—	—	0	0	—	—	0	0	208	0	208
Idaho	25.3	0	13,300	7.20	1,920	19.7	0	—	—	0	0	15,300	0	15,300
Illinois	1,410	0	4.25	0	—	259	0	—	—	11,300	0	12,900	0	12,900
Indiana	326	0	45.4	14.6	—	2,300	0	78.3	0	6,700	0	9,460	0	9,460
Iowa	79.8	0	1.08	27.1	—	11.7	0	30.3	0	2,530	0	2,680	0	2,680
Kansas	244	0	288	23.5	2.27	6.74	0	17.4	0	2,240	0	2,820	0	2,820
Kentucky	455	8.00	28.2	—	—	222	0	—	—	3,250	0	3,970	0	3,970
Louisiana	404	0	232	3.31	115	2,400	0	—	—	5,580	0	8,730	0	8,730
Maine	72.5	0	5.23	—	—	237	0	—	—	108	295	423	295	718

continued

Maryland	740	0	12.6	3.18	14.8	49.9	227	4.10	.02	377	6,260	1,200	6,490	7,690
Massachusetts	542	0	106	—	—	26.2	0	—	—	108	3,610	783	3,610	4,390
Michigan	896	0	73.2	1.15	—	589	0	581	0	7,710	0	9,260	0	9,260
Minnesota	171	0	36.6	0	—	97.8	0	—	—	2,260	0	3,150	0	3,150
Mississippi	40.4	0	99.1	—	49.8	124	0	—	—	318	148	632	148	781
Missouri	594	0	48.1	54.1	81.3	33.5	0	12.8	0	5,620	0	6,450	0	6,450
Montana	92.4	1.29	7,870	—	—	29.3	0	—	—	110	0	8,100	0	8,100
Nebraska	63.8	0	1,370	17.4	—	2.60	0	122	0	2,810	0	4,390	0	4,390
Nevada	478	0	1,540	—	—	5.00	0	—	—	24.7	0	2,050	0	2,050
New Hampshire	64.1	.16	4.25	—	13.1	37.9	0	6.72	0	235	761	362	761	1,120
New Jersey	650	0	117	0	0	66.2	0	104	0	648	3,390	1,590	3,390	4,980
New Mexico	33.8	0	1,630	—	—	1.67	0	—	—	45.0	0	1,710	0	1,710
New York	1,980	0	12.1	—	—	152	0	—	—	4,040	5,010	6,190	5,010	11,200
North Carolina	779	0	221	32.3	0	267	0	0	0	7,850	1,620	9,150	1,620	10,800
North Dakota	31.2	0	73.2	—	—	10.7	0	—	—	902	0	1,020	0	1,020
Ohio	966	2.71	17.8	17.1	0	645	0	35.5	0	8,590	0	10,300	0	10,300
Oklahoma	562	0	151	97.2	16.1	19.1	0	.23	0	143	0	990	0	990
Oregon	447	7.97	5,290	—	—	183	0	20.9	0	12.8	0	5,940	0	5,940
Pennsylvania	1,250	0	12.5	—	—	1,030	0	—	0	6,970	0	9,290	0	9,290
Rhode Island	102	0	2.99	—	—	2.09	0	—	—	2.40	290	110	290	400
South Carolina	462	0	162	—	—	514	0	—	—	5,700	0	6,840	0	6,840
South Dakota	39.1	.01	236	25.2	—	1.96	0	—	—	4.01	0	306	0	306
Tennessee	569	0	15.1	—	—	785	0	—	—	9,040	0	10,400	0	10,400
Texas	2,970	0	2,130	172	—	1,200	906	91.5	0	9,760	3,440	16,300	4,350	20,700
Utah	274	0	3,390	—	0	8.38	0	17.7	177	49.2	0	3,740	177	3,920

TABLE 3 (continued)
Surface-Water Withdrawals by Water-Use Category, 2000*

State	Public Supply Fresh	Domestic Fresh	Irrigation Fresh	Livestock Fresh	Aquaculture Fresh	Industrial Fresh	Industrial Saline	Mining Fresh	Mining Saline	Thermoelectric Power Fresh	Thermoelectric Power Saline	Total Fresh	Total Saline	Total Total
Vermont	40.6	.25	3.45	—	—	4.86	0	—	—	355	0	404	0	404
Virginia	650	0	22.8	—	—	365	53.3	—	—	3,850	3,580	4,880	3,640	8,520
Washington	552	.02	2,290	—	—	439	39.9	—	—	518	0	3,800	39.9	3,840
West Virginia	149	0.81	.02	—	—	958	0	—	—	3,950	0	5,060	0	5,060
Wisconsin	293	0	1.57	6.02	30.4	364	0	—	—	6,090	0	6,780	0	6,780
Wyoming	49.4	0	4,090	—	—	1.47	0	20.7	0	242	0	4,400	0	4,400
Puerto Rico	425	0	57.5	—	—	0	0	—	—	0	2,190	483	2,190	2,670
U.S. Virgin Islands	5.57	1.69	.21	—	—	3.12	0	—	—	0	136	10.6	136	147
Total	27,300	58.9	80,000	747	2,640	14,900	1,280	1,240	227	135,000	59,500	262,000	61,000	323,000

Source: Modified from table available at http://pubs.usgs.gov/circ/2004/circ1268/hdocs/table03.html, March 2007.
Note: Figures may not sum to totals because of independent rounding. All values are in million gallons per day. —, data not collected.
* See Figure 23.

TABLE 4
Groundwater Withdrawals by Water-Use Category, 2000*

State	Public Supply	Domestic	Irrigation	Livestock	Aquaculture	Industrial		Mining		Thermoelectric Power	Total		Total
	Fresh	Fresh	Fresh	Fresh	Fresh	Fresh	Saline	Fresh	Saline	Fresh	Fresh	Saline	Total
Alabama	281	78.9	14.5	—	8.93	56.0	0	—	—	0	440	0	440
Alaska	29.3	10.9	0.99	—	—	4.32	0	0.01	90.4	4.65	50.2	90.4	141
Arizona	469	28.9	2,750	—	—	19.8	0	81.2	8.17	74.3	3,420	8.17	3,430
Arkansas	132	28.5	6,510	—	187	67.0	0.08	0.21	0	2.92	6,920	0.08	6,920
California	2,800	257	11,600	182	158	183	0	21.0	152	3.23	15,200	152	15,400
Colorado	53.7	66.8	2,160	—	—	23.6	0	—	—	16.1	2,320	0	2,320
Connecticut	66.0	56.2	17.0	—	—	4.13	0	—	—	0.08	143	0	143
Delaware	45.0	13.3	35.6	3.70	0.07	17.0	0	—	—	0.47	115	0	115
District of Columbia	0	0	0	—	—	0	0	—	—	0	0	0	0
Florida	2,200	199	2,180	31.0	7.81	216	0	160	0	29.5	5,020	0	5,020
Georgia	278	110	750	1.66	7.70	290	0	7.75	0	1.03	1,450	0	1,450
Hawaii	243	4.82	171	—	—	14.5	0.85	—	—	0	433	0.85	434
Idaho	219	85.2	3,720	27.7	51.5	35.8	0	—	—	0	4,140	0	4,140
Illinois	353	135	150	37.6	—	132	0	—	—	5.75	813	0	813
Indiana	345	122	55.5	27.3	—	99.7	0	4.20	0	2.58	656	0	656
Iowa	303	33.20	20.4	81.8	—	226	0	2.49	0	11.9	679	0	679
Kansas	172	21.6	3,430	87.2	3.33	46.6	0	14.0	0	14.9	3,790	0	3,790

continued

TABLE 4 (continued)
Groundwater Withdrawals by Water-Use Category, 2000*

State	Public Supply Fresh	Domestic Fresh	Irrigation Fresh	Livestock Fresh	Aquaculture Fresh	Industrial Fresh	Industrial Saline	Mining Fresh	Mining Saline	Thermoelectric Power Fresh	Total Fresh	Total Saline	Total
Kentucky	71.0	19.5	1.14	—	—	95.2	0	—	0	2.71	189	0	189
Louisiana	349	41.2	791	4.03	128	285	0	—	—	28.4	1,630	0	1,630
Maine	29.6	35.7	0.61	—	—	9.90	0	—	—	4.92	80.8	0	80.80
Maryland	84.6	77.1	29.8	7.18	4.81	15.9	0	4.21	0	1.80	225	0	225
Massachusetts	197	42.2	19.7	—	—	10.7	0	—	—	0	269	0	269
Michigan	247	239	128	10.2	—	110	0	—	—	0	734	0	734
Minnesota	329	80.8	190	52.8	—	56.3	0	6.90	0	4.17	720	0	720
Mississippi	319	69.3	1,310	—	321	118	0	—	—	43.5	2,180	0	2,180
Missouri	278	53.6	1,380	18.3	2.01	29.2	0	4.10	0	12.2	1,780	0	1,780
Montana	56.1	17.3	83.0	—	—	31.9	0	—	—	0	188	0	188
Nebraska	266	48.4	7,420	76.0	—	35.5	0	5.64	4.55	6.87	7,860	4.55	7,860
Nevada	151	22.4	567	—	—	5.29	0	—	—	12.0	757	0	757
New Hampshire	33.0	40.9	0.50	—	3.12	6.95	0	0.08	0	0.71	85.2	0	85.2
New Jersey	400	79.7	22.8	1.68	6.46	65.3	0	6.12	0	2.24	584	0	584
New Mexico	262	31.4	1,230	—	—	8.80	0	—	—	11.4	1,540	0	1,540
New York	583	142	23.3	—	—	145	0	—	—	0	893	0	893
North Carolina	166	189	65.8	89.1	7.88	25.6	0	36.4	0	0.09	580	0	580
North Dakota	32.4	11.9	72.2	—	—	6.88	0	—	—	0	123	0	123

Ohio	500	132	13.9	8.20	1.36	162	0	53.1	0	7.57	878	0	878
Oklahoma	113	25.5	566	53.6	0.29	5.83	0	2.25	256	3.27	771	256	1,030
Oregon	118	68.3	792	—	—	12.1	0	—	—	2.47	993	0	993
Pennsylvania	212	132	1.38	—	—	155	0	162	0	3.98	666	0	666
Rhode Island	16.9	8.99	0.46	—	—	2.19	0	—	—	0	28.6	0	28.6
South Carolina	105	63.5	106	16.9	—	50.9	0	—	—	5.83	330	0	330
South Dakota	54.2	9.52	137	—	—	3.16	0	—	—	1.23	222	0	222
Tennessee	321	32.6	7.33	137	—	56.3	0	—	—	0	417	0	417
Texas	1,260	131	6,500	—	—	244	0.50	129	504	60.2	8,470	504	8,970
Utah	364	16.1	469	—	116	34.3	5.08	8.60	21.5	13.1	1,020	26.5	1,050
Vermont	19.5	20.7	0.33	—	—	2.05	0	—	—	0.66	43.2	0	43.2
Virginia	70.7	133	3.57	—	—	104	0	—	—	1.50	314	0	314
Washington	464	125	747	—	—	138	0	—	—	0.92	1,470	0	1,470
West Virginia	41.6	39.6	0.02	—	—	9.70	0	—	—	0	90.9	0	90.9
Wisconsin	330	96.3	195	60.3	39.8	83.0	0	—	—	8.99	813	0	813
Wyoming	57.2	6.57	413	—	—	4.31	0	58.8	222	1.13	541	222	763
Puerto Rico	88.5	.88	36.9	—	—	11.2	0	—	—	0	137	0	137
U.S. Virgin Islands	0.52	0	0.29	—	—	0.22	0	—	—	0	1.03	0	1.03
Total	16,000	3,530	56,900	1,010	1,060	3,570	6.51	767	1,260	409	83,300	1,260	84,500

Source: Modified from table available at http://pubs.usgs.gov/circ/2004/circ1268/htdocs/table04.html, March 2007.

* See Figure 22.

Note: Figures may not sum to totals because of independent rounding. All values are in million gallons per day. —, data not collected.

TABLE 5
Public Supply Water Withdrawals, 2000

| State | Population (in thousands) | | | Withdrawals (in million gallons per day) | | | Withdrawals (in thousand acre-feet per year) | | |
| | Total | Served by Public Supply | | By Source | | | By Source | | |
		Population	Population (in percent)	Groundwater	Surface Water	Total	Groundwater	Surface Water	Total
Alabama	4,450	3,580	80	281	553	834	315	620	935
Alaska	627	421	67	29.3	50.7	80.0	32.9	56.9	89.7
Arizona	5,130	4,870	95	469	613	1,080	526	688	1,210
Arkansas	2,670	2,320	87	132	289	421	148	324	472
California	33,900	30,100	89	2,800	3,320	6,120	3,140	3,730	6,860
Colorado	4,300	3,750	87	53.7	846	899	60.2	948	1,010
Connecticut	3,410	2,660	78	66.0	358	424	74.0	402	476
Delaware	784	617	79	45.0	49.8	94.9	50.5	55.9	106
District of Columbia	572	572	100	0	0	0	0	0	0
Florida	16,000	14,000	88	2,200	237	2,440	2,470	266	2,730
Georgia	8,190	6,730	82	278	968	1,250	311	1,090	1,400
Hawaii	1,210	1,140	94	243	7.60	250	272	8.52	281
Idaho	1,290	928	72	219	25.3	244	245	28.3	274
Illinois	12,400	10,900	88	353	1,410	1,760	396	1,580	1,970
Indiana	6,080	4,480	74	345	326	670	386	365	751
Iowa	2,930	2,410	83	303	79.8	383	340	89.5	429
Kansas	2,690	2,500	93	172	244	416	193	273	466

Kentucky	4,040	3,490	86	71.0	455	525	79.5	510	589
Louisiana	4,470	3,950	88	349	404	753	392	453	844
Maine	1,270	726	57	29.6	72.5	102	33.2	81.3	115
Maryland	5,300	4,360	82	84.6	740	824	94.8	829	924
Massachusetts	6,350	5,880	93	197	542	739	220	608	828
Michigan	9,940	7,170	72	247	896	1,140	277	1,000	1,280
Minnesota	4,920	3,770	77	329	171	500	369	192	561
Mississippi	2,840	2,190	77	319	40.4	359	357	45.3	402
Missouri	5,600	4,770	85	278	594	872	311	666	978
Montana	902	664	74	56.1	92.4	149	62.9	104	167
Nebraska	1,710	1,390	81	266	63.8	330	299	71.6	370
Nevada	2,000	1,870	94	151	478	629	169	536	705
New Hampshire	1,240	756	61	33.0	64.1	97.1	37.0	71.9	109
New Jersey	8,410	7,460	89	400	650	1,050	449	729	1,180
New Mexico	1,820	1,460	80	262	33.8	296	294	37.9	332
New York	19,000	17,100	90	583	1,980	2,570	653	2,220	2,880
North Carolina	8,050	5,350	66	166	779	945	186	873	1,060
North Dakota	642	493	77	32.4	31.2	63.6	36.3	35.0	71.3
Ohio	11,400	9,570	84	500	966	1,470	560	1,080	1,640
Oklahoma	3,450	3,150	91	113	562	675	127	631	757
Oregon	3,420	2,730	80	118	447	566	133	501	634
Pennsylvania	12,300	10,100	82	212	1,250	1,460	237	1,400	1,640
Rhode Island	1,050	922	88	16.9	102	119	19.0	115	134

continued

TABLE 5 (continued)
Public Supply Water Withdrawals, 2000

| State | Population (in thousands) | | | Withdrawals (in million gallons per day) | | | Withdrawals (in thousand acre-feet per year) | | |
| | | Served by Public Supply | | By Source | | | By Source | | |
	Total	Population	Population (in percent)	Groundwater	Surface Water	Total	Groundwater	Surface Water	Total
South Carolina	4,010	3,160	79	105	462	566	117	517	635
South Dakota	755	625	83	54.2	39.1	93.3	60.7	43.9	105
Tennessee	5,690	5,240	92	321	569	890	360	638	997
Texas	20,900	19,700	94	1,260	2,970	4,230	1,420	3,330	4,740
Utah	2,230	2,180	97	364	274	638	408	307	715
Vermont	609	362	59	19.5	40.6	60.1	21.8	45.6	67.4
Virginia	7,080	5,310	75	70.7	650	720	79.3	728	808
Washington	5,890	4,900	83	464	552	1,020	520	619	1,140
West Virginia	1,810	1,300	72	41.6	149	190	46.6	167	213
Wisconsin	5,360	3,620	67	330	293	623	370	329	699
Wyoming	494	406	82	57.2	49.4	107	64.1	55.3	119
Puerto Rico	3,810	3,800	100	88.5	425	513	99.2	476	576
U.S. Virgin Islands	109	53.4	49	0.52	5.57	6.09	0.58	6.24	6.83
Total	285,000	242,000	85	16,000	27,300	43,300	17,900	30,600	48,500

Source: Modified from table available at http://pubs.usgs.gov/circ/2004/circ1268/htdocs/table05.html, March 2007.
Note: Figures may not sum to totals because of independent rounding.

TABLE 6
Self-Supplied Domestic Water Withdrawals, 2000

| State | Population (in thousands) | | | | Withdrawals (in million gallons per day) | | | Withdrawals (in thousand acre-feet per year) | | |
| | Total | Served by Public Supply | Self-Supplied Domestic Population | Self-Supplied Domestic Population (in percent) | By Source | | | By Source | | |
					Groundwater	Surface Water	Total	Groundwater	Surface Water	Total
Alabama	4,450	3,580	868	20	78.9	0	78.9	88.4	0	88.4
Alaska	627	421	206	33	10.9	.25	11.2	12.2	.28	12.5
Arizona	5,130	4,870	265	5	28.9	0	28.9	32.4	0	32.4
Arkansas	2,670	2,320	351	13	28.5	0	28.5	31.9	0	31.9
California	33,900	30,100	3,810	11	257	28.6	286	288	32.0	320
Colorado	4,300	3,750	555	13	66.8	0	66.8	74.9	0	74.9
Connecticut	3,410	2,660	749	22	56.2	0	56.2	63.0	0	63.0
Delaware	784	617	166	21	13.3	0	13.3	14.9	0	14.9
District of Columbia	572	572	0	0	0	0	0	0	0	0
Florida	16,000	14,000	1,950	12	199	0	199	223	0	223
Georgia	8,190	6,730	1,450	18	110	0	110	123	0	123
Hawaii	1,210	1,140	72.9	6	4.82	7.22	12.0	5.40	8.09	13.5
Idaho	1,290	928	366	28	85.2	0	85.2	95.6	0	95.6
Illinois	12,400	10,900	1,500	12	135	0	135	152	0	152
Indiana	6,080	4,480	1,600	26	122	0	122	137	0	137

continued

TABLE 6 (continued)
Self-Supplied Domestic Water Withdrawals, 2000

State	Population (in thousands)		Self-Supplied Domestic		Withdrawals (in million gallons per day) By Source			Withdrawals (in thousand acre-feet per year) By Source		
	Total	Served by Public Supply	Population	Population (in percent)	Groundwater	Surface Water	Total	Groundwater	Surface Water	Total
Iowa	2,930	2,410	511	17	33.2	0	33.2	37.2	0	37.2
Kansas	2,690	2,500	193	7	21.6	0	21.6	24.2	0	24.2
Kentucky	4,040	3,490	552	14	19.5	8.00	27.5	21.9	8.97	30.8
Louisiana	4,470	3,950	523	12	41.2	0	41.2	46.2	0	46.2
Maine	1,270	726	549	43	35.7	0	35.7	40.0	0	40.0
Maryland	5,300	4,360	932	18	77.1	0	77.1	86.4	0	86.4
Massachusetts	6,350	5,880	473	7	42.2	0	42.2	47.2	0	47.2
Michigan	9,940	7,170	2,770	28	239	0	239	268	0	268
Minnesota	4,920	3,770	1,150	23	80.8	0	80.8	90.6	0	90.6
Mississippi	2,840	2,190	654	23	69.3	0	69.3	77.7	0	77.7
Missouri	5,600	4,770	824	15	53.6	0	53.6	60.1	0	60.1
Montana	902	664	238	26	17.3	1.29	18.6	19.4	1.45	20.8
Nebraska	1,710	1,390	324	19	48.4	0	48.4	54.3	0	54.3
Nevada	2,000	1,870	124	6	22.4	0	22.4	25.2	0	25.2
New Hampshire	1,240	756	479	39	40.9	.16	41.0	45.8	.18	46.0

New Jersey	8,410	7,460	952	11	79.7	0	79.7	89.3	0	89.3
New Mexico	1,820	1,460	360	20	31.4	0	31.4	35.2	0	35.2
New York	19,000	17,100	1,890	10	142	0	142	159	0	159
North Carolina	8,050	5,350	2,700	34	189	0	189	212	0	212
North Dakota	642	493	149	23	11.9	0	11.9	13.3	0	13.3
Ohio	11,400	9,570	1,790	16	132	2.71	134	148	3.04	151
Oklahoma	3,450	3,150	299	9	25.5	0	25.5	28.5	0	28.5
Oregon	3,420	2,730	692	20	68.3	7.97	76.2	76.5	8.93	85.5
Pennsylvania	12,300	10,100	2,190	18	132	0	132	148	0	148
Rhode Island	1,050	922	127	12	8.99	0	8.99	10.1	0	10.1
South Carolina	4,010	3,160	847	21	63.5	0	63.5	71.2	0	71.2
South Dakota	755	625	129	17	9.52	.01	9.53	10.7	.01	10.7
Tennessee	5,690	5,240	453	8	32.6	0	32.6	36.6	0	36.5
Texas	20,900	19,700	1,190	6	131	0	131	147	0	147
Utah	2,230	2,180	56.2	3	16.1	0	16.1	18.0	0	18.0
Vermont	609	362	247	41	20.7	.25	21.0	23.2	.28	23.5
Virginia	7,080	5,310	1,770	25	133	0	133	150	0	150
Washington	5,890	4,900	993	17	125	.02	125	140	.02	140
West Virginia	1,810	1,300	505	28	39.6	.81	40.4	44.4	.91	45.3
Wisconsin	5,360	3,620	1,750	33	96.3	0	96.3	108	0	108
Wyoming	494	406	87.5	18	6.57	0	6.57	7.36	0	7.36
Puerto Rico	3,810	3,800	12.8	0	.88	0	.88	.99	0	.99
U.S. Virgin Islands	109	53.4	55.2	51	0	1.69	1.69	0	1.89	1.89
TOTAL	285,000	242,000	43,500	15	3,530	58.9	3,590	3,960	66.1	4,030

Source: Modified from table available at http://pubs.usgs.gov/circ/2004/circ1268/htdocs/table06.html, March 2007.

(*Note:* Figures may not sum to totals because of independent rounding.)

TABLE 7
Irrigation Water Withdrawals, 2000

State	Irrigated Land (in thousand acres) By type of irrigation				Withdrawals (in million gallons per day) By source			Withdrawals (in thousand acre-feet per year) By source			Application Rate (in acre-feet per acre)
	Sprinkler	Micro-irrigation	Surface	Total	Ground-water	Surface Water	Total	Ground-water	Surface Water	Total	
Alabama	68.7	1.30	0	70.00	14.5	28.7	43.1	16.2	32.2	48.4	0.69
Alaska	2.43	0	0.07	2.50	0.99	0.02	1.01	1.11	0.02	1.13	0.45
Arizona	183	14.0	779	976	2,750	2,660	5,400	3,080	2,980	6,060	6.21
Arkansas	631	0	3,880	4,510	6,510	1,410	7,910	7,290	1,580	8,870	1.97
California	1,660	3,010	5,470	10,100	11,600	18,900	30,500	13,100	21,100	34,200	3.37
Colorado	1,190	1.16	2,220	3,400	2,160	9,260	11,400	2,420	10,400	12,800	3.76
Connecticut	20.6	0.39	0	21.0	17.0	13.4	30.4	19.0	15.0	34.0	1.62
Delaware	81.1	0.71	0	81.8	35.6	7.89	43.5	39.9	8.84	48.7	0.60
District of Columbia	0.32	0	0	0.32	0	0.18	0.18	0	0.20	0.20	0.63
Florida	515	704	839	2,060	2,180	2,110	4,290	2,450	2,370	4,810	2.34
Georgia	1,470	73.8	0	1,540	750	392	1,140	841	439	1,280	0.83
Hawaii	16.70	105	0	122	171	193	364	191	216	407	3.35
Idaho	2,440	4.70	1,300	3,750	3,720	13,300	17,100	4,170	15,000	19,100	5.10
Illinois	365	0	0	365	150	4.25	154	168	4.76	173	0.47
Indiana	250	0	0	250	55.5	45.4	101	62.2	51.0	113	0.45
Iowa	84.5	0	0	84.5	20.4	1.08	21.5	22.9	1.21	24.1	0.28
Kansas	2,660	2.14	647	3,310	3,430	288	3,710	3,840	323	4,160	1.26

continued

Kentucky	66.6	0	0	66.6	1.14	28.2	29.3	1.28	31.6	32.9	0.49
Louisiana	110	0	830	940	791	232	1,020	887	261	1,150	1.22
Maine	35.0	0.95	0.03	36.0	0.61	5.23	5.84	0.68	5.86	6.55	0.18
Maryland	57.3	3.32	0	60.6	29.8	12.6	42.4	33.4	14.1	47.6	0.78
Massachusetts	26.6	2.35	0	29.0	19.7	106	126	22.1	119	141	4.88
Michigan	401	8.67	4.87	415	128	73.2	201	144	82.0	226	0.54
Minnesota	546	0	26.9	573	190	36.6	227	213	41.1	254	0.44
Mississippi	455	0	966	1,420	1,310	99.1	1,410	1,470	111	1,580	1.11
Missouri	532	1.43	792	1,330	1,380	48.1	1,430	1,550	53.9	1,600	1.21
Montana	506	0	1,220	1,720	83.0	7,870	7,950	93.0	8,820	8,920	5.18
Nebraska	4,110	0	3,710	7,820	7,420	1,370	8,790	8,320	1,540	9,860	1.26
Nevada	192	0	456	647	567	1,540	2,110	635	1,730	2,360	3.65
New Hampshire	6.08	0	0	6.08	0.50	4.25	4.75	0.56	4.76	5.32	0.88
New Jersey	109	15.7	3.70	128	22.8	117	140	25.5	131	156	1.22
New Mexico	461	7.17	530	998	1,230	1,630	2,860	1,380	1,830	3,210	3.22
New York	70.0	8.73	1.84	80.6	23.3	12.1	35.5	26.1	13.6	39.8	0.49
North Carolina	193	3.70	0	196	65.8	221	287	73.8	248	322	1.64
North Dakota	200	0	26.7	227	72.2	73.2	145	80.9	82.1	163	0.72
Ohio	61.0	0	0	61.0	13.9	17.8	31.7	15.6	19.9	35.5	0.58
Oklahoma	392	1.50	113	507	566	151	718	635	170	804	1.59
Oregon	1,160	4.02	1,000	2,170	792	5,290	6,080	887	5,920	6,810	3.14
Pennsylvania	28.9	7.17	0	36.0	1.38	12.5	13.9	1.55	14.0	15.6	0.43
Rhode Island	4.48	0.29	0.05	4.82	0.46	2.99	3.45	0.52	3.35	3.87	0.80

TABLE 7 (continued)
Irrigation Water Withdrawals, 2000

State	Irrigated Land (in thousand acres) By type of irrigation				Withdrawals (in million gallons per day) By source			Withdrawals (in thousand acre-feet per year) By source			Application Rate (in acre-feet per acre)
	Sprinkler	Micro-irrigation	Surface	Total	Ground-water	Surface Water	Total	Ground-water	Surface Water	Total	
South Carolina	166	3.66	17.5	187	106	162	267	118	181	300	1.60
South Dakota	276	0	78.3	354	137	236	373	153	264	418	1.18
Tennessee	51.2	5.35	3.96	60.5	7.33	15.1	22.4	8.22	16.9	25.1	0.41
Texas	4,010	89.4	2,390	6,490	6,500	2,130	8,630	7,290	2,390	9,680	1.49
Utah	526	1.68	880	1,410	469	3,390	3,860	526	3,800	4,330	3.08
Vermont	4.95	0	0	4.95	0.33	3.45	3.78	0.37	3.87	4.24	0.86
Virginia	64.3	13.9	0	78.2	3.57	22.8	26.4	4.00	25.6	29.6	0.38
Washington	1,270	49.9	252	1,570	747	2,290	3,040	837	2,570	3,400	2.16
West Virginia	2.21	0	0.98	3.19	0.02	0.02	0.04	0.02	0.02	0.04	0.01
Wisconsin	355	0	0	355	195	1.57	196	218	1.76	220	0.62
Wyoming	190	4.73	964	1,160	413	4,090	4,500	463	4,580	5,050	4.36
Puerto Rico	15.5	33.0	5.35	53.8	36.9	57.5	94.5	41.4	64.5	106	1.97
U.S. Virgin Islands	0.20	0	0	0.20	0.29	0.21	0.50	0.33	0.24	0.56	2.80
Total	28,300	4,180	29,400	61,900	56,900	80,000	137,000	63,800	89,700	153,000	2.48

Source: Modified from table available at http://pubs.usgs.gov/circ/2004/circ1268/htdocs/table07.html, March 2007.
(Note: Figures may not sum to totals because of independent rounding.)

TABLE 8
Livestock Water Withdrawals, 2000

State	Withdrawals (in million gallons per day) By Source			Withdrawals (in thousand acre-feet per year) By Source		
	Ground-water	Surface Water	Total	Ground-water	Surface Water	Total
Alabama	—	—	—	—	—	—
Alaska	—	—	—	—	—	—
Arizona	—	—	—	—	—	—
Arkansas	—	—	—	—	—	—
California	182	227	409	204	255	458
Colorado	—	—	—	—	—	—
Connecticut	—	—	—	—	—	—
Delaware	3.70	0.22	3.92	4.15	0.25	4.39
District of Columbia	—	—	—	—	—	—
Florida	31.0	1.51	32.5	34.7	1.69	36.4
Georgia	1.66	17.7	19.4	1.86	19.9	21.7
Hawaii	—	—	—	—	—	—
Idaho	27.7	7.20	34.9	31.0	8.07	39.1
Illinois	37.6	0	37.6	42.1	0	42.1
Indiana	27.3	14.6	41.9	30.6	16.4	47.0
Iowa	81.8	27.1	109	91.8	30.4	122
Kansas	87.2	23.5	111	97.7	26.3	124
Kentucky	—	—	—	—	—	—
Louisiana	4.03	3.31	7.34	4.52	3.71	8.23
Maine	—	—	—	—	—	—
Maryland	7.18	3.18	10.4	8.05	3.56	11.6
Massachusetts	—	—	—	—	—	—
Michigan	10.2	1.15	11.3	11.4	1.29	12.7
Minnesota	52.8	0	52.8	59.2	0	59.2
Mississippi	—	—	—	—	—	—
Missouri	18.3	54.1	72.4	20.5	60.6	81.1
Montana	—	—	—	—	—	—
Nebraska	76.0	17.4	93.4	85.2	19.5	105
Nevada	—	—	—	—	—	—
New Hampshire	—	—	—	—	—	—

continued

TABLE 8 (continued)
Livestock Water Withdrawals, 2000

State	Withdrawals (in million gallons per day) By Source Ground-water	Surface Water	Total	Withdrawals (in thousand acre-feet per year) By Source Ground-water	Surface Water	Total
New Jersey	1.68	0	1.68	1.88	0	1.88
New Mexico	—	—	—	—	—	—
New York	—	—	—	—	—	—
North Carolina	89.1	32.3	121	99.9	36.2	136
North Dakota	—	—	—	—	—	—
Ohio	8.20	17.1	25.3	9.19	19.2	28.4
Oklahoma	53.6	97.2	151	60.0	109	169
Oregon	—	—	—	—	—	—
Pennsylvania	—	—	—	—	—	—
Rhode Island	—	—	—	—	—	—
South Carolina	—	—	—	—	—	—
South Dakota	16.9	25.2	42.0	18.9	28.2	47.1
Tennessee	—	—	—	—	—	—
Texas	137	172	308	153	192	346
Utah	—	—	—	—	—	—
Vermont	—	—	—	—	—	—
Virginia	—	—	—	—	—	—
Washington	—	—	—	—	—	—
West Virginia	—	—	—	—	—	—
Wisconsin	60.3	6.02	66.3	67.6	6.75	74.4
Wyoming	—	—	—	—	—	—
Puerto Rico	—	—	—	—	—	—
U.S. Virgin Islands	—	—	—	—	—	—
Total	1,010	747	1,760	1,140	838	1,980

Source: Modified from table available at http://pubs.usgs.gov/circ/2004/circ1268/htdocs/table08.html, March 2007.

Note: Figures may not sum to totals because of independent rounding. —, data not collected.

TABLE 9
Aquaculture Water Withdrawals, 2000

State	Withdrawals (in million gallons per day) By Source			Withdrawals (in thousand acre-feet per year) By Source		
	Ground-water	Surface Water	Total	Ground-water	Surface Water	Total
Alabama	8.93	1.44	10.4	10.0	1.61	11.6
Alaska	—	—	—	—	—	—
Arizona	—	—	—	—	—	—
Arkansas	187	10.4	198	210	11.6	222
California	158	380	537	177	426	603
Colorado	—	—	—	—	—	—
Connecticut	—	—	—	—	—	—
Delaware	0.07	0	0.07	0.08	0	0.08
District of Columbia	—	—	—	—	—	—
Florida	7.81	0.21	8.02	8.76	0.24	8.99
Georgia	7.70	7.72	15.4	8.63	8.65	17.3
Hawaii	—	—	—	—	—	—
Idaho	51.5	1,920	1,970	57.7	2,150	2,210
Illinois	—	—	—	—	—	—
Indiana	—	—	—	—	—	—
Iowa	—	—	—	—	—	—
Kansas	3.33	2.27	5.60	3.73	2.54	6.28
Kentucky	—	—	—	—	—	—
Louisiana	128	115	243	144	129	273
Maine	—	—	—	—	—	—
Maryland	4.81	14.8	19.6	5.39	16.6	22.0
Massachusetts	—	—	—	—	—	—
Michigan	—	—	—	—	—	—
Minnesota	—	—	—	—	—	—
Mississippi	321	49.8	371	360	55.9	416
Missouri	2.01	81.3	83.3	2.25	91.2	93.4
Montana	—	—	—	—	—	—
Nebraska	—	—	—	—	—	—
Nevada	—	—	—	—	—	—
New Hampshire	3.12	13.1	16.3	3.50	14.7	18.2

continued

TABLE 9 (continued)
Aquaculture Water Withdrawals, 2000

State	Withdrawals (in million gallons per day) By Source			Withdrawals (in thousand acre-feet per year) By Source		
	Ground-water	Surface Water	Total	Ground-water	Surface Water	Total
New Jersey	6.46	0	6.46	7.24	0	7.24
New Mexico	—	—	—	—	—	—
New York	—	—	—	—	—	—
North Carolina	7.88	0	7.88	8.83	0	8.83
North Dakota	—	—	—	—	—	—
Ohio	1.36	0	1.36	1.52	0	1.52
Oklahoma	0.29	16.1	16.4	0.33	18.1	18.4
Oregon	—	—	—	—	—	—
Pennsylvania	—	—	—	—	—	—
Rhode Island	—	—	—	—	—	—
South Carolina	—	—	—	—	—	—
South Dakota	—	—	—	—	—	—
Tennessee	—	—	—	—	—	—
Texas	—	—	—	—	—	—
Utah	116	0	116	130	0	130
Vermont	—	—	—	—	—	—
Virginia	—	—	—	—	—	—
Washington	—	—	—	—	—	—
West Virginia	—	—	—	—	—	—
Wisconsin	39.8	30.4	70.2	44.6	34.1	78.7
Wyoming	—	—	—	—	—	—
Puerto Rico	—	—	—	—	—	—
U.S. Virgin Islands	—	—	—	—	—	—
Total	1,060	2,640	3,700	1,180	2,960	4,150

Source: Modified from table available at http://pubs.usgs.gov/circ/2004/circ1268/htdocs/table09.html, March 2007.

Note: Figures may not sum to totals because of independent rounding. —, data not collected.

TABLE 10
Industrial Self-Supplied Water Withdrawals, 2000

	Withdrawals (in million gallons per day)									Withdrawals (in thousand acre-feet per year)		
	By Source and Type									By Type		
	Groundwater			Surface Water			Total					
State	Fresh	Saline	Total	Fresh	Saline	Total	Fresh	Saline	Total	Fresh	Saline	Total
Alabama	56.0	0	56.0	777	0	777	833	0	833	934	0	934
Alaska	4.32	0	4.32	3.80	3.86	7.66	8.12	3.86	12.0	9.10	4.33	13.4
Arizona	19.8	0	19.8	0	0	0	19.8	0	19.8	22.2	0	22.2
Arkansas	67.0	0.08	67.1	66.8	0	66.8	134	0.08	134	150	0.09	150
California	183	0	183	5.65	13.6	19.3	188	13.6	202	211	15.3	226
Colorado	23.6	0	23.6	96.4	0	96.4	120	0	120	135	0	135
Connecticut	4.13	0	4.13	6.61	0	6.61	10.7	0	10.7	12.0	0	12.0
Delaware	17.0	0	17.0	42.5	3.25	45.70	59.4	3.25	62.7	66.6	3.64	70.3
District of Columbia	0	0	0	0	0	0.0	0	0	0	0	0	0
Florida	216	0	216	74.7	1.18	75.9	291	1.18	292	326	1.32	328
Georgia	290	0	290	333	30.0	363	622	30.0	652	698	33.6	731
Hawaii	14.5	0.85	15.4	0	0	0	14.5	0.85	15.4	16.2	0.95	17.2
Idaho	35.8	0	35.8	19.7	0	19.7	55.5	0	55.5	62.2	0	62.2
Illinois	132	0	132	259	0	259	391	0	391	438	0	438
Indiana	99.7	0	99.7	2,300	0	2,300	2,400	0	2,400	2,690	0	2,690

continued

TABLE 10 (continued)
Industrial Self-Supplied Water Withdrawals, 2000

State	Withdrawals (in million gallons per day)									Withdrawals (in thousand acre-feet per year)		
	By Source and Type									By Type		
	Groundwater			Surface Water			Total					
	Fresh	Saline	Total	Fresh	Saline	Total	Fresh	Saline	Total	Fresh	Saline	Total
Iowa	226	0	226	11.7	0	11.7	237	0	237	266	0	266
Kansas	46.6	0	46.6	6.74	0	6.74	53.3	0	53.3	59.8	0	59.8
Kentucky	95.2	0	95.2	222	0	222	317	0	317	356	0	356
Louisiana	285	0	285	2,400	0	2,400	2,680	0	2,680	3,010	0	3,010
Maine	9.90	0	9.90	237	0	237	247	0	247	277	0	277
Maryland	15.9	0	15.9	49.9	227	277	65.8	227	292	73.8	254	328
Massachusetts	10.7	0	10.7	26.2	0	26.2	36.8	0	36.8	41.3	0	41.3
Michigan	110	0	110	589	0	589	698	0	698	782	0	782
Minnesota	56.3	0	56.3	97.8	0	97.8	154	0	154	173	0	173
Mississippi	118	0	118	124	0	124	242	0	242	271	0	271
Missouri	29.2	0	29.2	33.5	0	33.5	62.7	0	62.7	70.3	0	70.3
Montana	31.9	0	31.9	29.3	0	29.3	61.3	0	61.3	68.7	0	68.7
Nebraska	35.5	0	35.5	2.60	0	2.60	38.1	0	38.1	42.7	0	42.7
Nevada	5.29	0	5.29	5.00	0	5.00	10.3	0	10.3	11.5	0	11.5
New Hampshire	6.95	0	6.95	37.9	0	37.9	44.9	0	44.9	50.3	0	50.3
New Jersey	65.3	0	65.3	66.2	0	66.2	132	0	132	147	0	147
New Mexico	8.80	0	8.80	1.67	0	1.67	10.5	0	10.5	11.7	0	11.7

New York	145	0	145	152	0	152	297	0	297	333	0	333
North Carolina	25.6	0	25.6	267	0	267	293	0	293	329	0	329
North Dakota	6.88	0	6.88	10.7	0	10.7	17.6	0	17.6	19.7	0	19.7
Ohio	162	0	162	645	0	645	807	0	807	905	0	905
Oklahoma	6.83	0	6.83	19.1	0	19.1	25.9	0	25.9	29.1	0	29.1
Oregon	12.1	0	12.1	183	0	183.00	195	0	195	218	0	218
Pennsylvania	155	0	155	1,030	0	1,030	1,190	0	1,190	1,330	0	1,330
Rhode Island	2.19	0	2.09	2.09	0	2.09	4.28	0	4.28	4.80	0	4.80
South Carolina	50.9	0	50.9	514	0	514	565	0	565	633	0	633
South Dakota	3.16	0	3.16	1.96	0	1.96	5.12	0	5.12	5.74	0	5.74
Tennessee	56.3	0	56.3	785	0	785	842	0	842	944	0	944
Texas	244	0.50	244	1,200	906	2,110	1,450	907	2,350	1,620	1,020	2,640
Utah	34.3	5.08	39.4	8.38	0	8.38	42.7	5.08	47.8	47.8	5.69	53.5
Vermont	2.05	0	2.05	4.86	0	4.86	6.91	0	6.91	7.75	0	7.75
Virginia	104	0	104	365	53.3	419	470	53.3	523	526	59.7	586
Washington	138	0	138	439	39.9	479	577	39.9	617	647	44.7	692
West Virginia	9.70	0	9.70	958	0	958	968	0	968	1,090	0	1,090
Wisconsin	83.0	0	83.0	364	0	364	447	0	447	501	0	501
Wyoming	4.31	0	4.31	1.47	0	1.47	5.78	0	5.78	6.48	0	6.48
Puerto Rico	11.2	0	11.2	0	0	0	11.2	0	11.2	12.5	0	12.5
U.S. Virgin Islands	0.22	0	0.22	3.12	0	3.12	3.34	0	3.34	3.74	0	3.74
Total	3,570	6.51	3,580	14,900	1,280	16,200	18,500	1,280	19,700	20,700	1,440	22,100

Source: Modified from table available at http://pubs.usgs.gov/circ/2004/circ1268/htdocs/table10.html, March 2007.
Note: Figures may not sum to totals because of independent rounding. —, data not collected.

TABLE 11
Mining Water Withdrawals, 2000

State	Withdrawals (in million gallons per day) By Source and Type									Withdrawals (in thousand acre-feet per year) By Type		
	Groundwater			Surface Water			Total					
	Fresh	Saline	Total	Fresh	Saline	Total	Fresh	Saline	Total	Fresh	Saline	Total
Alabama	—	—	—	—	—	—	—	—	—	—	—	—
Alaska	0.01	90.4	90.4	27.4	49.5	76.9	27.4	140	167	30.7	157	188
Arizona	81.2	8.17	89.4	4.43	0	4.43	85.7	8.17	93.8	96.0	9.16	105
Arkansas	0.21	0	0.21	2.57	0	2.57	2.78	0	2.78	3.12	0	3.12
California	21.0	152	173	2.71	0.46	3.17	23.7	153	177	26.6	171	198
Colorado	—	—	—	—	—	—	—	—	—	—	—	—
Connecticut	—	—	—	—	—	—	—	—	—	—	—	—
Delaware	—	—	—	—	—	—	—	—	—	—	—	—
District of Columbia	—	—	—	—	—	—	—	—	—	—	—	—
Florida	160	0	160	57.8	0	57.8	217	0	217	244	0	244
Georgia	7.75	0	7.75	2.05	0	2.05	9.80	0	9.80	11.0	0	11.0
Hawaii	—	—	—	—	—	—	—	—	—	—	—	—
Idaho	—	—	—	—	—	—	—	—	—	—	—	—
Illinois	—	—	—	—	—	—	—	—	—	—	—	—
Indiana	4.20	0	4.20	78.3	0	78.3	82.5	0	82.5	92.5	0	92.5
Iowa	2.49	0	2.49	30.3	0	30.3	32.8	0	32.8	36.8	0	36.8
Kansas	14.0	0	14.0	17.4	0	17.4	31.4	0	31.4	35.2	0	35.2

State												
Kentucky	—	—	—	—	—	—	—	—	—	—	—	—
Louisiana	—	—	—	—	—	—	—	—	—	—	—	—
Maine	—	—	—	—	—	—	—	—	—	—	—	—
Maryland	4.21	0	4.21	4.10	0.02	4.12	8.31	0.02	8.33	9.32	0.02	9.34
Massachusetts	—	—	—	—	—	—	—	—	—	—	—	—
Michigan	6.90	0	6.90	581	0	581	588	0	588	659	0	659
Minnesota	—	—	—	—	—	—	—	—	—	—	—	—
Mississippi	—	—	—	—	—	—	—	—	—	—	—	—
Missouri	4.10	0	4.10	12.8	0	12.8	16.9	0	16.9	19.0	0	19.0
Montana	—	—	—	—	—	—	—	—	—	—	—	—
Nebraska	5.64	4.55	10.2	122	0	122	128	4.55	132	143	5.10	148
Nevada	—	—	—	—	—	—	—	—	—	—	—	—
New Hampshire	0.08	0	0.08	6.72	0	6.72	6.80	0	6.80	7.62	0	7.62
New Jersey	6.12	0	6.12	104	0	104	110	0	110	124	0	124
New Mexico	—	—	—	—	—	—	—	—	—	—	—	—
New York	—	—	—	—	—	—	—	—	—	—	—	—
North Carolina	36.4	0	36.4	0	0	0	36.4	0	36.4	40.8	0	40.8
North Dakota	—	—	—	—	—	—	—	—	—	—	—	—
Ohio	53.1	0	53.1	35.5	0	35.5	88.5	0	88.5	99.2	0	99.2
Oklahoma	2.25	256	258	0.23	0	0.23	2.48	256	258	2.78	287	290
Oregon	—	—	—	—	—	—	—	—	—	—	—	—
Pennsylvania	162	0	162	20.9	0	20.9	182	0	182	205	0	205
Rhode Island	—	—	—	—	—	—	—	—	—	—	—	—

continued

TABLE 11 (continued)
Mining Water Withdrawals, 2000

State	Withdrawals (in million gallons per day)									Withdrawals (in thousand acre-feet per year)		
	By Source and Type									By Type		
	Groundwater			Surface Water			Total					
	Fresh	Saline	Total	Fresh	Saline	Total	Fresh	Saline	Total	Fresh	Saline	Total
South Carolina	—	—	—	—	—	—	—	—	—	—	—	—
South Dakota	—	—	—	—	—	—	—	—	—	—	—	—
Tennessee	—	—	—	—	—	—	—	—	—	—	—	—
Texas	129	504	633	91.5	0	91.5	220	504	724	247	565	812
Utah	8.60	21.5	30.1	17.7	177	194	26.3	198	225	29.4	222	252
Vermont	—	—	—	—	—	—	—	—	—	—	—	—
Virginia	—	—	—	—	—	—	—	—	—	—	—	—
Washington	—	—	—	—	—	—	—	—	—	—	—	—
West Virginia	—	—	—	—	—	—	—	—	—	—	—	—
Wisconsin	—	—	—	—	—	—	—	—	—	—	—	—
Wyoming	58.8	222	280	20.7	0	20.7	79.5	222	301	89.1	248	338
Puerto Rico	—	—	—	—	—	—	—	—	—	—	—	—
U.S. Virgin Islands	—	—	—	—	—	—	—	—	—	—	—	—
Total	767	1,260	2,030	1,240	227	1,470	2,010	1,490	3,490	2,250	1,660	3,920

Source: Modified from table available at http://pubs.usgs.gov/circ/2004/circ1268/htdocs/table11.html, March 2007.

Note: Figures may not sum to totals because of independent rounding. —, data not collected.

TABLE 12
Thermoelectric Power Water Withdrawals, 2000

State	Withdrawals (in million gallons per day)							Withdrawals (in thousand acre-feet per year)		
	By Source and Type							By Type		
	Groundwater	Surface Water			Total					
	Fresh	Fresh	Saline	Total	Fresh	Saline	Total	Fresh	Saline	Total
Alabama	0	8,190	0	8,190	8,190	0	8,190	9,180	0	9,180
Alaska	4.65	28.9	0	28.9	33.6	0	33.6	37.6	0	37.6
Arizona	74.3	26.2	0	26.2	100	0	100	113	0	113
Arkansas	2.92	2,170	0	2,170	2,180	0	2,180	2,440	0	2,440
California	3.23	349	12,600	12,900	352	12,600	12,900	395	14,100	14,500
Colorado	16.1	122	0	122	138	0	138	155	0	155
Connecticut	0.08	186	3,440	3,630	187	3,440	3,630	209	3,860	4,070
Delaware	0.47	366	738	1,100	366	738	1,100	411	827	1,240
District of Columbia	0	9.69	0	9.69	9.69	0	9.69	10.9	0	10.9
Florida	29.5	629	12,000	12,600	658	12,000	12,600	738	13,400	14,100
Georgia	1.03	3,240	61.7	3,310	3,250	61.7	3,310	3,640	69.2	3,710
Hawaii	0	0	0	0	0	0	0	0	0	0
Idaho	0	0	0	0	0	0	0	0	0	0
Illinois	5.75	11,300	0	11,300	11,300	0	11,300	12,600	0	12,600
Indiana	2.58	6,700	0	6,700	6,700	0	6,700	7,510	0	7,510
Iowa	11.9	2,530	0	2,530	2,540	0	2,540	2,850	0	2,850

continued

TABLE 12 (continued)
Thermoelectric Power Water Withdrawals, 2000

State	Withdrawals (in million gallons per day)								Withdrawals (in thousand acre-feet per year)		
	By Source and Type								By Type		
	Groundwater	Surface Water			Total						
	Fresh	Fresh	Saline	Total	Fresh	Saline	Total		Fresh	Saline	Total
Kansas	14.9	2,240	0	2,240	2,260	0	2,260		2,530	0	2,530
Kentucky	2.71	3,250	0	3,250	3,260	0	3,260		3,650	0	3,650
Louisiana	28.4	5,580	0	5,580	5,610	0	5,610		6,290	0	6,290
Maine	4.92	108	295	403	113	295	408		127	330	457
Maryland	1.80	377	6,260	6,640	379	6,260	6,640		425	7,020	7,440
Massachusetts	0	108	3,610	3,720	108	3,610	3,720		121	4,050	4,170
Michigan	0	7,710	0	7,710	7,710	0	7,710		8,640	0	8,640
Minnesota	4.17	2,260	0	2,260	2,270	0	2,270		2,540	0	2,540
Mississippi	43.5	318	148	467	362	148	510		406	166	572
Missouri	12.2	5,620	0	5,620	5,640	0	5,640		6,320	0	6,320
Montana	0	110	0	110	110	0	110		123	0	123
Nebraska	6.87	2,810	0	2,810	2,820	0	2,820		3,160	0	3,160
Nevada	12.0	24.7	0	24.7	36.7	0	36.7		41.1	0	41.1
New Hampshire	0.71	235	761	997	236	761	997		265	854	1,120
New Jersey	2.24	648	3,390	4,040	650	3,390	4,040		729	3,800	4,530
New Mexico	11.4	45.0	0	45.0	56.4	0	56.4		63.2	0	63.2
New York	0	4,040	5,010	9,050	4,040	5,010	9,050		4,530	5,610	10,100

North Carolina	0.09	7,850	1,620	9,470	7,850	1,620	9,470	8,800	1,810	10,600
North Dakota	0	902	0	902	902	0	902	1,010	0	1,010
Ohio	7.57	8,590	0	8,590	8,590	0	8,590	9,630	0	9,630
Oklahoma	3.27	143	0	143	146	0	146	164	0	164
Oregon	2.47	12.8	0	12.8	15.3	0	15.3	17.2	0	17.2
Pennsylvania	3.98	6,970	0	6,970	6,980	0	6,980	7,820	0	7,820
Rhode Island	0	2.40	290	293	2.40	290	293	2.69	326	328
South Carolina	5.83	5,700	0	5,700	5,710	0	5,710	6,400	0	6,400
South Dakota	1.23	4.01	0	4.01	5.24	0	5.24	5.87	0	5.87
Tennessee	0	9,040	0	9,040	9,040	0	9,040	10,100	0	10,100
Texas	60.2	9,760	3,440	13,200	9,820	3,440	13,300	11,000	3,860	14,900
Utah	13.1	49.2	0	49.2	62.2	0	62.2	69.8	0	69.8
Vermont	0.66	355	0	355	355	0	355	398	0	398
Virginia	1.50	3,850	3,580	7,430	3,850	3,580	7,430	4,310	4,020	8,330
Washington	0.92	518	0	518	519	0	519	582	0	582
West Virginia	0	3,950	0	3,950	3,950	0	3,950	4,430	0	4,430
Wisconsin	8.99	6,090	0	6,090	6,090	0	6,090	6,830	0	6,830
Wyoming	1.13	242	0	242	243	0	243	273	0	273
Puerto Rico	0	0	2,190	2,190	0	2,190	2,190	0	2,460	2,460
U.S. Virgin Islands	0	0	136	136	0	136	136	0	153	153
Total	409	135,000	59,500	195,000	136,000	59,500	195,000	152,000	66,700	219,000

Source: Modified from table available at http://pubs.usgs.gov/circ/2004/circ1268/htdocs/table12.html, March 2007.

Note: Figures may not sum to totals because of independent rounding. —, data not collected.

TABLE 13
Priority Pollutants

Priority Pollutant	CAS Number	Freshwater CMC (acute) (µg/L)	Freshwater CCC (chronic) (µg/L)	Saltwater CMC (acute) (µg/L)	Saltwater CCC (chronic) (µg/L)	Human Health — Water + Organism (µg/L)	Human Health — Organism Only (µg/L)	FR Cite/Source
1 Antimony	7440360					5.6 B	640 B	65FR66443
2 Arsenic	7440382	340 A,D,K	150 A,D,K	69 A,D,bb	36 A,D,bb	0.018 C,M,S	0.14 C,M,S	65FR31682 57FR60848
3 Beryllium	7440417					Z		65FR31682
4 Cadmium	7440439	2.0 D,E,K,bb	0.25 D,E,K,bb	40 D,bb	8.8 D,bb	Z		EPA-822-R-01-001 65FR31682
5a Chromium (III)	16065831	570 D,E,K	74 D,E,K			Z Total		EPA820/B-96-001 65FR31682
5b Chromium (VI)	18540299	16 D,K	11 D,K	1,100 D,bb	50 D,bb	Z Total		65FR31682
6 Copper	7440508	13 D,E,K,cc	9.0 D,E,K,cc	4.8 D,cc,ff	3.1 D,cc,ff	1,300 U		65FR31682
7 Lead	7439921	65 D,E,bb,gg	2.5 D,E,bb,gg	210 D,bb	8.1 D,bb			65FR31682
8a Mercury	7439976	1.4 D,K,hh	0.77 D,K,hh	1.8 D,ee,hh	0.94 D,ee,hh			EPA823-R-01-001
8b Methylmercury	22967926					0.3 mg/kg J		
9 Nickel	7440020	470 D,E,K	52 D,E,K	74 D,bb	8.2 D,bb	610 B	4,600 B	65FR31682
10 Selenium	7782492	L,R,T	5.0 T	290 D,bb,dd	71 D,bb,dd	170 Z	4200	62FR42160 65FR31682 65FR66443
11 Silver	7440224	3.2 D,E,G		1.9 D,G				65FR31682
12 Thallium	7440280					0.24	0.47	68FR75510

No.	Name	CAS							Reference
13	Zinc	7440666	120 D,E,K	120 D,E,K	90 D,bb	81 D,bb	7,400 U	26,000 U	65FR31682; 65FR66443
14	Cyanide	57125	22 K,Q	5.2 K,Q	1 Q,bb	1 Q,bb	140 jj	140 jj	EPA820/B-96-001; 57FR60848; 68FR75510
15	Asbestos	1332214					7 million fibers/L I		57FR60848
16	2,3,7,8-TCDD (Dioxin)	1746016					5.0E-9 C	5.1E-9 C	65FR66443
17	Acrolein	107028					190	290	65FR66443
18	Acrylonitrile	107131					0.051 B,C	0.25 B,C	65FR66443
19	Benzene	71432					2.2 B,C	51 B,C	IRIS 01/19/00; 65FR66443
20	Bromoform	75252					4.3 B,C	140 B,C	65FR66443
21	Carbon Tetrachloride	56235					0.23 B,C	1.6 B,C	65FR66443
22	Chlorobenzene	108907					130 Z,U	1,600 U	68FR75510
23	Chlorodibromomethane	124481					0.40 B,C	13 B,C	65FR66443
24	Chloroethane	75003							
25	2-Chloroethylvinyl Ether	110758							
26	Chloroform	67663					5.7 C,P	470 C,P	62FR42160
27	Dichlorobromomethane	75274					0.55 B,C	17 B,C	65FR66443
28	1,1-Dichloroethane	75343							
29	1,2-Dichloroethane	107062					0.38 B,C	37 B,C	65FR66443
30	1,1-Dichloroethylene	75354					330	7,100	68FR75510
31	1,2-Dichloropropane	78875					0.50 B,C	15 B,C	65FR66443
32	1,3-Dichloropropene	542756					0.34 C	21 C	68FR75510
33	Ethylbenzene	100414					530	2,100	68FR75510
34	Methyl Bromide	74839					47 B	1,500 B	65FR66443
35	Methyl Chloride	74873							65FR31682

continued

TABLE 13 (continued)
Priority Pollutants

	Priority Pollutant	CAS Number	Freshwater		Saltwater		Human Health for the Consumption of		FR Cite/Source
			CMC (acute) (µg/L)	CCC (chronic) (µg/L)	CMC (acute) (µg/L)	CCC (chronic) (µg/L)	Water + Organism (µg/L)	Organism Only (µg/L)	
36	Methylene Chloride	75092					4.6 B,C	590 B,C	65FR66443
37	1,1,2,2-Tetrachloroethane	79345					0.17 B,C	4.0 B,C	65FR66443
38	Tetrachloroethylene	127184					0.69 C	3.3 C	65FR66443
39	Toluene	108883					1,300 Z	15,000	68FR75510
40	1,2-Trans-Dichloroethylene	156605					140 Z	10,000	68FR75510
41	1,1,1-Trichloroethane	71556					Z		65FR31682
42	1,1,2-Trichloroethane	79005					0.59 B,C	16 B,C	65FR66443
43	Trichloroethylene	79016					2.5 C	30 C	65FR66443
44	Vinyl Chloride	75014					0.025 C,kk	2.4 C,kk	68FR75510
45	2-Chlorophenol	95578					81 B,U	150 B,U	65FR66443
46	2,4-Dichlorophenol	120832					77 B,U	290 B,U	65FR66443
47	2,4-Dimethylphenol	105679					380 B	850 B,U	65FR66443
48	2-Methyl-4,6-Dinitrophenol	534521					13	280	65FR66443
49	2,4-Dinitrophenol	51285					69 B	5,300 B	65FR66443
50	2-Nitrophenol	88755							
51	4-Nitrophenol	100027							
52	3-Methyl-4-Chlorophenol	59507					U	U	
53	Pentachlorophenol	87865	19 F,K	15 F,K	13 bb	7.9 bb	0.27 B,C	3.0 B,C,H	65FR31682 65FR66443
54	Phenol	108952					21,000 B,U	1,700,000 B,U	65FR66443
55	2,4,6-Trichlorophenol	88062					1.4 B,C	2.4 B,C,U	65FR66443

No.	Compound	CAS			
56	Acenaphthene	83329	670 B,U	990 B,U	65FR66443
57	Acenaphthylene	208968			
58	Anthracene	120127	8,300 B	40,000 B	65FR66443
59	Benzidine	92875	0.000086 B,C	0.00020 B,C	65FR66443
60	Benzo(a) Anthracene	56553	0.0038 B,C	0.018 B,C	65FR66443
61	Benzo(a) Pyrene	50328	0.0038 B,C	0.018 B,C	65FR66443
62	Benzo(b) Fluoranthene	205992	0.0038 B,C	0.018 B,C	65FR66443
63	Benzo(ghi) Perylene	191242			
64	Benzo(k) Fluoranthene	207089	0.0038 B,C	0.018 B,C	65FR66443
65	Bis(2-Chloroethoxy) Methane	111911			
66	Bis(2-Chloroethyl) Ether	111444	0.030 B,C	0.53 B,C	65FR66443
67	Bis(2-Chloroisopropyl) Ether	108601	1,400 B	65,000 B	65FR66443
68	Bis(2-Ethylhexyl) Phthalate[X]	117817	1.2 B,C	2.2 B,C	65FR66443
69	4-Bromophenyl Phenyl Ether	101553			
70	Butylbenzyl Phthalate[w]	85687	1,500 B	1,900 B	65FR66443
71	2-Chloronaphthalene	91587	1,000 B	1,600 B	65FR66443
72	4-Chlorophenyl Phenyl Ether	7005723			
73	Chrysene	218019	0.0038 B,C	0.018 B,C	65FR66443
74	Dibenzo(a,h)Anthracene	53703	0.0038 B,C	0.018 B,C	65FR66443
75	1,2-Dichlorobenzene	95501	420	1,300	68FR75510
76	1,3-Dichlorobenzene	541731	320	960	65FR66443
77	1,4-Dichlorobenzene	106467	63	190	68FR75510
78	3,3'-Dichlorobenzidine	91941	0.021 B,C	0.028 B,C	65FR66443
79	Diethyl Phthalate[w]	84662	17,000 B	44,000 B	65FR66443
80	Dimethyl Phthalate[w]	131113	270,000	1,100,000	65FR66443

continued

TABLE 13 (continued)
Priority Pollutants

	Priority Pollutant	CAS Number	Freshwater		Saltwater		Human Health for the Consumption of		FR Cite/Source
			CMC (acute) (µg/L)	CCC (chronic) (µg/L)	CMC (acute) (µg/L)	CCC (chronic) (µg/L)	Water + Organism (µg/L)	Organism Only (µg/L)	
81	Di-n-Butyl Phthalate[w]	84742					2,000 B	4,500 B	65FR66443
82	2,4-Dinitrotoluene	121142					0.11 C	3.4 C	65FR66443
83	2,6-Dinitrotoluene	606202							
84	Di-n-Octyl Phthalate	117840							
85	1,2-Diphenylhydrazine	122667					0.036 B,C	0.20 B,C	65FR66443
86	Fluoranthene	206440					130 B	140 B	65FR66443
87	Fluorene	86737					1,100 B	5,300 B	65FR66443
88	Hexachlorobenzene	118741					0.00028 B,C	0.00029 B,C	65FR66443
89	Hexachlorobutadiene	87683					0.44 B,C	18 B,C	65FR66443
90	Hexachlorocyclopentadiene	77474					40 U	1,100 U	68FR75510
91	Hexachloroethane	67721					1.4 B,C	3.3 B,C	65FR66443
92	Ideno(1,2,3-cd)Pyrene	193395					0.0038 B,C	0.018 B,C	65FR66443
93	Isophorone	78591					35 B,C	960 B,C	65FR66443
94	Naphthalene	91203							
95	Nitrobenzene	98953					17 B	690 B,H,U	65FR66443
96	N-Nitrosodimethylamine	62759					0.00069 B,C	3.0 B,C	65FR66443
97	N-Nitrosodi-n-Propylamine	621647					0.0050 B,C	0.51 B,C	65FR66443
98	N-Nitrosodiphenylamine	86306					3.3 B,C	6.0 B,C	65FR66443
99	Phenanthrene	85018							
100	Pyrene	129000					830 B	4,000 B	65FR66443

		CAS					35	70	Reference
101	1,2,4-Trichlorobenzene	120821							68FR75510
102	Aldrin	309002	3.0 G		1.3 G		0.000049 B,C	0.000050 B,C	65FR31682 65FR66443
103	Alpha-BHC	319846					0.0026 B,C	0.0049 B,C	65FR66443
104	Beta-BHC	319857					0.0091 B,C	0.017 B,C	65FR66443
105	Gamma-BHC (Lindane)	58899	0.95 K		0.16 G		0.98	1.8	65FR31682 68FR75510
106	Delta-BHC	319868							
107	Chlordane	57749	2.4 G	0.0043 G,aa	0.09 G	0.004 G,aa	0.00080 B,C	0.00081 B,C	65FR31682 65FR66443
108	4,4'-DDT	50293	1.1 G,ii	0.001 G,aa,ii	0.13 G,ii	0.001 G,aa,ii	0.00022 B,C	0.00022 B,C	65FR31682 65FR66443
109	4,4'-DDE	72559					0.00022 B,C	0.00022 B,C	65FR66443
110	4,4'-DDD	72548					0.00031 B,C	0.00031 B,C	65FR66443
111	Dieldrin	60571	0.24 K	0.056 K,O	0.71 G	0.0019 G,aa	0.000052 B,C	0.000054 B,C	65FR31682 65FR66443
112	Alpha-Endosulfan	959988	0.22 G,Y	0.056 G,Y	0.034 G,Y	0.0087 G,Y	62 B	89 B	65FR31682 65FR66443
113	Beta-Endosulfan	33213659	0.22 G,Y	0.056 G,Y	0.034 G,Y	0.0087 G,Y	62 B	89 B	65FR31682 65FR66443
114	Endosulfan Sulfate	1031078	0.086 K	0.036 K,O	0.037 G	0.0023 G,aa	62 B	89 B	65FR66443
115	Endrin	72208	0.086 K	0.036 K,O	0.037 G	0.0023 G,aa	0.059	0.060	65FR31682 68FR75510
116	Endrin Aldehyde	7421934					0.29 B	0.30 B,H	65FR66443
117	Heptachlor	76448	0.52 G	0.0038 G,aa	0.053 G	0.0036 G,aa	0.000079 B,C	0.000079 B,C	65FR31682 65FR66443
118	Heptachlor Epoxide	1024573	0.52 G,V	0.0038 G,V,aa	0.053 G,V	0.0036 G,V,aa	0.000039 B,C	0.000039 B,C	65FR31682 65FR66443

continued

TABLE 13 (continued)
Priority Pollutants

Priority Pollutant	CAS Number	Freshwater CMC (acute) (µg/L)	Freshwater CCC (chronic) (µg/L)	Saltwater CMC (acute) (µg/L)	Saltwater CCC (chronic) (µg/L)	Human Health for the Consumption of Water + Organism (µg/L)	Human Health for the Consumption of Organism Only (µg/L)	FR Cite/Source
119 Polychlorinated Biphenyls PCBs:			0.014 N,aa		0.03 N,aa	0.000064 B,C,N	0.000064 B,C,N	65FR31682 65FR66443
120 Toxaphene	8001352	0.73	0.0002 aa	0.21	0.0002 aa	0.00028 B,C	0.00028 B,C	65FR31682 65FR66443

A: This recommended water quality criterion was derived from data for arsenic (III) but is applied here to total arsenic, which might imply that arsenic (III) and arsenic (V) are equally toxic to aquatic life and that their toxicities are additive. In the *arsenic criteria document* (PDF, 74 pp., 3.2M) (EPA 440/5-84-033, January 1985), Species Mean Acute Values are given for both arsenic (III) and arsenic (V) for five species and the ratios of the SMAVs for each species range from 0.6 to 1.7. Chronic values are available for both arsenic (III) and arsenic (V) for one species; for the fathead minnow, the chronic value for arsenic (V) is 0.29 times the chronic value for arsenic (III). No data are known to be available concerning whether the toxicities of the forms of arsenic to aquatic organisms are additive.

B: This criterion has been revised to reflect The Environmental Protection Agency's q1* or RfD, as contained in the Integrated Risk Information System (IRIS) as of May 17, 2002. The fish-tissue bioconcentration factor (BCF) from the 1980 Ambient Water Quality Criteria document was retained in each case.

C: This criterion is based on carcinogenicity of 10^{-6} risk. Alternate risk levels may be obtained by moving the decimal point (e.g., for a risk level of 10^{-5}, move the decimal point in the recommended criterion one place to the right).

D: Freshwater and saltwater criteria for metals are expressed in terms of the dissolved metal in the water column. The recommended water-quality criteria value was calculated by using the previous 304(a) aquatic life criteria expressed in terms of total recoverable metal, and multiplying it by a conversion factor (CF). The term "Conversion Factor" (CF) represents the recommended conversion factor for converting a metal criterion expressed as the total recoverable fraction in the water column to a criterion expressed as the dissolved fraction in the water column. (Conversion Factors for saltwater CCCs are not currently available. Conversion factors derived for saltwater CMCs have been used for both saltwater CMCs and CCCs). See "Office of Water Policy and Technical Guidance on Interpretation and Implementation of Aquatic Life Metals Criteria," (PDF, 49 pp., 3M) October 1, 1993, by Martha G. Prothro, Acting Assistant Administrator for Water, available from the Water Resource center and 40CFR§ 131.36(b)(1). Conversion Factors applied in the table can be found in Appendix A to the Preamble — Conversion Factors for Dissolved Metals.

E: The freshwater criterion for this metal is expressed as a function of hardness (mg/L) in the water column. The value given here corresponds to a hardness of 100 mg/L. Criteria values for other hardness may be calculated from the following: CMC (dissolved) = exp{m_A [ln(hardness)]+ b_A} (CF), or CCC (dissolved) = exp{m_C [ln (hardness)]+ b_C} (CF) and the parameters specified in Appendix B — Parameters for Calculating Freshwater Dissolved Metals Criteria That Are Hardness-Dependent.

F: Freshwater aquatic life values for pentachlorophenol are expressed as a function of pH, and are calculated as follows: CMC = exp(1.005(pH)-4.869); CCC = exp(1.005(pH)-5.134). Values displayed in table correspond to a pH of 7.8.

G: This criterion is based on 304(a) aquatic life criterion issued in 1980, and was issued in one of the following documents: *Aldrin/Dieldrin* (PDF, 153 pp., 7.3M) (EPA 440/5-80-019), *Chlordane* (PDF, 68 pp., 3.1M) (EPA 440/5-80-027), *DDT* (PDF, 175 pp., 8.3M) (EPA 440/5-80-038), *Endosulfan* (PDF, 155 pp., 7.3M) (EPA 440/5-80-046), *Endrin* (PDF, 103 pp., 4.6M) (EPA 440/5-80-047), *Heptachlor* (PDF, 114 pp., 5.4M) (EPA 440/5-80-052), *Hexachlorocyclohexane* (PDF, 109 pp., 4.8M) (EPA 440/5-80-054), *Silver* (EPA 440/5-80-071). The Minimum Data Requirements and derivation procedures were different in the 1980 Guidelines than in the 1985 Guidelines (PDF, 105 pp., 4.5M). For example, a "CMC" derived using the 1980 Guidelines was derived to be used as an instantaneous maximum. If assessment is to be done using an averaging period, the values given should be divided by 2 to obtain a value that is more comparable to a CMC derived using the 1985 Guidelines (PDF, 105 pp., 4.5M).

H: No criterion for protection of human health from consumption of aquatic organisms excluding water was presented in the 1980 criteria document or in the *1986 Quality Criteria for Water*. Nevertheless, sufficient information was presented in the 1980 document to allow the calculation of a criterion, even though the results of such a calculation were not shown in the document.

I: This criterion for asbestos is the Maximum Contaminant Level (MCL) developed under the Safe Drinking Water Act (SDWA).

J: This fish tissue residue criterion for methylmercury is based on a total fish consumption rate of 0.0175 kg/day.

K: This recommended criterion is based on a 304(a) aquatic life criterion that was issued in the *1995 Updates: Water Quality Criteria Documents for the Protection of Aquatic Life in Ambient Water* (EPA-820-B-96-001, September 1996). This value was derived using the GLI Guidelines (60FR15393-15399, March 23, 1995; 40CFR132 Appendix A); the difference between the 1985 Guidelines and the GLI Guidelines are explained on page iv of the 1995 Updates. None of the decisions concerning the derivation of this criterion were affected by any considerations that are specific to the Great Lakes.

L: The CMC = 1/[(f1/CMC1) + (f2/CMC2)] where f1 and f2 are the fractions of total selenium that are treated as selenite and selenate, respectively, and CMC1 and CMC2 are 185.9 g/l and 12.82 g/l, respectively.

M: EPA is currently reassessing the criteria for arsenic.

N: This criterion applies to total pcbs, (e.g., the sum of all congener or all isomer or homolog or Aroclor analyses.)

O: The derivation of the CCC for this pollutant (Endrin) did not consider exposure through the diet, which is probably important for aquatic life occupying upper trophic levels.

continued

TABLE 13 (continued)
Priority Pollutants

P: Although a new RfD is available in IRIS, the surface-water criteria will not be revised until the National Primary Drinking Water Regulations: Stage 2 Disinfectants and Disinfection Byproducts Rule (Stage 2 DBPR) is completed, since public comment on the relative source contribution (RSC) for chloroform is anticipated.

Q: This recommended water-quality criterion is expressed as g free cyanide (as CN)/L

R: This value for selenium was announced (61FR58444-58449, November 14, 1996) as a proposed GLI 303(c) aquatic-life criterion. EPA is currently working on this criterion and so this value might change substantially in the near future.

S: This recommended water-quality criterion for arsenic refers to the inorganic form only.

T: This recommended water-quality criterion for selenium is expressed in terms of total recoverable metal in the water column. It is scientifically acceptable to use the conversion factor (0.996 — CMC or 0.922 — CCC) that was used in the GLI to convert this to a value that is expressed in terms of dissolved metal.

U: The organoleptic effect criterion is more stringent than the value for priority toxic pollutants.

V: This value was derived from data for heptachlor and the criteria document provides insufficient data to estimate the relative toxicities of heptachlor and heptachlor epoxide.

W: Although EPA has not published a completed criteria document for butylbenzyl phthalate, it is EPA's understanding that sufficient data exist to allow calculation of aquatic criteria. It is anticipated that industry intends to publish in the peer-reviewed literature draft aquatic life criteria generated in accordance with EPA Guidelines. EPA will review such criteria for possible issuance as national WQC.

X: There is a full set of aquatic life toxicity data that show that DEHP is not toxic to aquatic organisms at or below its solubility limit.

Y: This value was derived from data for endosulfan and is most appropriately applied to the sum of alpha-endosulfan and beta-endosulfan.

Z: A more stringent MCL has been issued by EPA. Refer to drinking water regulations (40 CFR 141) or Safe Drinking Water Hotline (1-800-426-4791) for values.

aa: This criterion is based on a 304(a) aquatic life criterion issued in 1980 or 1986, and was issued in one of the following documents: *Aldrin/Dieldrin* (PDF, 153 pp., 7.3M) (EPA 440/5-80-019), *Chlordane* (PDF, 68 pp., 3.1M) (EPA 440/5-80-027), *DDT* (PDF, 175 pp., 8.3M) (EPA 440/5-80-038), *Endrin* (PDF, 103 pp., 4.6M) (EPA 440/5-80-047), *Heptachlor* (PDF, 114 pp., 5.4M) (EPA 440/5-80-052), Polychlorinated biphenyls (EPA 440/5-80-068), *Toxaphene* (EPA 440/5-86-006). This CCC is currently based on the Final Residue Value (FRV) procedure. Since the publication of the Great Lakes Aquatic Life Criteria Guidelines in 1995 (60FR15393-15399, March 23, 1995), the Agency no longer uses the Final Residue Value procedure for deriving CCCs for new or revised 304(a) aquatic life criteria. Therefore, the Agency anticipates that future revisions of this CCC will not be based on the FRV procedure.

bb: This water quality criterion is based on a 304(a) aquatic life criterion that was derived using the *1985 Guidelines* (PDF, 105 pp., 4.5M) (*Guidelines for Deriving Numerical National Water Quality Criteria for the Protection of Aquatic Organisms and Their Uses*, PB85-227049, January 1985) and was issued in one of the following criteria documents: *Arsenic* (PDF, 74 pp., 3.2M) (EPA 440/5-84-033), *Cadmium* (EPA-822-R-01-001), *Chromium* (EPA 440/5-84-029), *Copper* (PDF, 150 pp., 6.2M) (EPA 440/5-84-031), *Cyanide* (PDF, 67 pp., 2.7M) (EPA 440/5- 84-028), *Lead* (EPA 440/5-84-027), *Nickel* (EPA 440/5-86-004), *Pentachlorophenol* (EPA 440/5-86-009), *Toxaphene* (EPA 440/5-86-006), *Zinc* (EPA 440/5-87-003).

cc: When the concentration of dissolved organic carbon is elevated, copper is substantially less toxic and use of Water-Effect Ratios might be appropriate.

dd: The selenium criteria document (EPA 440/5-87-006, September 1987) provides that if selenium is as toxic to saltwater fishes in the field as it is to freshwater fishes in the field, the status of the fish community should be monitored whenever the concentration of selenium exceeds 5.0 g/L in salt water because the saltwater CCC does not take into account uptake via the food chain.

ee: This recommended water-quality criterion was derived on page 43 of the *mercury criteria document* (PDF, 144 pp., 6.4M) (EPA 440/5-84-026, January 1985). The saltwater CCC of 0.025 ug/L given on page 23 of the criteria document is based on the Final Residue Value procedure in the 1985 Guidelines. Since the publication of the Great Lakes Aquatic Life Criteria Guidelines in 1995 (60FR15393-15399, March 23, 1995), the Agency no longer uses the Final Residue Value procedure for deriving CCCs for new or revised 304(a) aquatic life criteria.

ff: This recommended water quality criterion was derived in *Ambient Water Quality Criteria Saltwater Copper Addendum* (Draft, April 14, 1995) and was promulgated in the Interim final National Toxics Rule (60FR22228-222237, May 4, 1995).

gg: The EPA is actively working on this criterion and so this recommended water quality criterion may change substantially in the near future.

hh: This recommended water quality criterion was derived from data for inorganic mercury (II) but is applied here to total mercury. If a substantial portion of the mercury in the water column is methylmercury, this criterion will probably be under protective. In addition, even though inorganic mercury is converted to methylmercury and methylmercury bioaccumulates to a great extent, this criterion does not account for uptake via the food chain because sufficient data were not available when the criterion was derived.

ii: This criterion applies to DDT and its metabolites (i.e., the total concentration of DDT and its metabolites should not exceed this value).

jj: This recommended water quality criterion is expressed as total cyanide, even though the IRIS RFD we used to derive the criterion is based on free cyanide. The multiple forms of cyanide that are present in ambient water have significant differences in toxicity due to their differing abilities to liberate the CN-moiety. Some complex cyanides require even more extreme conditions than refluxing with sulfuric acid to liberate the CN-moiety. Thus, these complex cyanides are expected to have little or no "bioavailability" to humans. If a substantial fraction of the cyanide present in a water body is present in a complexed form (e.g., $Fe_4[Fe(CN)_6]_3$), this criterion may be overconservative.

kk: This recommended water quality criterion was derived using the cancer slope factor of 1.4 (LMS exposure from birth).

Source: Modified from table available at http://www.epa.gov/waterscience/criteria/wqcriteria.html, March 2007.

TABLE 14
Different Methods of Measuring Land Subsidence

Method	Component Displacement	Resolution (millimeters)*	Spatial Density (Samples/Survey)**	Spatial Scale (Elements)
Spirit level	vertical	0.1–1	10-100	line-network
Geodimeter	horizontal	1	10–100	line-network
Borehole extensometer	vertical	0.01–0.1	1–3	point
Horizontal extensometer				
Tape	horizontal	0.3	1–10	line-array
Invar wire	horizontal	0.0001	1	line
Quartz tube	horizontal	0.00001	1	line
GPS	vertical horizontal	20 5	10–100	network
InSAR	range	5–10	100,000–10,000,000	map pixel3***

Note: Under ideal conditions, it is possible to resolve changes in elevation on the order of 0.2 to 0.4 inches (5 to 10 mm) at the scale of one pixel. Interferograms, formed from patterns of interference between the phase components of two radar scans made from nearly the same antenna position (viewing angle) but at different times, have demonstrated dramatic potential for high-density spatial mapping of ground-surface displacements associated with tectonic (Massonnet and others, 1993; Zebker and others, 1994) and volcanic strains (Massonnet and others, 1995; Rosen and others, 1996; Wicks and others, 1998). InSAR has also recently been used to map localized crustal deformation and land subsidence associated with geothermal fields in Imperial Valley, California (Massonnet and others, 1997), Long Valley, California (W. Thatcher, USGS, written communication, 1997), and Iceland (Vadon and Sigmundsson, 1997), and with oil and gas fields in the Central Valley, California (Fielding and others, 1998). InSAR has also been used to map regional-scale land subsidence caused by aquifer-system compaction in the Antelope Valley, California (Galloway and others, 1998), Las Vegas Valley, Nevada (Amelung and others, 1999), and Santa Clara Valley, California (Ikehara and others, 1998).

Source: Table and explanation modified from http://pubs.usgs.gov/fs/fs-051-00/, September 2007.

* Measurement resolution attainable under optimum conditions. Values are given in metric units to conform with standard geodetic guidelines. (One inch is equal to 25.4 millimeters and 1 foot is equal to 304.8 millimeters.)

** Number of measurements generally attainable under good conditions to define the spatial extent of land subsidence at the scale of the survey.

*** A pixel on an InSAR displacement map is typically 40 to 80 meters square on the ground.

TABLE 15
Interstate Compacts Related to Water Law

Atlantic Salmon Compact

Interstate Water Apportionment

Alabama-Coosa-Tallapoosa (ACT) River Basin Compact
Animas-La Plata Project Compact
Apalachicola-Chattahoochee-Flint (ACF) River Basin Compact
Arkansas River Basin Compact
Arkansas River Compact of 1949
Arkansas River Compact of 1965
Bear River Compact
Belle Fourche River Compact
Big Blue River Compact
California-Nevada Interstate Compact
Canadian River Compact
Colorado River Compact*
Columbia River Compact
Columbia River Gorge Compact
Connecticut River Compact
Costilla Creek Compact
Interstate Compact for Jurisdiction on the Colorado River
Interstate Public Water Supply Compact
Klamath River Compact
La Plata River Compact
Nebraska-South Dakota-Wyoming Water Compact
Oregon-California Goose Lake Interstate Compact
Pecos River Compact
Red River Compact
Republican River Compact
Rio Grande Interstate Compact
Sabine River Compact
Saco Watershed Compact
Snake River Compact
South Platte River Compact
Upper Colorado River Basin Compact
Upper Niobrara River Compact
Yellowstone River Compact

Water-Pollution Control

Bi-State Metropolitan Development District Compact
New England Interstate Water Pollution Control Compact
New Hampshire-Vermont Interstate Sewage and Waste Disposal Facilities Compact
Ohio River Valley Water Sanitation Compact
Red River of the North
Tennessee River Basin Water Pollution Control Compact
Tri-State Sanitation Compact

continued

TABLE 15 (continued)
Interstate Compacts Related to Water Law

Water Resources and Flood Control

Connecticut River Valley Flood Control Compact
Delaware River Basin Compact
Great Lakes Basin Compact
Kansas-Missouri Flood Prevention and Control Compact
Merrimack River Flood Control Compact
Missouri River Barge Traffic Compact (1984)
Missouri River Barge Traffic Compact
Potomac Valley Compact
Susquehanna River Basin Compact
Thames River Flood Control Compact
Wheeling Creek Watershed Protection & Flood Prevention Compact

Source: List available in part at http://www.fws.gov/laws/laws_digest/compact.html, March 2007.
*   Recent drought conditions have led to another agreement in December 2007, now awaiting congressional approval.

TABLE 16
Riparian States

1. Alabama	17. Minnesota
2. Arkansas	18. Missouri
3. Connecticut	19. New Hampshire
4. Delaware	20. New Jersey
5. Florida	21. New York
6. Georgia	22. North Carolina
7. Hawaii *	23. Ohio
8. Illinois	24. Pennsylvania
9. Indiana	25. Rhode Island
10. Iowa	26. South Carolina
11. Kentucky	27. Tennessee
12. Louisiana *	28. Vermont
13. Maine	29. Virginia
14. Maryland	30. West Virginia
15. Massachusetts	31. Wisconsin
16. Michigan	

*   Though usually listed as a riparian state, Louisiana uses a civil code, based on French and Spanish law, that is somewhat unique in the United States. See §§ 13.1–13.4. The Hawaiian water law is heavily influenced by the island's feudal common law history. See §§ 13.5–13.9. Hawaii has a comprehensive water code, which is quite unique in the United States. Water is viewed as being owned by the whole for the common good, and it is unclear the extent to which water rights can be transferred in Hawaii. Water is said to be appurtenant to the land, meaning for the benefit of certain specific parcels.

TABLE 17
Prior Appropriation States

1. Alaska	6. Nevada
2. Arizona	7. New Mexico
3. Colorado	8. Utah
4. Idaho	9. Wyoming
5. Montana	

TABLE 18
Hybrid System States

1. California	6. Oklahoma
2. Kansas	7. Oregon
3. Mississippi	8. South Dakota
4. Nebraska	9. Texas
5. North Dakota	10. Washington

TABLE 19
Different Water Rights in the United States

1. **Riparian Rights**—Rights of landowners whose land borders the body of water. These rights ordinarily include the right to use the surface of the water and to have the body flow in its natural, free-flowing capacity. Even appropriation states generally recognize some riparian rights to use the surface of the water. Some appropriation states, known as hybrid states, officially recognize riparian rights that predated their appropriation system, though only a few recognize new riparian rights. Most riparian jurisdictions have moved toward a reasonable use policy, to make some water available for consumptive uses. Nearly all recognize some right to consume small quantities of water for domestic use. Some riparian jurisdictions allow nonriparian uses to the extent that riparians are not harmed by the use, while others consider any use for nonriparian land, even by a riparian owner, an appropriation and therefore not allowed under the riparian system.[1]

2. **Appropriative Rights**—Rights granted in the West, by states to a specific priority among water users and a limited allotment of water, for beneficial use, usually by permit. These rights ordinarily exist independent of any riparian land ownership and may exist for surface or groundwater, though not all appropriation states apply the same standards to each. The appropriation system is largely a response to the circumstances of the West, where water remains scarce in many places and much of the riparian land is or was originally owned by the federal government. The need for water to irrigate and operate mines and other industries required a different system than the eastern riparian common law.[2]

3. **Absolute Ownership Rights in Groundwater**—Some states recognize absolute ownership of all that is beneath the soil and in the air above a landowner's property. This rule has ordinarily been tempered some to deal with damage to other's property from overpumping, and most jurisdictions apply at least a reasonable use requirement today. Most jurisdictions no longer use the absolute ownership doctrine and have moved to a permit system similar to surface water (see appropriative rights).[3] Some states use what is known as the "reasonable use doctrine," which allows a riparian user of groundwater to take reasonable amounts for beneficial use. This is also sometimes called the "American Rule." Similar but slightly different is the "correlative rights doctrine," which states that all overlying landowners have an equal right to use a fair portion of the whole.

4. **Native American Rights (also termed *Winters* rights)**—These rights are a form of federal reserved rights (see below) that are usually guaranteed by contract and may be implied in the contract, rather than specifically stated. They guarantee enough water to make the land on a reservation useful for living. Unlike other rights, these rights are not lost by nonuse. They also predate most other rights in priority.[4]

5. **Pueblo Rights**—These rights come from Spanish law, which predates the Treaty of Guadalupe, whereby the United States acquired much of its western territory. These rights belong to cities and towns that existed prior to the treaty and settled near certain water supplies. They guarantee not only enough water to meet past and present needs of the city (previously called pueblo) but also the future needs. Another variation of these rights applies to the Pueblo Indians, who because of their fee interest in their land, meaning they can sell it, cannot have *Winters* rights. The Pueblo rights are said to predate all others and operate much as *Winters* rights would. These rights apply to all traditional native lands, for so long as the Indians own the land.[5]

6. **Rights in a Water Company's Stock**—Water stock is usually equated with a right to the amount of water guaranteed by the stock. It operates similar to the appropriation rights above, with a water corporation operating as a go between for the state and the irrigators.[6]

TABLE 19 (continued)
Different Water Rights in the United States

7. **Public Use**—Most states guarantee certain rights of public use of waters within the state. Federal law ordinarily allows use of navigable waterways for transportation and other purposes. State laws often expand this right to bodies not covered by the federal definition. Public use generally applies to such uses of the surface water, as the body is suitable for, including boating, swimming, hunting, or fishing. Some states treat all water in the state as owned by the state in trust for the public.[7]

8. **Eminent Domain**—Government has the right to confiscate private land for public benefit with just compensation to the landowner.[8]

9. **Hawaiian Rights**—Water law on the Hawaiian Islands traces its roots to pre-colonial days, where rights were allocated for the common good, by the monarch. Water is still viewed as owned by the whole for the common good and it is unclear, the extent to which water may be transferred there. [9]

10. **Louisiana Civil Code**—Louisiana traces its code to its early French and Spanish settlements. Some of this law has been carried forward and exists still today by way of treaty and custom. While often considered a riparian state, the civil code does restrict and regulate water uses.[10]

11. **Reserved Rights**—These are rights the federal government sometimes holds for use on its lands for public purposes (see Native American water rights, above). These purposes may also include military enclaves or national parks.[11]

12. **Prescriptive Rights**—These are rights that exist in only a handful of states, based on adverse use for a prescriptive period of time, usually established by statute. Prescription regarding water rights is difficult to prove because of the nature of water. It is difficult to use water in a manner that would place the actual holder of a water right on notice. Most states seem to view water rights that are not used, as reverting back to the state and since prescription against the state is not allowed, adverse users must perfect their rights through the permit or appropriations system.[12]

[1] See §§ 6.2–6.18.
[2] See §§ 7.1–7.24.
[3] See § 8.14.
[4] See §§ 9.1–9.16.
[5] See § 7.26.
[6] See §§ 15.4–15.7.
[7] See §§ 3.1–3.6, 3.8.
[8] See §§ 3.10–3.12.
[9] See §§ 13.5–13.9.
[10] See §§ 13.1–13.4.
[11] See § 4.2.
[12] See § 3.7.

TABLE 20
Western Rights for In-Stream Flow Appropriations

State	Ownership	Authorization and Date	New Appropriation	Transfers	Beneficial Uses
Alaska	Public or Private	Statute—1980	Yes, by reservation	Yes	Protection of Fish and Wildlife habitat; Recreation and parks; Navigation; Sanitation and water quality
Arizona	Public or Limited Private[1]	Statute—1941[2]	Yes	Yes[3]	Wildlife; Fish; Recreation
California	Public or Private	Statute—1991	No[4]	Yes	Wetland habitat; Fish and Wildlife; Recreation; Water Quality
Colorado	Colorado Water Conservation Board	Statute—1973	Yes	Yes	"To preserve the natural environment," but to date only streams supporting fisheries have been protected
Idaho	Public or Limited Private[5]	Statute—1974	Yes, by reservation	Yes, temporary[6]	Fish and Wildlife habitat; Aquatic life; Recreation; Aesthetic beauty; Navigation; Transportation; Water Quality
Montana	Public or Limited Private[7]	Statute—1969	Yes, by reservation	Yes	Fisheries; Water Quality; Other uses that benefit the appropriator, other persons, or the public[8]
Nevada	Public or Private	Case Law—1988	Yes	Yes	Wildlife; Recreation[9]
New Mexico[10]	Public or Private	Case Law—1998	No	Yes	Fish and Wildlife Habitat; Recreation; (note: instream flow in itself is not a recognized beneficial use)
Oregon	Oregon Water Resource Department	Statute—1915	Yes[11]	Yes	Recreation; Conservation; Fish and Wildlife; Ecological Values; Pollution Abatement; Navigation
Washington	Wash. Dept of Ecology	Statute—2003	Yes	Yes	Fish and Wildlife; water quality; or other instream resources; hydropower and federal programs
Utah	Divisions of Wildlife Resources and Parks and Recreation	Statute—1986	No	Yes	Propagation of Fish; Public Recreation; Preservation or Enhancement of the Natural Stream Environment
Wyoming	State of Wyoming	Statute—1986	Yes	Yes	Only Fisheries

1 Ownership in the private sector is limited to public-interest groups.

2 Legislation in 1941 and 1962 established wildlife and fish and then recreation as beneficial uses, but case law in 1976 actually established in-stream flow appropriations.

3 Transfers are legally allowed, but have not yet occurred. However, there have been temporary leases.

4 The State Water Board can require bypass flows for new consumptive uses, but these conditions do not constitute in-stream flow rights.

5 Private ownership is only possible on a temporary basis through the water banks or other leases.

6 Transfers are limited to temporary transfers of storage rights through water banks.

7 Private holdership can only be established through transfers.

8 Beneficial use is at the discretion of the DNRC. To date, in-stream flows have been for fisheries and water quality, but the law does not limit the program to these uses.

9 Beneficial use is determined on a case-by-case basis so in-stream flow uses are not necessarily limited to wildlife and recreation.

10 New Mexico does not have a legislated in-stream flow program and in-stream flow is not a recognized beneficial use. Case law has begun their in-stream flow program.

11 Only the Department of Fish and Wildlife, the Department of Environmental Quality, and the Department of Parks and Recreation may apply for new appropriations for in-stream flow. Although these departments apply, the right is held in trust by the Water Resources Department.

Source: Modified from chart at http://www.blm.gov/nstc/WaterLaws/stateflowsummary.html, September 2007 (Washington State was not included in the original chart).

TABLE 21
Specified Beneficial Uses

	Agriculture	Conservation	Domestic	Fish and Wildlife	Frost Protection	Ground Recharge	Industrial	Manufacturing	Mining	Municipal	Navigation	Parks	Pollution Control	Power	Recreational	Stock Watering	Transportation	Water Quality
Alaska	√	√	√				√	√	√	√	√			√	√		√	√
Arizona	√	√	√			√			√	√				√	√	√		
California	√	√	√				√		√	√				√	√	√		√
Colorado	√	√	√				√			√					√			
Idaho																		
Kansas	√	√					√			√				√	√			
Montana	√	√	√				√		√	√				√	√	√		
Nebraska																		
Nevada	√	√	√						√						√			
New Mexico																		
North Dakota	√	√	√				√			√					√	√		
Oklahoma	√		√				√			√					√			
Oregon	√	√	√				√		√	√			√	√	√			
South Dakota																		
Texas	√		√	√	√		√		√	√		√		√	√	√		
Utah	√																√	
Washington	√	√	√	√	√		√		√					√	√	√		
Wyoming	√	√								√								

Source: Based on chart in David H. Getches, *Water Law*, 3rd Ed. at 98 (1997).

Note: The following states had no comprehensive definition for beneficial use as of 1997: Idaho, Nebraska, Nevada, Oklahoma, Utah, and Wyoming. Beneficial use is defined by case law in relation to the land from which water is taken in New Mexico.* A South Dakota statute defines beneficial use as "any use of water within or outside the state that is reasonable and useful and beneficial to the appropriator" and is consistent with the interests of the public.**

* Erickson v. McLean, 62 N.M. 264, 308 P.2d 983 (1957).

** S.D. Cod. Laws § 46-1-6 (3).

TABLE 22
Precipitation Size and Speed

	Intensity inches/hour (cm/hour)	Median Diameter (millimeters)	Velocity of Fall feet/second (meters/ second)	Drops per Second per Square Foot (square meter)
Fog	0.005 (0.013)	0.01	0.01 (0.003)	6,264,000 (67,425,000)
Mist	.002 (.005)	.1	.7 (.21)	2,510 (27,000)
Drizzle	.01 (.025)	.96	13.5 (4.1)	14(151)
Light rain	.04 (1.02)	1.24	15.7 (4.8)	26 (280)
Moderate rain	.15 (.38)	1.60	18.7 (5.7)	46 (495)
Heavy rain	.60 (1.52)	2.05	22.0 (76.)	46 (495)
Excessive rain	1.60 (4.06)	2.40	24.0 (7.3)	76 (818)
Cloudburst	4.00 (10.2)	2.85	25.9 (7.9)	113 (1,220)

Source: Lull, H.W., 1959, *Soil Compaction on Forest and Range Lands,* U.S. Dept. of Agriculture, Forestry Service, Misc. Publication No.768.

Available at http://ga.water.usgs.gov/edu/watercycleprecipitation.html, January 2008.

TABLE 23
Water Equivalents Table

1 gallon	8.34 pounds
	231 cubic inches
	0.134 acre-feet
	3.785 liters
	3.785 cubic decimeters (dm³)
	0.003785 square meter (m²)
gallon per minute	0.06309 liter per second (L/s)
gallon per day	0.003785 cubic meter per day (m³/d)
gallon per day per square mile (gal/d/ml²)	0.001461 cubic meter per day per square kilometer (m³/d/km²)
1 million gallons	3.07 acre-feet
	3.785 cubic meter (m³)
1 million gallons per day (mgd)	1120 acre-feet per year
	1.55 cubic feet per second
	694.4 gallons per minute (gpm)
	0.04381 cubic meter per second (m³/s)
	1,461 cubic meter per day per square kilometer (m³/d/km²)
	3.785 cubic meter per day (m³/d)
1 cubic foot	7.48 gallons
	62.4 pounds
1 cubic foot per second (cfs)	646,317 gallons per day
	448.8 gallons per minute
	1.98 acre-feet per day
	38.4 miner's inches (CO)
	40 miner's inches (AZ, CA, MT, NV, OR)
	50 miner's inches (ID, NE, NM, ND, SD, UT)
cubic mile (mi³)	4.168 cubic kilometers (km³)
acre-foot (acre-ft)	1,233 cubic meter (m³)
	0.001233 cubic hectometer (hm³)
acre-foot per day (acre-ft/d)	0.01427 cubic meter per second (m³/s)
acre-foot per year (acre-ft/yr)	1,233 cubic meter per year (m³/yr)
	0.001233 cubic hectometer per year (hm³/yr)
1 miner's inch	.02-.028 c.f.s. (depending on state)
1 acre-foot (af)	325,851 gallons
	43,560 cubic feet
1 inch of rain	27,200 gallons per acre
	113 tons per acre
inch per hour (in/h)	meter per hour (m/h)
inch per year (in/yr)	25.4 millimeter per year (mm/yr)
mile perhour (mi/h)	1.609 kilometer per hour (km/h)
foot (ft)	0.3048 meter (m)
cubic foot per second (cfs)	0.02832 cubic meter per second (m3/s)
mile (mi)	1.609 kilometer (km)
acre	0.405 hectare (ha)
acre-foot (af)	1233 cubic meter (m3)

Appendix 4

BLM Fact Sheets on Western States Water Law

The following fact sheets were taken from the BLM (Bureau of Land Management) Web page and are dated August 15, 2001. Links to the fact sheets can be found at http://www.blm.gov/nstc/WaterLaws/abstract1.html (accessed September 2007). These fact sheets originally contained contact information, which has been deleted here, since similar information was already included in the section "Western State Water Departments and Permitting Agencies." The BLM provides fact sheets for the following western states: Alaska, Arizona, California, Colorado, Idaho, Montana, Nevada, New Mexico, Oregon, Utah, and Wyoming. A new fact sheet was created by the author of this text for Washington, based on the BLM format, and is not available from BLM.

ALASKA

WATER RIGHTS SYSTEM

Alaska water law is based on the doctrine of prior appropriation. Historically, there have been riparian rights in Alaska, but when the Alaska Water Use Act was passed in 1966, all riparian rights were converted to prior appropriation rights. Water for public water supplies may be granted as a preferred use in Alaska. This means that a prior appropriation water right is not absolute but may be subject to changes to meet public needs for domestic water use. If this occurs, the water right holder must be compensated for the loss. The state's water law is contained in the Alaska Water Use Act, Alaska Statute 46.15. Water rights are regulated by Alaska Administrative Code 11 AC 93.

RESPONSIBLE AGENCY

The Alaska Department of Natural Resources, Division of Mining, Land, and Water (the Division), administers water rights in Alaska. This agency is responsible for the appropriation and distribution of surface and ground water in the state.

APPLICATION PROCESS

The only way to establish a new water right in Alaska is to file an application to appropriate water. The types of applications that can be filed in Alaska can be seen in Appendix 1. The applicant is required to submit the application along with a filing fee to the Division of Mining, Land, and Water, at which time the application is indexed into a tracking system. The date when the application is filed is the priority date for the water right. Pending applications in Alaska are adjudicated in the order in which they are received. Public notice of an application is required in the following instances: If the proposed appropriation is over 5,000 gallons per day; if it comes from an anadromous fish stream (one in which fish migrate upstream from the sea to breed), or if the water source has a high level of competition. If notice is required, certified mailings are sent to current appropriators that may be affected by the new application, and to the Departments of Fish and Game and Environmental Conservation. In addition, legal notices are published in a local newspaper or post office for 15 days. Objections to the proposed appropriation can be directed to the division, and all objections are addressed in writing prior to the issuance of a permit.

When approving or rejecting an application, the division considers whether: rights of other appropriators will be affected; the proposed means of diversion are adequate; the proposed use of water is beneficial; and if the proposed appropriation is in the public interest (see Appendix 2 for criteria to assess the public Interest). After these considerations, the division issues a decision. If the applicant or objecting parties disagrees with the decision, an appeal can be requested. The appeal must be received within thirty days from the receipt of notification, and the division then holds a hearing on the objection(s).

When a permit is approved, a specific time period (usually two to five years) is granted within which to develop the project. Once the system is fully developed, the total amount of beneficially used water is established, and all permit conditions have been met, a Certificate of Appropriation is granted.

POINT OF DIVERSION AND CHANGE OF USE PROCEDURES

A Certificate of Appropriation can be amended to change the quantity of water, the legal description for the point of diversion, the type of use, the depth of taking, or to add take points. The division reviews the proposed change to determine the impact on other water users. If approved, a one-year permit is issued to make the change. If no objections to the change are filed within that year, the change becomes final.

STATE-RECOGNIZED BENEFICIAL USES

Alaska defines "beneficial use" to mean a use of water for the benefit of the appropriator, other persons, or the public, that is reasonable and consistent with the public interest, including, but not limited to:

Agriculture
Domestic

Fish and Wildlife
Fish and Shellfish Processing
Industrial
Irrigation Manufacturing
Mining
Navigation and Transportation
Power
Public
Recreation Uses
Water Quality

GROUNDWATER

Groundwater and surface water in Alaska are treated the same. They are considered conjunctive, and the administration and regulation of groundwater does not differ from surface water, except in one location. In the Critical Water Management Area around Juneau, there are additional regulations on ground water use relating to salt-water intrusion in the aquifier.

WATER RIGHTS

There are no restrictions in Alaska as to who can hold a water right. State law says any "person" can hold a water right and "person" is defined as "an individual, partnership, association, public or private corporation, state agency, political sub-division of the state, and the United States." A water right in Alaska is attached to the land where the water is being used. If the land is sold, the water right transfers with the land, unless a severance application is approved by the division. Water rights in Alaska can be transferred from one owner to another by being bought and sold or traded. The transfer of a water right, however, must be approved through the filing of a change application with the Division of Mining, Land, and Water. The approval criterion for a change application is that the change cannot harm another water user and it must be in the public interest.

A water right in Alaska can be lost by abandonment or forfeiture. Abandonment occurs when a water right holder voluntarily relinquishes his/her water right by submitting a notarized statement to the Division. A water right is lost by forfeiture if it is not used for five consecutive years. Water lost through abandonment or forfeiture reverts back to the state and is subject to future appropriation.

ADJUDICATIONS

In 1986, the Alaska Water Use Act was amended to establish procedures for basin wide adjudications in order to clarify water rights. Procedures were established for both administrative and judicial adjudications. Administrative adjudications are conducted by the Division of Mining, Land, and Water and results in a judicial decree, which is then submitted to the courts. A judicial adjudication involves federal reserved water rights. Although procedures for these adjudications have been established (and can be found in 11AC 93 0400), they have never been used. Alaska has yet to have a basin wide adjudication.

NUMBER OF ONGOING ADJUDICATIONS

There has never been a basin-wide adjudication in Alaska.

INSTREAM FLOWS

An instream flow amendment was added to Alaska's Water Use Act in 1980. This amendment allowed for the new appropriation of instream flows through reservations. An instream flow reservation may be established on any stream or body of water in Alaska that is not fully appropriated. Upon receiving an Application for Reservation of Water, the Division must establish that there is a need for the reservation, that there will be no adverse impacts on other water-right holders, and that the right is in the public interest. An assessment is also made to confirm that water is available for the reservation. Instream flow reservations differ from consumptive water rights, because they are subject to additional burdens of proof of beneficial use. An instream flow right is reviewed every ten years to determine if the reservation is providing a beneficial use. Depending upon the findings of the review, the instream flow reservation may be extended, restricted, or revoked.

The 1980 amendment also established a means for transferring a water right to an instream flow reservation. In order to do this, an application must be filed with the State Water Commission. A one-year permit is granted to allow other water users to object to the transfer to instream flow. If approved, the reservation retains the priority date of the original water right and becomes an established instream flow reservation subject to review every ten years.

RECOGNIZED BENEFICIAL USES FOR INSTREAM FLOW

In Alaska, permissible instream uses include:

Protection of fish and wildlife habitat, migration, and propagation
Recreation and parks
Navigation and transportation
Sanitation and water quality
Holdership of instream flow water rights:

Any "person" may apply for and hold an instream flow reservation. A "person" refers to any private individual, organization, or government agency as defined above.

QUANTIFICATION REQUIREMENTS AND PROCEDURES

In Alaska, there are no standard quantification requirements or procedures for the establishment of an instream flow right. In order to establish an instream flow there must be a justifiable quantification based upon the particular beneficial use. There is not, however, a standard method or procedure that must be used.

FEDERAL RESERVED WATER RIGHTS

Federal reserved water rights are included in basin-wide adjudications if the federal government consents to have its federal reserved water rights administratively adjudicated by the state. Forty-nine percent of Alaska is federal lands (of which 26.1 million acres are BLM-reserved land) and may have extensive federal reserved water rights.

Federal reserved water rights in Alaska are different from state appropriated water rights. They:

May apply to both instream and out-of-stream water uses

May be created without actual diversion or beneficial use

Are not lost by nonuse

Have priority dates established as the date the land was withdrawn

Are for the minimum amount of water reasonably necessary to satisfy both existing and foreseeable future uses of water for the primary purposes for which the land is withdrawn

Because most federal reserved water rights are not quantified, the Division does not know how much water is needed or used for the primary purposes of federal land withdrawals in Alaska. Although procedures have been established for the adjudication of federal reserved water rights, this process has not yet taken place. Because of the controversy surrounding the Alaska National Wildlife Refuge (ANWR), a temporary moratorium was placed on the processing of federal reserved water rights. That moratorium has been lifted, but no federal reserved water rights applications have been processed since the lifting.

BLM-SPECIFIC INFORMATION

The Division of Mining, Land, and Water does not require water rights applicants to have the necessary rights-of-way approval from the BLM approved prior to approving an application.

The BLM is required to pay filing fees in Alaska. The fee for an instream flow application is $500 per application.

The Bureau of Land Management in Alaska is applying for and holds federal reserved water rights for Wild and Scenic Rivers. The BLM State Office has submitted eight applications of which one is perfected and seven are pending. Apart from Wild and Scenic Rivers, the BLM does not have (and cannot apply for at this time) any other federal reserved water rights.

The relationship between the BLM and the State of Alaska (Division of Mining, Land, and Water) regarding water rights is tenuous at best. There is a good working relationship between individuals in both offices, but policy differences frustrate meaningful cooperation. Alaska is a strong proponent of states' rights and has conflicted with federal agencies over federal reserved water rights. In some cases, they do not recognize reserved rights to which federal agencies feel they are entitled. The state has even delayed in processing federal applications. These circumstances have affected the BLM to some extent but are mainly being played out between the state and the National Fish and Wildlife Service.

APPENDIX 1: TYPES OF APPLICATIONS

Application for Reservation of Water
Application for Water Rights
Application for Temporary Water Use
Change of Address for Water Rights
Change of Property Ownership for Water Rights
Notice of Relinquishment of Water Rights Form
Request for Water Right Permit Extension
Statement of Beneficial Use of Water

APPENDIX 2: CRITERIA TO ASSESS THE PUBLIC INTEREST

In determining the public interest, the Division of Mining, Land, and Water shall consider:

The benefit to the applicant resulting from the proposed appropriation
The effect of the economic activity resulting from the proposed appropriation
The effect on fish and game resources and on public recreational opportunities
The effect on public health
The effect of loss of alternate uses of water that might be made within a reason-
    able time if not precluded or hindered by the proposed appropriation
Harm to other persons resulting from the proposed appropriation
The intent and ability of the applicant to complete the appropriation
The effect upon access to navigable or public water

Accessed at http://www.blm.gov/nstc/WaterLaws/alaska.html, September 2007.

ARIZONA

WATER RIGHTS SYSTEM

Arizona's water law is based on the doctrine of prior appropriation, but it is administered based on a bifurcated system where surface water is regulated separately from ground water. There are basically four categories of water supplies available in Arizona: Colorado River water, surface water other than Colorado River water, ground water, and effluent. Each water supply is managed in a different manner. Colorado River water is allocated through the law of the river and Arizona's water banking program, surface water rights are based on "first in time, first in right," and groundwater rights vary depending on location. The Arizona water code is located in Title 45 of the Arizona Revised Statutes.

Responsible Agency

The Arizona Department of Water Resources (ADWR) is responsible for ensuring that dependable, long-term water supplies are available for Arizona. The ADWR oversees the use of surface and groundwater resources, administers state water laws (except those related to water quality), explores methods of augmenting water supplies to meet future demands, and works to develop public policies that promote conservation and distribution of water.

Application Process

In order to appropriate surface water, one must file an application with the Department of Water Resources (The types of applications available can be found in Appendix 1.). The application must describe the source of the water, the location of the proposed diversion, the proposed place of use, the beneficial use, and the proposed quantity and periods of use. Upon confirmation of completeness and correctness, the ADWR provides a public notice and there is an opportunity for public protest. Protests must allege that the proposed allocation will impair a prior water right, will be contrary to public interest, or will pose a threat to public safety. If the application is protested the ADWR can, but is not required to, hold a public hearing. After the protest period and any hearing, the ADWR may either grant or reject the application. If the application is approved, a permit is granted. In general a permit is granted if the application does not conflict with vested rights, is not a threat to public safety, and is not contrary to the interests and welfare of the public. The issuance of a permit allows the permittee five years to complete the necessary construction and to put the water to beneficial use. Upon putting the water to beneficial use, the water right is perfected and the permittee is granted a Certificate of Water Right.

In order to withdrawal groundwater inside an active management area (see below ground water discussion), an application must be filed with the ADWR. The permit process is the same for groundwater as it is for surface water. Once a permit is issued, the permittee can withdraw a specific amount of water, from a specific location, for a specified purpose. The groundwater withdrawal permit is limited in the duration of use, but the applicant may apply to renew the permit.

Point of Diversion and Change of Use Procedures

A point of diversion is required for all consumptive uses in Arizona, but may be changed through an application to the ADWR. A water right holder may also change the use of surface water, and this can be done in two ways. A new use may be added to the certificate while retaining the existing use, or the existing use can be changed to a new use. In order to add a new use, a permit must be applied for and obtained to appropriate the water for the new use. The application will be processed in the same manner as any other permit to appropriate surface water and the priority date of the new water right will be the date the application was filed. In order to change the existing use to a new use, the requirements depend upon on the use. If the existing use is for irrigation, domestic, or municipal use, the use may not be changed without the approval of the ADWR. If the existing use is for any other use, the ADWR must be notified of the change in use, but approval is not required. When a change in use has been effected, the new use retains the same priority date as the original use.

STATE RECOGNIZED BENEFICIAL USES

The following beneficial uses are recognized in Arizona:

Domestic
Municipal
Irrigation
Stockwatering
Power Mining
Recreation
Wildlife and Fish
Groundwater Recharge

GROUNDWATER

The separate administration of surface water and groundwater is a defining characteristic of water management in Arizona. The legal separation of these two types of waters requires a water manager to determine what type of water is at issue before determining which law is applicable. Historically, Arizonans have been pumping ground water faster than it is replaced naturally—a situation called overdraft. Because of the significant problems due to overdraft, the Arizona Ground Water Management Code (Code) was passed in 1980. The Code has three primary goals. The first is to control the severe overdraft currently occurring in many parts of the state. The second goal is to provide a means to allocate the state's limited ground water resources. The third goal is to augment Arizona's groundwater through water supply development.

To accomplish these goals, a comprehensive management framework was established within the Arizona Department of Water Resources. This management framework consists of three levels of water management to respond to different groundwater conditions. The lowest level of management includes general provisions that apply statewide. The next level applies to Irrigation Non-Expansion Areas (INAs). The highest level of management, with the most extensive provisions, is applied to Active Management Areas (AMAs) where groundwater overdraft is most severe. The boundaries of AMAs and INAs are generally defined by groundwater basins. There are currently five designated AMAs in Arizona and they are the areas surrounding Phoenix, Tucson, Pinal, Santa Cruz, and Prescott. INAs were established in rural farming areas where the groundwater overdraft problems are less severe. There are currently three INAs: Douglas, Joseph City, and the Harquahala INA. New AMAs and INAs can be designated by the ADWR, if necessary, to protect the water supply or on the basis of a public vote held by local residents of an area.

Outside of AMAs and INAs, groundwater may be withdrawn and used for reasonable and beneficial use without a permit. Use of this groundwater, however, does require the filing of a notice of intent to drill with ADWR. Within AMAs, ground-water use requires a permit. Groundwater withdrawal permits (which allow for new use of water) are limited to certain specified activities. Arizona groundwater law requires certain criteria to be met for each type of withdrawal before a permit can be issued. In addition to rights granted through permits, three other types of ground-water withdrawal rights exist within AMAs. The first is grandfathered ground water rights. These rights are based on historic use of groundwater for five years prior to the designation of the AMA. Most grandfathered rights are appurtenant to the land, but some are not and may be purchased or leased from the owner. Withdrawal rights can also be granted to municipal water providers, private water companies, and irrigation districts within AMAs, enabling them to provide service to their customers. Finally, small domestic wells are exempt from the regulations within an AMA. Users of small domestic wells may withdraw groundwater for non-irrigation purposes without a permit.

Groundwater use and management in each AMA is coordinated by a Ground Water Users Advisory Council appointed by the governor. These councils develop water conservation strategies within the AMA. The requirement of each AMA is to achieve a "safe yield," which occurs when the amount of water consumed from the aquifer equals the amount of water recharged to the aquifer.

WATER RIGHTS

Water rights in Arizona can be held by any legal entity. There are no restrictions on who can hold water rights, thus the owner can be an individual, group of individu-als (related or not), corporations, government agencies, etc. A surface water right is considered to be attached to the land, and therefore, may not be transferred without approval. The owner of a right must apply to the ADWR to sever and transfer the use of a water right to a new location. If the water right was granted for domestic, municipal, or irrigation use, the holder must be granted approval from the ADWR before changing the use of the water.

An owner of a water right may voluntarily abandon the right, or the right may be found to have been forfeited if no use is made of the water for five consecutive years. Water that is abandoned or forfeited reverts to the public and becomes available for new appropriation.

ADJUDICATIONS

General stream adjudications in Arizona are State Superior Court determinations of the status of all rights to surface water. These determinations are based upon state law and federal claims to water within the river system. The Department of Water Resources serves as the technical advisor to the State Superior Court and provides administrative assistance. Adjudications quantify and prioritize surface water rights within the watersheds. The goals of the adjudications are to assess all uses in priority and quantity for improved water management and to integrate federal reserved rights in the state allocation system.

Ongoing Adjudications

There are currently two adjudications pending in Arizona: the Gila River and the Little Colorado River.

Instream Flows

The Arizona Legislature amended its water codes to add wildlife and fish in 1941 and recreation in 1962 as uses for which "any person" could appropriate water. In 1976, the Arizona Court of Appeals, in *McClellan v. Jantzen* found that these amendments constituted instream appropriation and these rights could be held without a diversion.

Rights for instream flow can be obtained through new appropriation. The State has developed a process for instream flow appropriation that requires an applicant to collect at least one year of flow data in order to submit an application. In addition, the applicant must submit a report of the flow measurements and conclusions of expected benefits. Upon submitting this data, a temporary permit is issued with the requirement that an additional four years of flow data be collected. Once this data is collected, an instream flow right may be issued. The filing date of the application establishes the priority date as it does for other appropriation applications.

It is still unclear whether existing water rights can be transferred to instream flow rights. The law states that water rights may be "transferred for use for . . . wildlife purposes, including fish" (ARS 45-172), but no instream flow transfers have been attempted. There have been several temporary leases of stored water to augment stream flows, but these leases occurred outside of the transfer process.

Recognized Beneficial Uses for Instream Flow

Arizona recognizes stream flow maintenance to support wildlife, fish, and recreation as appropriate beneficial uses for instream flow.

Holdership of Instream Flow Water Rights

Both public agencies and private organizations may hold instream flow rights. Although not legislatively bound to do so, the Arizona Department of Water Resources has so far limited the ownership of instream flow rights in the private sector to public interest groups such as the Nature Conservancy.

Quantification Requirements and Procedures

For a more detailed description of Arizona's instream flow program as well as the quantification and monitoring methods and procedures, see "A Guide to Filing Applications for Instream Flow Water Rights in Arizona," Arizona Department of Water Resources, November, 1997.

Available at http://www.blm.gov/nstc/WaterLaws/arizona.html, September 2007.

CALIFORNIA

WATER RIGHTS SYSTEM

California's system of water rights is referred to as a "dual system" in which both the riparian doctrine and the prior appropriation doctrine apply to water rights. There is also a separate doctrinal basis for ground water, as well as pueblo rights, so a more accurate classification of California's system would be a "plural system." Water rights in California are use rights. All waters are the property of the state. A water right in California is a property right allowing the use of water, but it does not involve ownership of the water. California's water law is contained in the California Code of Regulations, Title 23, and can found at http://www.calregs.com/.

Riparian rights result from the ownership of land bordering a surface water source (a stream, lake, or pond). As a class, these rights are senior to most appropriative rights, and riparian landowners may use natural flows directly for beneficial purposes on riparian lands without applying for a permit (see Appendix 1 for Attributes of Riparian Rights).

Appropriative rights are acquired by putting surface water to beneficial use. Prior to 1914, appropriative rights could be claimed by simply diverting and using the water, posting a notice of appropriation at the point of diversion, and recording a copy of the notice with the County Recorder. Since 1914, the acquisition of appropriative rights has required an application through the State Water Board.

In addition to riparian and appropriative rights, California recognizes pueblo rights. These rights are derived from Spanish law whereby Spanish or Mexican pueblos could claim water rights. As a result, pueblo rights are paramount to the beneficial use of all needed, naturally occurring surface and subsurface water from the entire watershed of the stream flowing through the original pueblo. Water use under a pueblo right must occur within the modern city limits, and excess water may not be sold outside the city. The quantity of water available for use under a pueblo right increases with population and with extensions of city limits. In general pueblo rights are limited to use of water for ordinary municipal purposes.

Responsible Agency

Responsibility for water in California is shared among several agencies. The State Water Resources Control Board (State Water Board) is responsible for the water rights and water quality functions of the state. They have the jurisdiction to issue permits and licenses for appropriation from surface and underground streams. The board also has the authority to declare watercourses fully appropriated. The California courts have jurisdiction over the use of percolating ground water, riparian use of surface waters, and the appropriate use of surface waters initiated prior to 1914. The Department of Water Resources is responsible for planning the use of state water supplies, and develops, in consultation with the California Water Commission, rules and regulations for this purpose.

Application Process

Any entity intending to appropriate water is required to file an application for a water right permit (or a use registration for small scale domestic use) with the State Water Board. A list of available applications can be seen in Appendix 2. A permit is not required from riparian right holders, ground water users, users of purchased waters, or those who use water from a spring or standing pool lacking a natural outlet on the land they are located. Once the application or registration has been accepted, a priority is established in relation to other appropriators. For domestic registration, the State Water Board provides a Certificate of Registration, which establishes general conditions under which the diversion may be made. When an application for a water right permit is filed, public notice is given to interested parties. This indicates an opportunity to file protests against the proposed application. If differences cannot be resolved, either a field investigation (for small applications requesting 3cf or 200 acre-feet per year) or a State Water Board hearing is conducted.

An application for a new water appropriation is approved if it is determined to be for a useful or beneficial purpose and if water is available for appropriation. In evaluating an application, the Board considers the relative benefits derived from the beneficial uses, possible water pollution, and water quality. If a permit is approved, it may be approved in full or it may be subject to specified conditions. A decision or order from the State Water Board is reviewable by the Superior Court. Once the State Water Board issues a permit, the use and diversion of water is authorized (see Appendix 3 for a summary of the steps to obtain a permit).

Once the permittee completes the necessary works, the water is put to full beneficial use, and all terms and conditions are met, a license is issued. The license is the final confirmation of an appropriative right and it remains in effect as long as the license conditions are met and the water is put to beneficial use.

The time frame involved in obtaining a license in California is highly variable. Permit decisions are required to be reached within six months on accepted applications for nonprotested projects, which do not require extensive environmental review. Applications with unique requirements for environmental review and/or require protest resolution, may extend the time frame by months and even years.

POINT OF DIVERSION AND CHANGE OF USE PROCEDURES

In 1928, the California Constitution was amended to require reasonable diversion and use in the exercise of all water rights. The only exception to the point of diversion requirement is for instream flow rights. The State Water Board and the courts have concurrent jurisdiction to apply and enforce diversion and use requirements. The holder of an appropriative right may change the point of diversion, place of use, or purpose of use, so long as other rights are not injured by the change. In order to change an attribute of a water right in California, a change application must be filed with and approved by the State Water Board. Change applications follow the application process described above.

STATE RECOGNIZED BENEFICIAL USES

Beneficial uses in California include the following:

Aquaculture—Raising fish or other aquatic organisms not for release to other waters

Domestic—Water used by homes, resorts, or campgrounds, including water for household animals, lawns, and shrubs

Fire Protection—Water to extinguish fires

Fish and Wildlife—Enhancement of fish and wildlife resources, including raising fish or other organisms for scientific study or release to other waters of the state

Frost Protection—Sprinkling to protect crops from frost damage

Heat Control—Sprinkling to protect crops from heat

Industrial Use—Water needs of commerce, trade, or industry

Irrigation—Agricultural water needs

Mining—Hydraulicking, drilling, and concentrator table use

Municipal—City and town water supplies

Power—Generating hydroelectric and hydromechanical power

Recreation—Boating, swimming, and fishing

Stockwatering—Commercial livestock water needs

Water Quality Control—Protecting and improving waters that are put to beneficial use.

GROUNDWATER

The vast majority of California's groundwater is unregulated. The state does not have a comprehensive groundwater permit process to regulate ground water withdrawal.

There are three legally recognized classifications of groundwater in California: subterranean streams, underflow of surface waters, and percolating groundwater. Subterranean streams and underflow of surface waters are subject to the laws of surface waters and are regulated by the State Water Board. Percolating groundwater, on the other hand, has few regulation requirements.

Percolating groundwater has two subclassifications: overlying land use and surplus groundwater. Landowners overlying percolating groundwater may use it on an equal and correlative basis. This means that all property owners above a common aquifer possess a shared right to reasonable use of the groundwater aquifer. These rights are similar to riparian rights and since they are correlative, a user cannot take unlimited quantities without regard to the needs of other users. Surplus groundwater may be appropriated for use on nonoverlying lands, provided such use will not create overdraft conditions. A permit is not required to use percolating groundwater of either classification, but the appropriation of surplus groundwater is subordinate to the correlative rights of overlying users.

WATER RIGHTS

Water rights in California can be held by any legal entity. There are no restrictions on who can hold water rights, thus the owner can be an individual, related individuals, non-related individuals, trusts, corporations, government agencies, etc. Water rights are considered real property (they can be owned separately from the land on which the water is used or diverted) and can be transferred from one owner to another, both temporarily or permanently. Any transfer (sale, lease, or exchange) is subject to approval by the State Water Board through the application process discussed above. Approval is granted upon finding that the transfer would not result in injury to any other water right and would not unreasonably affect fish, wildlife, or other instream beneficial use.

An appropriative water right in California can be maintained only by continuous beneficial use, and can be lost by five or more continuous years of nonuse. Riparian rights, on the other hand, cannot be lost through nonuse. Appropriative rights can also be lost through abandonment, but to constitute abandonment of an appropriative right, there must be the intent not to resume the beneficial use of the water right. As a result, abandonment is always voluntary. The rights to waters lost through abandonment or nonuse revert to the public, but only after notice has been given and a public hearing is held.

ADJUDICATIONS

In California, adjudication can be initiated through the court or through statutory procedures. Court initiated adjudication occurs when a water right lawsuit is filed in court (all surface and ground water rights may be included in this procedure). In the case of a court initiated adjudication, the court often asks the State Water Board to act a referee and to conduct an investigation and report back. Statutory adjudications result when one or more entities claim a right from a specific source and file a petition with the State Water Board. The statutory procedure can be used to determine all rights to any body of water including percolating groundwater. The result of a statutory adjudication is a decree that integrates all rights on the water source and sets quantity, season, priority, etc.

Ongoing Adjudications

As of 2000, sixteen basins in California had been adjudicated.

Instream Flows

In 1991, California adopted changes to its water laws that permitted the transfer of existing consumptive water rights to the purpose of instream flow. These transfers can be made for the purposes of enhancing wetlands habitats, enhancing fish and wildlife resources, or increasing recreation on the water. California law allows transfers to be either permanent or temporary changes in use; therefore instream flow rights can be both purchased and leased. New instream flow rights retain the priority date of the original right.

California state law does not permit new appropriations of water for instream flow. The State Water Board may attach conditions requiring bypass flows to new consumptive use appropriations, but these conditions do not constitute newly appropriated instream flow rights. When a new water use permit application is submitted, the State Water Board must notify the Department of Game an Fish, which has the authority to recommend amounts of water necessary to preserve fish, wildlife, and recreation in the affected stream. The board then considers these recommendations and may set instream flow requirements as conditions for the new permit. In this way, current flows can be protected even though new appropriations for instream flow rights are prohibited.

Recognized Beneficial Uses for Instream Flow

Recognized beneficial uses of instream flow in California include enhancing wetlands habitat, enhancing fish and wildlife resources, increasing recreation on the water, and protecting water quality.

Holdership of Instream Flow Water Rights

Under California law, any "person" (public or private) may hold an instream flow right, as long as that right was established through a legal transfer.

BLM-Specific Information

The application process in California has proven to be expensive for the BLM. For the appropriation process, the BLM pays $1,050, which includes the following: $100 application fee, $850 environmental filing fee, and $100 upon issuance of the permit. Since 1991, water right applicants have been required to pay an $850 environmental filing fee to the California Department of Fish and Game (CDFG) with each application. This is a concern for the BLM because the CDFG's review is redundant to the

BLM's NEPA process. The 1991 memo introducing the fee states that "these fees are not intended to reimburse costs specifically identifiable to individual projects, but rather to offset a relative portion of the cumulative effect of all projects." Therefore the BLM cannot request a waiver of this fee. In terms of other water rights applicants, the BLM is required to approve the necessary right-of-ways prior to the approval of the application by the state.

The BLM is not currently involved in any of California's adjudications. In the past, however, the BLM has been involved in the Eagle Lake and Alturas adjudications.

Regarding federal reserved water rights, the BLM California holds a number of PWRs. In order to assert a PWR 107 or other PWR, the BLM provides notice to the State of California. In the past fifteen years, there have been relatively few PWR assertions in California and the extent of unasserted PWRs is unknown. There are a number of PWRs that are included in the Master Title Plans on BLM lands, and these probably originated from assertions made before the early 1980's. The BLM does not have any federal reserved water rights on Wild and Scenic Rivers or on wilderness areas in California.

The relationship between the BLM and the State of California is very close and cooperative. The staff of the Division of Water Rights have been especially helpful to the BLM in interpreting the details involved in each particular water right decision. The staff has a practical mindset and helps the BLM achieve their goals. The BLM also commonly assists the state in their capacity surveys for the BLM reservoirs, which are moving from permit to license. This has expedited the process. The state has also been quite receptive to suggestions from the BLM for streamlining some of the water right reporting requirements.

APPENDIX 1: ATTRIBUTES OF RIPARIAN RIGHTS

Riparian rights are of equal priority.

Unless adjudicated, the right is not quantified; rather, it extends to the amount of water that can be reasonably and beneficially used on the riparian parcel.

Riparian rights are correlative. During times of water shortage, the riparian proprietors share the shortage.

Water may be used only upon that portion of the riparian parcel, which is within the watershed of the water source.

The riparian right does not extend to seasonal storage of water.

The riparian right is part of the riparian land and cannot be transferred for use on other lands.

The riparian rights remains with the land when riparian lands are sold.

When riparian lands are subdivided, parcels which are severed from the adjacent water source lose their riparian rights, unless the rights are reserved.

A riparian right is not lost by nonuse.

APPENDIX 2: TYPES OF APPLICATIONS

Water Right Application Form
Environmental Information Form

Notice of Assignment Form
Agent Assignment Request Form
Application Protest Form
Cancellation of Application Form
Registration Form
Notice of Assignment Form
Complaint Form
Answer to Complaint Form
Petition for Extension of Time Form
Petition for Correction Form
Petition for Change Form
Petition for Change in Distribution of Storage Form
Petition for Protest Form
Notice of Assignment Form
Request for Revocation Form
Petition for Temporary Permit Form
Petition for Temporary Urgency Change Form
Temporary Transfer
Long term Transfer
Wastewater Change Petition Form

APPENDIX 3: STEPS TO OBTAIN A PERMIT

Step Board's Role Applicant's Role

File Application

If you need assistance, Board engineers will help you prepare application forms, small project maps, and other documents. Incomplete applications won't be accepted. You prepare an application, which meets specific requirements, including a filing fee.

Acceptance of Application

Board notifies you within 30 days that either your application is incomplete or that it has been accepted. Acceptance of your application establishes your priority as the date of filing. Unless you are granted an extension, you must provide any additional information requested by the Board within 60 days of notification. If not, your application may be canceled.

Environmental Review

Your proposed project is assessed to determine to what extent it could alter the environment. You assume cost for preparation of any required environmental studies.

Public Notice

The Board will send you a public notice describing your proposed project. Copies of the notice are also sent to known interested parties and to post offices in the area of your project for posting. For small projects, you must post the notice for 40 consecutive days in two conspicuous places near your project site. For large projects, you must publish the notice in a newspaper at least once a week for three consecutive weeks.

Protests

During the noticing period, the Board may receive protests against your proposed project from interested individuals or groups. If protests are filed against your application, you must respond to them in writing and attempt to reach agreements so that protests can be withdrawn.

Hearings

If protests cannot otherwise be resolved, you and the protestant present your cases at a field investigation or during a hearing conducted by the Board. The Board issues a decision on protested applications based on information gathered at the field investigation or on evidence presented during the hearing. You prepare testimony and exhibits for presentation at the hearing and cooperate with the Board and protestant toward reaching a satisfactory resolution.

Permit Issuance

A water right permit is issued when protests, if any, are resolved or dismissed, or when the Board approves the application by decision following a hearing. In addition, a permit fee must be paid. During this phase, the Board determines whether water conservation measures are needed. Prior to issuance of a permit, you must submit a permit fee as directed by the Board. If water conservation measures are required, they will be included as a condition of your permit.

Available at http://www.blm.gov/nstc/WaterLaws/california.html, September 2007.

COLORADO

WATER RIGHTS SYSTEM

Colorado water law is based upon the doctrine of prior appropriation or "first in time — first in right," and the priority date is established by the date the water was first put to a beneficial use. Colorado water law is contained in the State Constitution Article XVI sections 5 and 6 and in the Colorado Revised Statutes, sections 37, articles 80 through 92.

RESPONSIBLE AGENCY

There is not a single state agency in Colorado responsible for issuing water rights. Water rights in Colorado are established through a water court system. There are seven water courts, one for each major river basin, which adjudicate water rights throughout the state. Each water court has an appointed water judge and water referee who hear all water related matters within their jurisdiction. The State Engineer administers and distributes the state's waters. The State Engineer is also responsible for issuing and denying permits to construct wells and divert groundwater, but these permits do not constitute rights to groundwater. The Colorado Ground Water

Commission (Commission) is a regulatory and an adjudicatory body authorized to manage and control designated groundwater resources. Finally, the Colorado Water Conservation Board (CWCB) oversees conservation and development in the state and is responsible for the state's instream flow program.

Application Process

Water rights in Colorado are established through a water courts system. Every water right application must go through the water courts, and must be handled by an attorney. Therefore, Colorado has a very large attorney workload relating to water rights.

In order to obtain a right to either surface or groundwater, an application must be filed with one of the seven water courts in the state. A list of applications can be seen in Appendix 1. The application must be filed in the division in which the diversion is located. Once an application is filed with the appropriate court, a summary (or the application in full) is published in "the resume" (publication in the resume is considered proper notice to all water rights holders). The resume contains all applications filed with a particular court each month. All applications are also published in a local newspaper.

Upon publication in the resume and paper, a statement of opposition can be filed by any person. Oppositions must be filed within a forty-five day period following publication. Any statement of opposition must outline the reasons why an application should not be granted or should be amended. At the end of the month following the month of publication of the application, the water referee examines the application and the statements of opposition. The referee consults with the division engineer and within thirty days, the engineer files a written report containing the recommendations. This report is sent to the applicant, who must then mail copies to all parties in the case. The referee can then either approve or disapprove (in whole or in part) the application. If no protest is filed before the twentieth day following the mailing of the referee's ruling, the ruling is signed by the judge and entered as a decree of the court.

Protests to the referee's ruling, however, can be filed with the court. If a protest is filed, a hearing is held before the water judge. Applications can also be referred to the water judge directly by the referee and engineer. When a matter goes to the water judge, a trial is set and the case proceeds before the water judge who either grants or denies the water right based upon factual issues in the case and how they relate to statutory and case law criteria. A granted water right is considered a "decreed water right."

Water rights in Colorado (both surface and groundwater) can be either absolute or conditional. An absolute right is water that has been diverted and put to beneficial use. A conditional right is a right that will be developed in the future. A conditional right maintains its priority until the project is complete. In order to maintain a conditional water right, the owner must file an application for a finding of reasonable diligence every six years with the water court. The applicant must prove that he or she has been diligently pursuing completion of the project. Upon completion, the owner of a conditional right may file for an absolute water right, and that absolute water right contains the appropriation date for which the conditional right was awarded.

The time frame to obtain a water right in Colorado varies depending upon the caseload of the specific water court.

POINT OF DIVERSION AND CHANGE OF USE PROCEDURES

Appropriations of water are made when an individual physically takes the water from a stream and transports it to another location for beneficial use. The use of water directly from a stream, such as by wildlife or livestock drinking, is considered a diversion in Colorado. A point of diversion is required for all water rights in Colorado except for instream flow. Instream flow rights, however, can only be held by the CWCB.

The point of diversion, location of use, and type of use of a water right can be changed through an application with the appropriate water court. In order to change a water right, the applicant must provide evidence that the change will not injure the vested water rights of other users.

STATE RECOGNIZED BENEFICIAL USES

Beneficial use in Colorado is statutorily defined as "the use of that amount of water that is reasonable and appropriate under reasonably efficient practices to accomplish without waste the purpose for which the appropriation is lawfully made." Specific uses are not described, but previous categories have included:

Aesthetics and Preservation of Natural Environments
Augmentation
Commercial
Domestic
Fire Protection
Fishery
Geothermal
Groundwater Recharge
Industrial
Irrigation
Livestock
Minimum Flow
Municipal
Power
Recreation
Silvicultural
Snowmaking
Wildlife Watering
Wildlife Habitat

GROUNDWATER

A modified form of prior appropriation governs the establishment and administration of groundwater rights in Colorado. Colorado groundwater use is governed by the

Ground Water Management Act of 1965, which was adopted to allow the full economic development of water resources while protecting the rights of senior appropriators. Colorado considers all water within the state to be tributary to a surface stream, unless the water applicant can prove otherwise in water court. The test for establishing a nontributary source of water is very rigorous. The proposed diversion cannot deplete surface streams more than 1/10 of 1 percent of the proposed diversion volume in any single year for up to 100 years. When a nontributary aquifer is established by law, the water in the aquifer is allocated based on the percentage of land owned on the surface above the aquifer. If the applicant cannot establish nontributary groundwater, then the use of groundwater falls under the prior appropriation system and water rights must be obtained through the court system described above.

In addition to the application process through the courts, groundwater in Colorado is subject to further restrictions administered by the State Engineer. By law, every new well in the state that diverts groundwater must have a well permit. Exempt Well Permits, however, can be obtained for wells that pump less that 15 gallons per minute. For these wells, the state will give well permits that are exempt from the priority system. In order to obtain a permit, a person must file an application for approval of a permit with the State Engineer. A permit must be obtained from the state engineer prior to any utilization of groundwater, but the permit does not constitute a water right to the groundwater. A ground water right can only be obtained through the formal application to a water court. The water court, however, cannot grant a groundwater right until the State Engineer has issued a permit.

In addition to issuing permits, the State Engineer also provides staff assistance and technical support to the Colorado Groundwater Commission. The Commission is responsible for the management of designated basins located primarily in the eastern plains. The Commission's duties are to administer groundwater rights, work towards water conservation, and to protect vested water rights. The commission also establishes pumping levels in the designated basins that will not deplete ground water supplies at an excessive rate. Currently, the Commission has established eight designated basins (Kiowa-Bijou, Southern High Plains, Upper Black Squirrel Creek, Lost Creek, Camp Creek, Upper Big Sandy, Upper Crow Creek, and Northern High Plains). Within each basin, Groundwater Management Districts (GWMDs) can be formed. The GWMDs are authorized to adopt additional rules and regulations to help administer groundwater within their district. There are currently 13 Groundwater Management Districts within the basins.

Water Rights

A water right in Colorado can be held by any legal entity. In other words, a water right can be held by an individual, group of individuals, organization, corporation, government agency, etc. The only restriction to who can hold a water right concerns instream flow rights which can only be held by the CWCB.

Water rights in Colorado are considered real property and can be bought, sold, and leased to other entities. Although water is considered to be the property of the state, a property right exists in the priority to the use of water. The transfer of a water requires filing a change of water right application with the appropriate water court. As with a change of use or point of diversion application, the applicant must provide evidence that the transfer will not injure the vested water rights of other users.

A conditional water right can be considered abandoned if the holder fails to show diligence to complete the necessary project. Any water right can be considered abandoned if it is not used for a period of ten years. Abandonment, however, must include the finding of an intent to abandon and, as a result, water rights in Colorado cannot be forfeited without proof of intent.

ADJUDICATIONS

Water rights in Colorado are adjudicated by the district water courts. Colorado has a process of individual adjudications where each right is adjudicated as it is approved. There are no general or basin wide adjudications in Colorado.

ONGOING ADJUDICATIONS:

Colorado does not have any ongoing general adjudications. Each of the seven water courts, however, have ongoing adjudications for all water rights within their jurisdiction.

INSTREAM FLOWS

In 1973, Colorado adopted legislation that recognized the maintenance of instream flows as a beneficial use of water. This legislation said that instream flow could be used "to preserve the natural environment to a reasonable degree," and it removed the requirement of a diversion to appropriate water. This established Colorado's instream flow program, and the CWCB has the exclusive responsibility for the protection of instream flows. In 1986, the instream flow legislation was amended to authorize the CWCB to acquire water rights for instream flows by methods other than appropriation. The CWCB is now allowed to acquire senior water rights through lease, purchase, or donation.

The CWCB is the only entity that may hold instream flow rights. It can apply for new appropriations through the state water courts. In order to do this, the board must ensure that a natural environment exists and will be preserved by the water available for appropriation and it must analyze the extent of the benefits of the water. The public has an opportunity to review and comment on the recommendations. The board then submits an application to the state water court. If granted the priority date for the instream right becomes the application date.

In addition to new appropriations, the CWCB can acquire water rights from other entities for instream flow. An existing consumptive right can be obtained by the board (through purchase, lease, or donation) and changed to an instream flow right. The CWCB is required to request recommendations for instream flow from the state Division of Wildlife, the Division of Parks and Outdoor Recreation, and from the U.S. Departments of Agriculture and Interior.

Recognized Beneficial Uses for Instream Flow

Instream flows in Colorado must be used to preserve the natural environment. Although the law authorizes a wide range of uses for instream flow, to date, the CWCB has acted only to protect streams that support fisheries.

Holdership of Instream Flow Water Rights

The CWCB is the only entity that can hold an instream flow right. Other entities, however, can acquire an existing right and transfer it to the board for instream flow.

Quantification Requirements and Procedures

In order to quantify an instream flow water right, the CWCB requires a multiple cross-section survey using the R2Cross methodology, averaging the survey results, and providing a written quantification recommendation to the board.

BLM-Specific Information

A water right applicant in Colorado does not have to have an approved right-of-way from the BLM in order to obtain an approved application. The BLM can challenge the applicant on land access issues in water court, and they can argue in court that the applicant does not have land access. If the applicant cannot prove that land access is available, the water court will dismiss the case.

The BLM is required to pay filing fees in Colorado. Filing fees are $45 for opposition, $91 for water rights application, and $150 for change of water right. In addition, the BLM must pay to have an application published in a local newspaper. This cost can range from $30 to $900, depending upon the size of the water right application and the number of newspapers.

Besides Colorado's instream flow program, the BLM can deny or condition rights-of-way in order to protect instream flows. The BLM establishes agreements with the owners of water diversions and reservoirs to protect stream flows. In addition, the BLM can designate areas as "Areas of Critical Environmental Concern," which makes the process for opposing water usage applications easier.

With regard to federal reserved water rights, the BLM has no designated Wild and Scenic Rivers in Colorado. Several rivers on BLM land have been studied and determined to be suitable, but no designations have been made. The BLM has some designated wilderness areas in Colorado, but the legislation that created them expressly stated that no reserved water rights were created. The BLM has completed the adjudication of all of its Public Water Reserves and holds approximately 1,400.

The BLM has an excellent relationship with the Colorado state government on water rights issues. However, the BLM is very disappointed with the implementation of the state's instream flow program. The legislation that authorizes the program is very broad, and enables the CWCB to protect a wide range of water-dependent values. The CWCB to date, however, has acted only to protect cold-water fisheries, and in a few cases, warm water fisheries and riparian values (but only when the

CWCB was placed under extreme pressure to do so). Colorado's Governor Owens is appointing increasingly conservative members to the board, meaning that the BLM's disappointment with the board is likely to increase.

APPENDIX 1: TYPES OF APPLICATIONS

Application for surface water right
Application for groundwater right
Motion to Intervene—A legal motion
Application for change in water right
Application for approval of plan for augmentation
Statement of Opposition—A legal motion

Available at http://www.blm.gov/nstc/WaterLaws/colorado.html, September 2007.

IDAHO

WATER RIGHTS SYSTEM

The doctrine of prior appropriation or "first in time—first in right" is the basis for administering water rights in Idaho. The constitution and statutes of Idaho declare all the waters of the state, when flowing in their natural channels, including the waters of all natural springs and lakes, within the boundaries of the state, and groundwaters of the state, to be public waters. A water right is the right to divert the public waters and put them to a beneficial use, in accordance with one's priority date. Idaho's water laws are contained in Idaho Code, Title 42, and can be found at: http://www3.state.id.us/idstat/TOC/42FTOC.html

RESPONSIBLE AGENCY

The Idaho Department of Water Resources (IDWR) is the agency responsible for the allocation of surface and groundwater within the state. The IDWR is also responsible for assisting the courts in the adjudication of water rights, processing change applications, and enforcing the state's water laws. In order to accomplish these tasks, as well as coordinate the management of the state's water, the IDWR has divided the state into more than 50 administrative basins. In addition to the IDWR, the Idaho Water Resource Board, an eight-member board appointed by the Governor and confirmed by the state senate, assists in the management of the state's water. The board provides guidance to the IDWR, is responsible for administering certain water programs, and is responsible for applying for and holding new appropriations for instream flow rights.

APPLICATION PROCESS

Since May 20, 1971, the only one way to establish a water right is by following the application/permit/license procedure discussed below. Prior to May 20, 1971, rights to surface waters were established by simply diverting water and putting it to beneficial use. These water rights are called "beneficial use," "historic use" or "constitutional" water rights. The priority date for a water right established by this method is the date water was first put to beneficial use.

Today, a new water right must be established by filing an application with the IDWR. Small domestic uses of groundwater and instream livestock water, however, are exempt from the permit application process. The application, which must quantify and describe the new use, is to be filed with the IDWR, and the filing date establishes the priority date of the water right. Upon receipt of the application, the IDWR notifies to the public by publishing a notice in a local newspaper for two consecutive weeks. For large applications, a notice is also published in a major newspaper in each region of the state. Protests can be filed against the application for a period of thirty days after the final legal notice appears. Protests are accepted from water users and any other entity concerned about the application. Efforts are made to resolve the dispute informally, but if this cannot be achieved, a hearing is scheduled. Hearings are held in accordance with the Idaho Administrative Procedure Act. An application can only be approved if it meets the following criteria:

The new use will not damage existing water rights.
The water supply is sufficient for the purpose of the new use.
The application is made in good faith and is not speculative.
The applicant has sufficient resources to complete the project.
The new use does not conflict with local public interests.
The project is consistent with the conservation of water in Idaho.
A portion of the Snake River Basin is held in trust by the State for the Idaho
    Power Co.

Applications in this area are subject to additional criteria (see Appendix 1). Once a decision on the application is made by the director of the IDWR, any dissatisfied party may appeal the decision. Appeals are handled by a judicial review, which is based on the record created in the administrative hearing. Once an application is approved, it is called a "permit." Upon receipt of a permit, the permittee has up to five years to submit proof of beneficial water use. Upon receipt of proof, the IDWR conducts an investigation and then issues a license. A license issued by the state is evidence of a water right. The types of applications that can be filed in Idaho can be seen in Appendix 2.

The time frame to obtain a water right in Idaho is extremely variable. In the best-case scenario, the minimum time it could take for an application to be approved and a permit to be issued is forty-five days. If the application is protested or there are other complications, however, it can take much longer. Even after a permit is issued, a licensed water right is not obtained until the permittee submits proof of beneficial use. This time-frame depends upon the work involved, but must be completed within five years.

Point of Diversion and Change of Use Procedures

A diversion is generally required to establish a water right. The Idaho Water Resources Board, however, is authorized to acquire water rights without diversions for instream flow. A water right may also be acquired without a diversion to water livestock directly from the stream. These rights are called "instream livestock" water rights.

The place of use, period of use, purpose of use, or the point of diversion can be changed by filing a change application with the IDWR. The change procedures are similar to those for an application for new appropriation, but the decision criteria are different. The IDWR may approve a proposed change if it:

Will not result in injury to the rights of other water users.
Does not constitute an enlargement of the original water right.
Is in the local public interest.
Is consistent with the conservation of water resources within Idaho.
The IDWR may deny the application or approve the change in whole, in part, or approve it subject to conditions necessary to meet the four criteria described above.

STATE RECOGNIZED BENEFICIAL USES

Aesthetics
Aquatic Life
Commercial
Cooling
Domestic
Fire Protection
Fish Propagation
Groundwater Recharge
Industrial
Irrigation
Manufacturing
Mining
Municipal
Navigation and Transportation
Power
Recreational Use
Stock watering
Water Quality Control
Wildlife

GROUNDWATER

The application for a groundwater right follows the same application/ permit/ license process as that for surface water. Prior to March 25, 1963, rights to groundwater could be established simply by putting the water to beneficial use. Today, the only exception to the application process for groundwater is a "beneficial use" right for domestic purposes. "Domestic purpose" is limited mainly to single-family domestic purposes, but is defined by statute as "water for homes, organization camps, public campgrounds, livestock and for any other purpose in connection therewith, including irrigation of up to one-half acre of land, if the total use is not in excess of 13,000 gallons per day, or any other uses if the total use does not exceed a diversion rate of 0.04 cubic feet per second and a diversion volume of 2,500 gallons per day." The exemption from the application process for domestic purposes does not included water for "multiple ownership subdivisions, mobile home parks, commercial or business establishments."

Idaho policy states that groundwater is to be managed to allow full economic development while protecting prior right holders. The pumpage from an aquifer is to be limited by the IDWR to prevent its mining. Low-temperature geothermal water (85 to 212 degrees F) is classified as groundwater and managed accordingly. Geothermal sources greater than 212 degrees F are managed under the geothermal statutes.

WATER RIGHTS

Any entity can hold a water right in Idaho. The water right can be in the name of an individual, group of individuals, organization, corporation, government agency, etc. The Idaho water board, however, is the only entity that may apply for and hold instream flow water rights. The holder of a water right in Idaho is considered to have established a real property right to that water, much like property rights for land. The constitution and statutes of the state of Idaho protect water rights as private property rights, and those rights can be bought and sold. Idaho has a thriving water market. Water rights can be transferred directly between individual buyers on a permanent basis. This requires filing change of owner and change of use applications with the IDWR. Water rights can also be transferred on a temporary basis through Idaho's water banking program. Idaho water banks are operated by the Water Resource Board and help to facilitate temporary water transfers. If a water right holder has excess water, that water can be deposited in the water bank. An entity that needs water may then rent that water on a one-year basis paying the water right holder a fixed price depending upon the purpose and location of use. Water banks are set up according to water districts, and priority is given to irrigation.

A water right can be lost in Idaho by abandonment or forfeiture. Abandonment requires proof of intent, where as forfeiture occurs if the water right is not used for five consecutive years. Water rights lost through abandonment or forfeiture revert back to the state for further appropriation.

ADJUDICATIONS

The state's district courts are responsible for the general adjudication of Idaho's watersheds. An adjudication is a court action for the determination of existing water rights that results in a decree that confirms and defines each water right. The application/permit/ license procedure described above is for the purpose of establishing new water rights. Adjudications in Idaho involve both surface and groundwater.

When an adjudication of a particular source is commenced, the IDWR is required to notify the water users of the commencement of the adjudication, and notify them that they are required to file a "notice of claim" to a water right with the IDWR. The IDWR then investigates the notices of claim and prepares a report that is filed with the court. Claimants of water rights are notified of the filing of the report and objections may be filed with the court by anyone who disagrees with the findings. If no objection is filed to a water right described in the report, the court decrees the water right as described in the report. If an objection is filed to a water right, the court then determines the right after a hearing and decrees the water right. Because water rights in Idaho could be establish without a permit until 1971, there are many unrecorded, yet valid, water rights in the state. A general adjudication of the Snake River Basin in Idaho is currently ongoing.

Although a "notice of claim" is required in an adjudication, there is another type of claim that may be filed with the IDWR. A "statutory claim" is filed with the IDWR to make a record of an existing beneficial use right. In 1978 a statute was enacted requiring persons with beneficial use rights (other than water rights used solely for domestic purposes as defined above) to record their water rights with the IDWR. The purpose of the statute was to provide some means to record of water rights for which there were previously no records. However, these records are merely affidavits of the water users, and do not result in a license, decree, or other confirmation of the water right.

ONGOING ADJUDICATIONS

The Snake River basin adjudication is the only general adjudication currently being conducted in Idaho. This adjudication began in 1987, and is one of the largest general adjudications in the country. Geographically, it involves thirty-eight of the forty-four counties in Idaho and accounts for abut 87 percent of the state's water rights.

INSTREAM FLOWS

The instream flow program in Idaho is complex and evolving. Instream flows were first recognized in the state in 1974 through legislation that established instream flow as a beneficial use. In that same year the Idaho Supreme Court (in *State Department of Parks v. Idaho Water Resources Department*) confirmed that an appropriation of water does not require a physical diversion. Idaho's instream flow program was further developed in 1978 when the state legislature adopted the Minimum Stream Flow Act. This act allows the Idaho Water Resources Board to apply for and hold minimum stream flow rights through new appropriation. Idaho law is clear in the establishment of instream flow through new appropriation. The Water Resource Board is responsible for filing an application with the IDWR. The application and processing procedures are similar to those for other appropriations except that a hearing must always be held. The decision criteria used by the IDWR are as follows:

Is the requested flow the minimum needed for the purpose requested?
Will the requested flow interfere with any existing water right?
Is it in the public as opposed to a private interest?
Is it necessary for the purpose requested (does it meet the beneficial use)?
Can the flow be maintained?

All instream right applications must be reviewed by the state legislature, which has the authority to accept, reject, or amend the approved application.

The law regarding the establishment of instream flow rights through transfers is less clear. Instream flows can be established through water right transfers, but are limited to temporary transfers of storage rights. Storage water rights can be leased on an annual basis through the state's water banking program. This method has been used effectively in the past by the federal government to augment stream flows for salmon. The depositor of the water, however, cannot specify their preferred intended use. In other words, an individual cannot deposit water in a water bank and state that the use is to be for instream flow. The renter of the water determines the use and water banks give preference to irrigation. The one exception to this is in the Lenhi River basin. In this basin, an entity can deposit water in the water bank for the express purpose of instream flow. This exception is legislatively stipulate and resulted from endangered species concerns in the basin. Entities also cannot transfer water rights to the water board to be held in trust for instream flow. In theory a water right could be gifted to the board and the board could then apply to have it transferred to an instream flow right. Although permissible, this has not been attempted and it is not certain that the Water Resource Board would have the political will to change the use to instream flow or that the process would not be challenged in court.

RECOGNIZED BENEFICIAL USES FOR INSTREAM FLOW

State law requires that instream flow rights be the "minimum flow of water required to protect the fish and wildlife habitat, aquatic life, recreation, aesthetic beauty, navigation, transportation, or water quality of a stream in the public interest" (Idaho Code 42-1502(f)).

HOLDERSHIP OF INSTREAM FLOW WATER RIGHTS

The Idaho Water Resources Board is the only entity that can apply for and hold new appropriations for instream flow water rights. Private ownership is possible on a temporary basis through the water banks.

QUANTIFICATION REQUIREMENTS AND PROCEDURES:

As with any water right in Idaho, the holder is limited to the minimum amount necessary for the beneficial use. When applying for a new instream flow right, the board must quantify the minimum amount necessary for the beneficial use indicated. In practice, however, the Water Resources Board most often applies for all unappropriated water in a stream segment in order to protect aesthetic beauty and preserve the natural habitat.

FEDERAL RESERVED WATER RIGHTS:

Federal reserved water rights in Idaho are handled through the general adjudication process. The federal government must make a claim during adjudication and participate in the adjudication process. Only federal reserved rights in the Snake River

Basin are currently being adjudicated. Federal reserved rights outside of this basin will not be adjudicated until the Snake River Basin is completed.

APPENDIX 1: CRITERIA USED TO EVALUATE WATER ALLOCATIONS IN THE SNAKE RIVER BASIN

In areas of the Snake River Basin held in trust for the Idaho Power Co., the director of the IDWR must consider, in addition to his normal considerations, whether or not the new use of water will significantly reduce the flows available for power generation. If it is determined that a significant reduction will occur, the director must consider the following criteria:

The potential benefits that the proposed use would provide to the state and local economies.

The economic impact the proposed use would have upon electric utility rates in Idaho, as well as the availability and cost of alternative energy sources, to ameliorate any impact

The promotion of the farming tradition.

The promotion of full economic and multiple use development of Idaho water resources.

In the Snake River Basin above the Murphy gauge, whether the proposed development conforms to a staged development policy of up to 20,000 acres per year or 80,000 acres in a four-year period.

APPENDIX 2: TYPES OF APPLICATIONS: WATER RIGHT/WATER BANK FORMS

Application for Permit
Assignment of Application for Permit
Assignment of Permit
Application for Amendment of a Permit
Request for Extension of time for Proof of Beneficial Use
Proof of Beneficial Use
Application for Temporary Approval of Water Appropriation
Application for Transfer of Water Right
Temporary Change Application
Application for Extension of Time to Avoid Forfeiture of a Water Right
Application for Exchange of Water
Notice of Change of Water Right Ownership
Application to Sell or Lease a Water Right to the Water Supply Bank
Application to Rent Water From the Water Supply Bank
Notice of Protest
Notice of Instream Diversion Stockwater Use of Water
Affidavit for Water Rights to Be Used for Power Purposes
Certified Water Rights Examiner Application

Available at http://www.blm.gov/nstc/WaterLaws/idaho.html, September 2007.

MONTANA

WATER RIGHTS SYSTEM

Water rights in Montana are guided by the prior appropriation doctrine. Montana law establishes that the state's water resources are the property of the State of Montana and are to be used for the benefit of the people. Montana has closed some of its river basins to certain types of new water appropriations due to water availability problems, over appropriation, and a concern for protecting existing water rights. Montana water law authorizes the closure of basins to certain new appropriations through the adoption of administrative rules and the negotiation of reserved water right compacts.

Montana water law is contained in the Montana Water Use Act (Title 85, Chapter 2, MCA) of 1973. The act (effective July 1, 1973) changed the water rights administration significantly in the following ways.

All water rights existing prior to July 1, 1973, are to be finalized through a statewide adjudication process in state courts.

A permit system was established for obtaining water rights for new or additional water developments.

An authorization system was established for changing water rights.

A centralized records system was established (prior to 1973, water rights were recorded, but not consistently, in county courthouses throughout the state).

A system was provided to reserve water for future consumptive uses and to maintain minimum instream flows for water quality, fish, and wildlife.

RESPONSIBLE AGENCY

Authority for water rights decisions is shared by the district court (including the water court) and the Water Resources Division of the Montana Department of Natural Resources and Conservation (DNRC). The Montana Water Court, a division of the district court, is in charge of general stream adjudications for all pre-July 1, 1973, water rights. The Water Resources Division within the DNRC is responsible for the administration, control, and regulation of water appropriated after June 30, 1973.

APPLICATION PROCESS

New appropriation of water or a new diversion, withdrawal, impoundment, or distribution requires the filing of an Application for Beneficial Water Use Permit. This form requests information describing the intended use, place of use, point of diversion, source of supply, amount of water to be used, diversion facilities, and other particulars of the proposed appropriation. An Application for Beneficial Water Use Permit is also required before appropriating groundwater of more than 35 gallons per minute and 10 acre-feet per year. Permits are not required, however, for groundwater uses of less than thirty-five gallons per minute, but a Notice of Completion must be filed in order to acquire the water right.

Upon receipt of an application, the regional office reviews and investigates the application. Upon completion of the review, the DNRC publishes a notice in a newspaper and contacts any potentially affected water users. Objections to the application can then be made, and if they cannot be resolved, a hearing examiner considers the case through an administrative hearing. An environmental review is also made to determine whether the proposed project will have significant environmental impacts and whether an environmental impact statement is needed.

The following criteria are considered when a new appropriation of water is requested in Montana:

Is water physically available at the proposed point of diversion in the amount that the applicant seeks to appropriate?

Can water reasonably be considered legally available during the period in which the applicant seeks to appropriate and in the amount requested?

Will the water rights of a prior appropriator under an existing water right, a certificate, a permit, or state water reservation be adversely affected?

Are the proposed means of diversion, construction, and operation of the appropriation works adequate?

Is the proposed use of water a beneficial use?

Does the applicant have possessory interest, or the written consent of the person with the possessory interest, in the property where the water is to be put to beneficial use?

If a valid objection pertaining to water quality is received, an applicant must also prove that:

The water quality of a prior appropriator will not be adversely affected,

The proposed use will be in accordance with the established classification of water for the source of supply pursuant to 75-5-301(1), MCA, or

The proposed use will not adversely affect the ability of a discharge permit holder to satisfy effluent limitations in accordance with Title 75, Chapter 5, Part 4.

If the application is approved by the DNRC, the applicant receives a permit. Once a permit is received, the permittee then must construct the project, divert the water, and put the water to the intended use as outlined in the permit. When this is finished, the permittee must provide the DNRC with a certified statement describing how the appropriation has been completed. This includes submitting a Project Completion Notice for Permitted Water Development to the DNRC before the deadline specified in the permit or any authorized extension of time. After the project is completed, the DNRC will review the project completion notice and determine whether the project was completed in accordance with the permit. The DNRC will then issue a Certificate of Water Right as long as the project has been completed, the water has been used according to the terms of the permit, and the basin in which the permit lies has been adjudicated and the final decree issued. The priority date of a certificate becomes the date of the original permit. The types of applications that can be filed in

Montana can be seen in Appendix 1. The estimated processing time for an application that is correct and complete is 210 days.

POINT OF DIVERSION AND CHANGE OF USE PROCEDURES:

A holder of a water right, permit, certificate, or water reservation may change the point of diversion, place of use, purpose of use, and place of storage by obtaining prior approval from the DNRC. In order to do this, a person must submit an Application for Change of Appropriation Water Right to DNRC and include information on the water right to be changed and the proposed change. An application for change follows the same general process for notice and hearing as outlined above. Upon completion of the change, the appropriator must file a Project Completion Notice for Change of a Water Right notifying the DNRC that the authorized change is completed.

STATE RECOGNIZED BENEFICIAL USES

Beneficial use in Montana means "a use of water for the benefit of the appropriator, other persons, or the public." Recognized uses have previously included, but are not limited to:

Agriculture
Commercial
Domestic
De-watering
Erosion Control
Fire Protection
Fish
Fish Raceways
Geothermal
Industrial
Irrigation
Mining
Municipal
Navigation
Power
Pollution Abatement
Recreation Uses
Sediment Control
Storage
Stock water
Waterfowl
Water Leased
Wildlife

GROUNDWATER

Groundwater use regulations are different within controlled groundwater areas than outside of these designated areas. Controlled Groundwater Areas may be proposed by the DNRC on its own motion, by petition of a state or local public health agency, or through a petition signed by at least 20 or one-fourth (whichever is less) of groundwater users where the petitioners feel a controlled groundwater area is necessary. One or more of the following criteria must be met in order for the DNRC to declare an area a Controlled Groundwater Area:

Groundwater withdrawals are in excess of recharge to the aquifer.

Excessive groundwater withdrawals are very likely to occur in the near future because withdrawals have consistently increased in the area.

There are significant disputes within the area concerning priority of rights, amounts of water being used, or priority of type of use.

Groundwater levels or pressures are declining or have declined excessively.

Excessive groundwater withdrawals would cause contaminant migration.

Groundwater withdrawals adversely affecting groundwater quality are occurring or are likely to occur.

Water quality within the groundwater area is not suited for a specific beneficial use.

When the DNRC is considering the designation of a Controlled Groundwater Area, it will notify concerned parties and hold public hearings to gather comments and information. After notice and public hearing, the DNRC will issue an order. If the order declares a permanent or temporary Controlled Groundwater Area, the order will contain the specific control provisions. See Appendix 1 for more information on Montana's Controlled Groundwater Areas.

Nine Controlled Groundwater Areas have been designated in Montana:

The South Pine Controlled Groundwater Area
The Larson Creek Controlled Groundwater Area
The Hayes Creek Controlled Groundwater Area
The Warm Springs Ponds Controlled Groundwater Area
The Rocker Controlled Groundwater Area
The Bozeman Solvent Site Controlled Groundwater Area
The Old Butte Landfill/Clark Tailings Controlled Groundwater Area
The Idaho Pole Company Site Controlled Groundwater Area
The North Hills Controlled Groundwater Area

Outside of Controlled Groundwater Areas, a permit to appropriate water is required before any development can begin, and obtaining this permit involves the application process described above. A person does not, however, need to apply for a permit to develop a well or a groundwater spring with an anticipated use of 35 gallons per minute or less, not to exceed 10 acre-feet per year. In this instance, the first step is to drill the well or develop the spring. A Well Log Report is completed by the driller and sent to the DNRC within 60 days. After the development is put to use, the owner submits a Notice of Completion of Groundwater Development to the DNRC. The priority date of the water right is the date the DNRC receives the Notice of Completion. A person must have exclusive property rights in the groundwater development works or written consent from the person with the property rights. A Certificate of Water Right will then be issued to the owner for the specified use.

WATER RIGHTS

A water right in Montana can be held by an individual, group of individuals, organization, corporation, government agency, etc. In Montana, water rights are attached to the piece of land on which they are used. If a piece of land is transferred, any water right attached to that land passes along with it unless specifically stated otherwise. A water right may be severed from the land and sold or retained independently from the land. If the land is sold but the water right is retained, the DNRC does not need to be notified. If, however, the water right alone is transferred to a new owner, an ownership update must be filed with the department. In either case, for the water right to be used again elsewhere, the owner must file an Application to Change a Water Right to change the water right's place of use.

A water right under a permit can be abandoned if it is not used and there is an intent to abandon. If an appropriator ceases to use all or part of an appropriation with the intention to abandon, the right is considered abandoned. In addition, a right is considered to be abandoned if it is not used for ten consecutive years (even if there is not evidence of intent to abandon).

ADJUDICATIONS

In 1979, the Montana Legislature passed a bill amending the adjudication procedures originally established by the Montana Water Use Act. Rather than adjudicating existing water rights one basin at a time, the legislature opted for a comprehensive general adjudication of the entire state's 85 drainage basins. Existing water rights are those that originated before July 1, 1973.

Montana is divided into four water divisions, and the Water Court presides over each division for the purpose of adjudicating existing water rights. The Reserved Water Rights Compact Commission (RWRCC) was created to negotiate compacts with federal agencies and Indian tribes to quantify their reserved water rights in Montana. These negotiated compacts are incorporated into Montana's adjudications.

The Montana Supreme Court has issued an order requiring every person claiming ownership of an existing water right to have filed a statement of claim for that right with the DNRC by January 1, 1982. Stockwater and domestic claims for ground water or instream flow, however, were exempted from this process, though such claims could be filed voluntarily. Existing water rights that were not filed by the deadline are considered to have been abandoned.

The DNRC's role in the adjudication process is to provide technical assistance to the Water Court. Prior to the issuance of a decree, the DNRC examines each claim for completeness, accuracy, and reasonableness. After all claims in a basin are examined, the DNRC issues a summary report to the Water Court, which is available to the public. The court uses this report in preparing the decree for the basin.

Notice of issuance of every temporary preliminary or preliminary decree is given to all parties who may be affected by the decree, along with notice of the time period for objecting to the rights or compacts, or both, in the decree. Water users are encouraged to review the decree and file objections if they feel their claims, or claims belonging to others in the basin, are in error or contain incorrect information. Following the expiration of a decree's objection period, each party whose claim received an objection must be given notice of the filing of that objection. This notice triggers a 60-day counter objection period. A water judge or water master hears all objections and counter objections. After all objections are resolved, the water judge issues a final decree. On the basis of the final decree, the DNRC will issue a Certificate of Water Right to each person decreed an existing water right.

Existing water right claims for livestock and domestic uses from instream flows or ground water sources are exempt from the adjudication process. If claims are not filed, exempt rights are placed in the DNRC's central records for notice purposes. The owner must submit a completed Notice of Water Right. The filing of this notice does not constitute recognition of a water right. The burden of proof of these water rights remains with the owners. Once the water right is entered into the records, the owner will receive notice of any actions on the source of supply that may affect the water right.

NUMBER OF ONGOING ADJUDICATIONS

The status of adjudications on Montana's basins can be found at http://www.dnrc. state.mt.us/wrd/WaterRights/adjStat.htm

INSTREAM FLOWS

Montana's instream flow program began in 1969 when the state enacted legislation allowing the Department of Fish, Wildlife, and Parks the right to appropriate water on twelve trout streams. In 1973, the state replaced this legislation with a reservation system which allowed state and federal agencies to request a reservation for minimum flows on any stream. In 1989, further legislation was enacted which allowed the Department of Fish, Wildlife, and Parks to lease water rights for instream flows (but on a limited number of stream reaches). The legislation has been modified several times, changing the number of stream reaches upon which leases could be held. In 1995, the water leasing program was expanded to allow individuals and private groups to lease water rights for instream use.

Today, instream flow rights in Montana can be established through new appropriations or through water transfers. New appropriations for instream flow can be established through the water reservations system. Under the reservation statute, the state, any sub-division of the state (including municipalities, conservation districts, and any other state agency), or a federal agency may apply to the DNRC for instream flow use. These applications for minimum flow reservation go through the same application process described above, but the DNRC reviews the right every ten years and may extend, condition, or revoke the reservation. Priority dates for the reservations, as with other applications, are determined by the application date.

Instream flows can also be maintained in Montana through water transfers. There are three ways to convert an existing consumptive use water right to instream use. A person may lease all or a portion of a water right to the Montana Department of Fish, Wildlife & Parks (FWP), lease the water right to another party interested in holding the right for the fishery, or convert the water right to an instream use. Any conversion to an instream use requires a temporary change authorization from the DNRC and must benefit fisheries.

RECOGNIZED BENEFICIAL USES FOR INSTREAM FLOW

Beneficial uses for instream flows are vaguely defined in Montana. State law indicates that a beneficial use can be any use that benefits the appropriator, other persons, or the public. This leaves the decision of what constitutes a beneficial instream flow use to the discretion of the DNRC. Most instream flow uses to date have been to benefit fisheries and to maintain water quality, but instream flow uses are not necessarily limited to these uses.

HOLDERSHIP OF INSTREAM FLOW WATER RIGHTS

Federal agencies and any political subdivision of the state may apply for and hold instream flow reservations (from new appropriations). With some restrictions, private or public entities may lease water rights for instream flow. "Any person" may also lease a water right for instream flow. Montana statute defines "person" as "an individual, association, partnership, corporation, state agency, political subdivision, the United States or any agency of the United States, or any other entity." The Montana Department of Fish, Wildlife, and Parks is authorized to lease rights for instream flow, but only on a certain number of stream reaches (the FWP is currently authorized to hold leases on 40 reaches).

FEDERAL RESERVED WATER RIGHTS

A Reserved Water Rights Compact Commission has been established in Montan to negotiate compacts with federal agencies and Native American tribes in an effort to quantify federal reserved rights. Negotiated compacts will be incorporated into the statewide general stream adjudications. When negotiations fail to produce compacts, federal reserved water rights will then be determined through the state adjudication process.

To date, the RWRCC has negotiated and the Montana Legislature has ratified these compacts:

Assiniboine and Sioux Tribes of the Fort Peck Indian Reservation-State of
    Montana
Northern Cheyenne Tribe-State of Montana

United States National Park Service-State of Montana (Big Hole National
  Battlefield, Glacier National Park, Yellowstone National Park, Bighorn Canyon
  National Recreation Area, and Little Bighorn Battlefield National Monument)
Chippewa Cree Tribe of the Rocky Boy's Reservation-State of Montana
United States Bureau of Land Management-State of Montana (2 Units—Bear
  Trap Canyon Public Recreation Site and Upper Missouri National Wild and
  Scenic River)
United States Fish and Wildlife Service-State of Montana (2 Units—Benton
  Lake National Wildlife Refuge and Black Coulee National Wildlife Refuge)
Crow Tribe-State of Montana

BLM-Specific Information

There are several aspects of the application process that are particularly relevant to
the BLM's work in Montana. The capability to proceed with the construction of small
stockwater ponds without first applying for a permit has greatly facilitated the BLM's
ability to meet the need for flexibility within the range improvement program. The
21-month average processing time for a regular water permit can sometimes frustrate
projects. In cases where objections are received to BLM permit applications, it is
becoming more difficult for the BLM to provide the expert testimony required to
support the application because the BLM in Montana has fewer hydrologists and soil
scientists and engineers available to provide such testimony. Additionally, many of
the BLM specialists do not have sufficient training in appearing as an expert witness
to allow them to adequately prepare for this role.

The following information applies to BLM right-of-way approvals and filing fees.
The water right application form requires a signature by the landowner if different
from the applicant. This signature is accepted by the state as proof that the necessary
permissions have been secured. There has never been any request for proof of final
right-of-way approval. The BLM pays the same filing fees for new appropriations
(post-1973) as all other applicants. In the statewide general adjudication legislation,
the legislature set the filing fees at $40.00 per claim up to a maximum of $400.00
per claimant per water division. The BLM also paid these fees at the maximum per
division. New appropriation fees generally amount to less than $2000.00 per year in
recent years. This is mostly due to the decrease in the number of applications due to
basin closures, increased difficulty in securing new permits, and decrease in project
dollars for range improvements.

The BLM is very active in Montana's general statewide adjudications. Montana is
adjudicating all water rights that existed prior to July 1, 1973, and the BLM has filed
more than 22,000 claims. The BLM state office in Montana routinely defends these
claims, as well as objects to other claims, which adversely impact the BLM's water
uses and programs.

The BLM has been active in protecting natural flows in Montana. Besides the
state's instream flow program (discussed above), BLM has used several other methods
to protect natural flows. Protection of instream flows was a critical issue in the compact negotiations for the Upper Missouri National Wild and Scenic River and the
Bear Trap Canyon Recreation Site. Montana also has initiated several state-based

reservation proceedings to allow users to reserve instream flows. The BLM has secured these instream flow reservations on 31 stream reaches in the Upper Missouri River Basin to protect riparian habitat and flows for threatened and endangered species (west slope cutthroat and/or Arctic grayling). The BLM has also held discussions with the Bureau of Reclamation concerning their ability to augment instream flows from storage facilities in Montana.

Negotiations over the BLM's federal reserved water rights are progressing in Montana. The BLM has reached a compact with the State of Montana for the Upper Missouri National Wild and Scenic River and the Bear Trap Canyon Recreation Site. The State has declined to attempt a compact for the various PWR107 water sources, but will handle them through the adjudication process. The BLM is currently working through BLM and USDI approvals to begin negotiations for the Upper Missouri River Breaks National Monument reserved water rights.

The BLM and the State of Montana have enjoyed a good and cooperative working relationship regarding water rights. The BLM is an active participants on many water related joint working groups to deal with water rights and water quality issues in Montana. Montana is in the early stages of a new state administration resulting from the last election and BLM has yet to see whether that will change this relationship (although no problems are expected).

Appendix 1: Types of Applications

Montana Water Right Forms include:

> Application for Beneficial Water Use Permit
> Notice of Completion of Groundwater Development
> Well Log Report
> Application for Provisional Permit for Completed Stockwater Pit or Reservoir
> Application for Change of Appropriation Water Right
> Application for Extension of Time
> DNRC Water Right Ownership Update
> Water Right Dispute Options
> Objection to Application
> Notice of Completion of Permitted Water Development
> Notice of Completion of Change of Appropriation Water Right

Available at http://www.blm.gov/nstc/WaterLaws/montana.html, September 2007.

NEVADA

Water Rights System

Nevada water law is founded on the doctrine of prior appropriation, or "first in time—first in right." Nevada law explicitly states that all waters of Nevada are public property, and a water right is a right to put that water to beneficial use. Beneficial use is the basis of a water right in Nevada. Nevada water law is set forth in the Nevada Revised Statutes, Chapters 532 through 538.

RESPONSIBLE AGENCY

The Nevada Water Resources Division, headed by the State Engineer, is responsible for the administration and enforcement of Nevada's water law. This includes overseeing the appropriation, distribution, and management of the state's surface and groundwater.

APPLICATION PROCESS

The only way to establish a new water right in Nevada is to file an application to appropriate water with the State Engineer (an application to change existing rights requires a similar process). A list of applications that can be filed in Nevada can be seen in Appendix 1. Following the filing, the application is reviewed for completeness and compliance with required procedures. A legal notice, including the point of diversion (if applicable), is then prepared and advertised in a local newspaper for four consecutive weeks. Following the advertisement, there is a thirty day protest period. Any interested person may file a protest with the State Engineer. The protest should set forth the grounds on which the protest is being submitted and whether the protestant seeks denial of the application or conditional approval. If an application is protested, a formal hearing may be held in which the applicant and protestant presents their evidence to the State Engineer. The hearings are formal and all testimony is sworn and recorded.

The State Engineer considers the following three criteria when approving or rejecting an application:

Is there unappropriated water in the source?
Will the proposed use impair existing rights?
Is the proposed use detrimental to the public interest?

The State Engineer may also consider water quality issues, and he may place conditions upon the approved application to protect any interests.

Approved applications are granted a specific time period within which to develop the beneficial use of the water. Once the water has been put to beneficial use, the applicant is required to file proof with the State Engineer. The proof must detail the quantity of water used, the extent of uses, the exact location of the point of diversion, and other related information. Once proof has been filed, the State Engineer issues a certificate of appropriation and the water right is "certified" or "perfected." Any party disagreeing with the decision of the State Engineer may appeal to the district court of the county in which the decision applies.

Vested rights are rights that do not have to go through the application process. Vested rights to surface water are those rights for which the work to establish beneficial use was initiated prior to March 1, 1905 (the date of adoption of Nevada's water law). Vested rights from underground sources are those rights initiated prior to

March 22, 1913, for artesian water and prior to March 22, 1939 for percolating water. The extent of all vested rights on a water source is determined through the adjudication process (see below).

Obtaining a water right in Nevada can take as little as a few months or as long as many years. It takes a minimum of two months to provide notice and allow for protest of an application. If there are no complications, the State Engineer can approve the application. However, if the application is protested or contains complications that need investigating, the State Engineer can take much longer to approve the application. Once the application has been approved, it is up to the permittee to complete the necessary work and file proofs which will result in the perfected water right. The time frame to obtain a perfected right once the application has been approved is dependant upon the work involved.

POINT OF DIVERSION AND CHANGE OF USE PROCEDURES

In Nevada, a diversion is not a necessary component of a water right. The basis of a water right is beneficial use, and if the requested beneficial use necessitates a point of diversion, then it is required and must be specified in the application. Beneficial uses which do not necessitate a point of diversion may be granted, as is the case for instream flow rights.

The point of diversion, place of use, and purpose of use on a water right may be changed by filing a change application with the State Engineer. However, any change may not impair existing rights or be detrimental to the public interest. The process for approving a change application is similar to the application process discussed above.

STATE RECOGNIZED BENEFICIAL USES

Beneficial use is the basis, the measure, and the limit of the right to use water. In Nevada, beneficial uses are determined on a case-by-case basis. The following have been accepted as beneficial uses, but recognized beneficial uses are not limited to these categories:

Commercial
Construction
Drilling
Industrial
Irrigation
Milling
Mining Municipal
Power
Recreation
Stockwatering
Storage
Wildlife

GROUNDWATER

The process for obtaining a groundwater right is similar to that for surface water (see above). New groundwater rights, however, may be restricted in Nevada if they will cause interference with preexisting wells. The State Engineer also has the authority to designate certain preferred uses when making groundwater appropriations, thus prior appropriation is not the strict doctrine for ground water use. In addition, domestic uses of groundwater (defined as water for one house), are exempt from the permitting process.

The general groundwater policy of the State Engineer is to limit water withdrawals from a basin to the average annual recharge for that basin. However, in basins where an outside source of supply is assured, the State Engineer may allow withdraws in excess of the annual recharge. To do this, the State Engineer designates the basin and issues temporary permits subject to revocation at a later date when water becomes available from an outside source. There are currently two designated basins in Nevada, the Las Vegas Artesian Basin and the Colorado River Basin. In these basins, "temporary revocable permits" have been issued, and they may be revoked when Colorado River water becomes available.

WATER RIGHTS

Nevada law states that any "person" may appropriate water for beneficial use. A "person" may be an individual, group of individuals, organization, corporation, government agency, etc. Water rights in Nevada are considered real property and are protected as such. As a result, a water right can be conveyed or transferred. Water rights, however, are appurtenant to the land and are conveyed by deed with the land unless the seller specifically reserves the water right in the deed. When transferring ownership of a water right, a Report of Conveyance must be filed with the State Engineer.

A water right in Nevada can be lost only by abandonment. Abandonment is determined by the intent of the water user to stop using a water right and it does not have a statutory time period. Until recently, water rights could be lost by forfeiture, which occurred if a right was not used for five consecutive years. This, however, has changed and water rights can now only be lost through voluntary abandonment. Water lost through abandonment reverts back to the public and is available for future appropriation.

ADJUDICATIONS

The adjudication process in Nevada focuses on verifying and quantifying pre-statutory water rights, Native American Indian water rights, and federal reserved water rights. An adjudication of surface water claims, other than claims of Native American Indian or federal reserved rights, involves vested rights established before the enactment of Nevada's statutory water law in 1905. An adjudication of ground water claims involves vested rights established before 1913 for artesian groundwater and 1939 for claims to percolating groundwater. An adjudication is initiated by the State Engineer, either upon petition by a water user or by his own initiative. Claimants in an adjudication must file a proof of claim and pay a filing fee. The State Engineer determines the validity of claims through hearings and field investigations. A notice that an adjudication is proceeding must be published for a period of four consecutive weeks in a newspaper of general circulation within the boundaries of the stream system. Upon completion of the adjudication, the State Engineer produces an Order of Determination. The order is submitted to the court where it then goes through further hearings and is subject to objections at the judicial level. After judicial review, the court enters the final decree affirming or modifying the Order of Determination. This decree is final and conclusive and describes the limit and extent of all rights. The adjudication process can be summarized in the following ten steps:

One or more water users on a stream system may petition the State Engineer to begin adjudication proceedings. In the absence of a petition, the State Engineer may initiate the proceedings.

The State Engineer investigates facts and conditions concerning the stream system and determines if he will enter an order granting the petition.

If the petition is granted the State Engineer notifies all claimants and has a Notice of Order and Proceedings published for four weeks in a newspaper nearest the stream system.

The next step in the process is the filing of proofs and title reports by the claimants according to the schedule published in the notice of order for taking proofs.

From the evidence submitted during the period for taking proofs, a preliminary order of determination is prepared by the State Engineer. The preliminary order allocates the waters of the stream system to claimants having valid vested rights.

All evidence submitted during the period for taking proofs and used in preparing the preliminary order is subject to inspection in the office of the State Engineer by any claimant, for a period of 20 days or more.

The preliminary order of determination is subject to objections by any of the claimants, and if objections are filed a hearing is held before the State Engineer.

Next, an order of determination is prepared by the State Engineer and is submitted to all claimants and to the district court having jurisdiction. All evidence and maps are also forwarded to the district court.

Any claimant may file an exception to the order of determination and be heard before the district judge at a hearing.

The district judge then enters findings of fact, conclusions of law and the decree, which determines the water rights on the stream system.

ONGOING ADJUDICATIONS

Decreed and ongoing adjudications are listed at http://water.nv.gov/Adjudications/adj-listing.htm.

INSTREAM FLOWS

Although no statutory law protects instream flows in Nevada, judicial determination has recognized it as a beneficial use. Nevada's instream flow program is based on a court decree in 1988 involving the Bureau of Land Management. The dispute in the case (*Nevada v. Morros*) was over whether or not the BLM could apply for and hold a water right for recreational and wildlife purposes on a lake within its jurisdiction. Opponents argued that a right could not be granted because the use did not involve a physical diversion. The Nevada Supreme Court found that a physical diversion was not necessary to establish a water right. The decision upheld the right to appropriate water for instream flow under state law for fish, wildlife, and recreation.

Instream flow rights in Nevada can be established either through new appropriation or through a water right transfer. Transfers can be a temporary or permanent change from the original use to an instream flow right. Historically transfers have been commonly used in Nevada to establish instream flow through the purchase of existing rights to provide water for state and federal wildlife refuges. Applications to establish an instream flow water right, either from a transfer or through a new appropriation, must go through the application process discussed above.

RECOGNIZED BENEFICIAL USES FOR INSTREAM FLOW

Nevada has recognized wildlife (including fish) and recreation as beneficial uses for instream flow. Since beneficial uses are determined on a case by case basis, however, uses for instream flow are not necessarily limited to these categories.

HOLDERSHIP OF INSTREAM FLOW WATER RIGHTS

According to state law "any person" may appropriate water for beneficial use. "Any person" includes individuals, private organizations, and government agencies. It appears that any entity which can hold a water right in Nevada is permitted to hold an instream flow right.

BLM-SPECIFIC INFORMATION

The BLM in Nevada is currently involved in the Walker River Adjudication. This process is just beginning and the state office will be working on this issue in the future. The BLM has filed the paperwork to assert federal claims in this basin, but no further progress has been made on the issue.

Federal reserved water rights for the BLM in Nevada are primarily limited to Public Water Reserves (PWR). There are no Wild and Scenic Rivers on BLM land in Nevada. In addition, the ten new wilderness areas in Nevada explicitly excluded federal reserved water rights. The BLM has a large number of Public Water Reserve (PWR) 107s. In order to assert these rights, the BLM files a notification with the state engineer's office and pays a notification fee of $50. PWRs other than 107s are sight specific and Nevada has approximately 110 of these.

The relationship between the BLM and the State of Nevada can be characterized as strained. After years of threatening, the BLM is enforcing trespassing laws related to grazing fees. This has angered States' rights advocates and has increased tensions between the Bureau and the state. In addition, the BLM has been involved in a state supreme court case over the BLM's right to hold stockwater rights. This case was recently decided in the BLM's favor. The BLM can now hold stockwater rights solely or jointly. This decision has settled the case, but has further served to stress the relationship. As the BLM asserts its right to hold stockwater rights and to enforce the payment of grazing fees, the relationship between the State of Nevada and the BLM could further deteriorate.

APPENDIX 1: TYPES OF APPLICATIONS

Water Rights Related Forms:

Water Right Application—New
Water Right Application—Change
Water Right Application—Temporary
Water Right Application—Environmental
Proof of Completion of Work
Proof of Beneficial Use
Resumption of Use
Extension of Time
Extension of Time to Prevent Forfeiture
Proof of Use for Stockwater or Wildlife
Proof of Appropriation for Irrigation
Protest Form

Available at http://www.blm.gov/nstc/WaterLaws/nevada.html, September 2007.

NEW MEXICO

WATER RIGHTS SYSTEM

New Mexico's water law is based on the doctrine of prior appropriation or "first in time—first in right." All waters in New Mexico are declared to be public and subject to appropriation for beneficial use. There are five basic components of a water right in New Mexico: Point of diversion (or constructed work), place of use, purpose of use, owner, and quantity. Although these factors are statutorily required, past court decisions, legal opinions, and the discretion of the state engineer allow flexibility in the interpretation of these basic requirements. The state's water law is presently in force in New Mexico Statutes Chapter 72.

RESPONSIBLE AGENCY AND LAW

The State Engineer, appointed by the Governor and confirmed by the State Senate, has broad authority over the supervision, appropriation and distribution of New Mexico's surface and groundwater. This office is responsible for supervision, measurement, appropriation, and distribution of the state's water. The State Engineer performs these duties according to state statute and according to the adjudication of the courts.

APPLICATION PROCESS

Apart from water rights acquired before 1907 and small scale stockwatering (10 acre-feet or less), a permit from the State Engineer is required to appropriate water, change the point of diversion, change the location of wells in declared basins, divert or store water, or change the place or purpose of water use. The types of applications and their associated fees can be seen in Appendix 1. An application for a new appropriation or a change in an existing water right must be advertised once a week for three consecutive weeks in a local newspaper. Those believing that their water rights would be impaired by the granting of the application may file a protest. Protests may also be filed on the basis that granting the application will adversely affect public welfare, or would be contrary to the conservation of water within the state. The protest must be filed within ten days of the last publication notice of the application. If a settlement cannot be reached, the applicant can request a hearing before the State Engineer (or the appointed hearing examiner). The burden of proof in the hearing is on the applicant, and appeals go to the district court.

When considering an application for permit, the State Engineer considers the following: The existence of unappropriated waters; if the application will impair existing water right; whether granting the application would be contrary to the conservation of water within the state; and if the application will be detrimental to the public welfare. The State Engineer can then issue a permit either in whole, in part, or conditioned to ensure non-impairment of water rights.

Once a permit is approved, the permittee must complete the work necessary to put the water to the intended use. Upon completion of the work, the State Engineer issues a certificate, which quantifies the right and describes the point(s) of diversion, place of use, and purpose of use. After full use of the water is made, the permittee must file proof of application of water to beneficial use, and upon inspection the State Engineer issues a license to appropriate water. The license defines the extent and conditions of the use.

There is a new requisite in New Mexico that prior to someone obtaining a water right involving the use of public lands, the person must prove they actually have a permit to use the public lands. This requirement is described in section 72-12-1 of New Mexico's water right's code.

The time frame involved in obtaining a water right in New Mexico is extremely variable. If an application is not complex and is not protested, it takes a minimum of 3 months to obtain approval. If, however, the application is protested, hearings are held, and complexities are involved. The State Engineer can then take much longer (in some cases decades) to reach a decision.

POINT OF DIVERSION AND CHANGE PROCEDURES

Statutory law states that beneficial use in New Mexico requires a diversion of water from its natural path to a place where that water produces revenue or sustains human life. Court rulings, however, have found that this requirement does not apply to all beneficial uses. As we will see below in the instream flow discussion, recreational use, for example, does not require a point of diversion.

One attribute of a water right in New Mexico is the right to change the point of diversion, the place of use, and the purpose of use. These changes, however, may not impair any other water right, may not be contrary to the conservation of water, and may not be detrimental to the public welfare. In addition, a change in diversion, place, or purpose may not increase consumptive use. Any such changes in surface or ground water require the filing of an appropriate application with and approval by the State Engineer.

STATE-RECOGNIZED BENEFICIAL USES

The State of New Mexico does not have an official list of approved beneficial uses. The recognition of a beneficial use is at the discretion of the State Engineer. According to state statute, a beneficial use in New Mexico requires a diversion of water from its natural path to a place where that water will produce revenue or sustain human life. Recent court decisions, however, have changed this allowing for beneficial uses without a diversion requirement. Therefore, the State Engineer has broad authority in considering what constitutes beneficial use in New Mexico. Recognized beneficial uses in the past have included:

Agriculture
Commercial
Domestic
Industrial
Recreational Uses
State Conservation Goals
Stockwatering

GROUNDWATER

The New Mexico groundwater code was enacted in 1931. Groundwater procedures closely parallel those for surface water, with several important differences. A permit to drill a well and appropriate water is not required in areas outside of declared "underground water basins." Within underground water basins, however, use is regulated by the State Engineer. The State Engineer has the authority to establish these basins when regulation is necessary to protect prior appropriations, ensure water is put to beneficial use, and to maintain orderly development of the state's water resources. There are currently 33 declared underground water basins throughout New Mexico.

Under New Mexico groundwater law, only well drillers licensed by the State may drill or alter wells (with a diameter larger than $2^3/8$ inches) within the boundaries of declared underground water basins. The State Engineer is required to issue permits within declared underground water basins in certain instances (see Appendix 4).

WATER RIGHTS

Water rights in New Mexico can be held by any entity accept by the State Engineer. In other words, rights can be held solely, jointly, collectively, or in the name of a corporation, organization, or government agency. All water appropriated for irrigation (unless otherwise stated) is appurtenant to the land upon which it is used and it cannot be transferred to other lands or used for other beneficial purposes unless the water right is separated from the land. A water right can be severed from the land through an application to the State Engineer.

Water rights in New Mexico can be transferred from one entity to another, but a change application must be filed and approved by the State Engineer. Water rights in New Mexico are considered real property and they may be bought or sold. A water right can be conveyed as part of a piece of property or separate (as long as that water right has been severed from the land by an approved application through the state engineer).

A water right in New Mexico can be lost by forfeiture. When all or any part of appropriated water is not put to beneficial use for a period of four consecutive years, the State Engineer issues a notice of nonuse. If the failure to beneficially use the water persists for one more year, the unused water is forfeited and becomes part of the public domain. Forfeiture does not occur, however, if the reason for nonuse is beyond the control of the owner.

ADJUDICATIONS

New Mexico has adjudicated water rights since 1907. Adjudication is through a program of hydrographic surveys and suits. The State Engineer is required to conduct surveys of every stream system in the state. During a survey, data is collected to help the court determine the amount of water to be awarded to each claimant. In an adjudication suit, each claimant has an opportunity to present evidence of water right to the court. The completion of adjudication results in a court decree outlining the priority, amount, purpose (determination of use), periods, and place of water use.

ONGOING ADJUDICATIONS

Currently there are ten ongoing adjudications in New Mexico.

INSTREAM FLOWS

New Mexico's instream flow program is complex, unclear, and continually evolving. New Mexico does not have a legislated instream flow program, and instream flow is not a recognized beneficial use. Recent case law, however, has allowed the development of an instream flow program in New Mexico. In 1998, the New Mexico Attorney General issued a legal opinion concluding that the transfer of a consumptive water right to an instream flow right is allowable under state law. The legal opinion determined that instream uses such as recreation and fish and wildlife habitat are beneficial uses, and that transfers of existing water rights to instream flows are not expressly prohibited. Prior to this opinion, New Mexico was the only state that did not recognize instream flow as a beneficial use.

The Attorney General's opinion is based upon case law. The New Mexico Supreme Court first recognized instream flows as a beneficial use in *State Game Commission v. Red River Valley Co.* in 1945. In that decision, the court ruled that "beneficial use," in relationship to unappropriated water included recreation and fishing. In 1972, the court further held that a diversion was necessary to establish agricultural water rights (*Reynolds v. Miranda*). Based upon these rulings the Attorney General found that recreational use is a beneficial use and that a diversion was not necessary to establish a water right other than for agricultural use.

The 1998 Attorney General's opinion is limited to the context of transferring existing water rights. The opinion notes that new appropriations of water for instream flow are not subject to this precedent. Although the opinion concludes that there are no legal barriers to the transfer of existing water rights to an instream flow right, the State Engineer still has the responsibility for approving such a transfer. Transfers are subject to the application process outlined above, and the State Engineer's office has further indicated that it will require any instream flow right to be conditioned upon gauging throughout the protected stream reach.

RECOGNIZED BENEFICIAL USES FOR INSTREAM FLOW

Instream flow in itself is not recognized as a beneficial use. It appears, however, that water can be dedicated to instream flow for the purpose of recreation and fish and wildlife habitat.

HOLDERSHIP OF INSTREAM FLOW WATER RIGHTS

The Attorney General's opinion does not explicitly address the issue of ownership of instream flow rights. It may be assumed that since ownership of other types of water rights are not limited, instream flow rights could be held by a public or private entity. Current law, however, is unclear and continues to develop in this area.

QUANTIFICATION REQUIREMENTS AND PROCEDURES

Since instream flow is not statutorily regulated, there are no explicit quantification requirements in New Mexico. Approval of water transfers to instream flow is subject to the approval of the State Engineer. The State Engineer has the authority to place restrictions on the approval of an instream flow, therefore quantification requirements are currently at the discretion of the State Engineer.

BLM-SPECIFIC INFORMATION

The New Mexico office of the State Engineer has not required a water applicant to have the necessary right-of-way approved from the BLM prior to the approval of the application. The new requirements that an applicant prove they have a permit to use public lands prior to obtaining a water right for use on that land, however, changes this precedent.

The BLM is required to pay any applicable fees. A list of these fees can be seen in Appendix 1.

The BLM is involved in all ten of New Mexico's ongoing adjudications.

With regard to federal reserved water rights, the BLM holds reserved rights on the Red River Wild and Scenic River. This reserved right was acquired through the adjudication of the Red River basin. The BLM has also applied for reserved rights on the Rio Chama Wild and Scenic River, but that adjudication is ongoing. In addition to Wild and Scenic River reserved rights, the BLM holds numerous Public Water Reserve 107 reserved water rights.

The New Mexico office of the State Engineer has recently threatened to deny federal agencies any claim to stockwater rights. The rational is the agencies do not own livestock and therefore do not put the water to beneficial use. There is, however, a history of federal agencies being granted stockwater claims in New Mexico.

The relationship between the New Mexico office of the State Engineer and the Bureau of Land Management can be characterized as business-like. With the recent requirement for proof of permits on public lands, there stands to be more inter-action between the two agencies. In the past, however, interaction has been limited to the BLM's proposed applications and to the BLM's protest of other applications. The new requirement for proof of permit could potentially lead to more cooperative interaction. The BLM is in the process of coordinating action within the agency for the approval of these permits and there is consideration of making the State Engineer office a cooperative partner.

APPENDIX 1: TYPES OF APPLICATIONS

Groundwater Rights Applications

Declaration of Water Right $1.00
Application to Appropriate (Domestic, Stock) $5.00
Application to Change Location Domestic Well $5.00
Application to Repair or Deepen Domestic Well $5.00
Application to Appropriate Irrigation, Municipal, Industrial, or Commercial Use $25.00
Application for Supplemental Well $25.00
Application to Change Place or Purpose of Use $25.00
Application to Change Location of Well and Place and/or Purpose of Use $50.00
Application for Extension of Time $25.00
Proof of Completion of Well and Proof of Beneficial Use $25.00
Application to Change Point of Diversion and Place and/or Purpose of Use from Surface to Ground Water $50.00

Surface Water Rights Applications

Declaration of Water right $1.00
Declaration of Livestock Dam $1.00
Application to Appropriate $25.00
Application to Change Point of Diversion $25.00
Application to Change Place and/or Purpose of Use $50.00
Application to Change Point of Diversion and Place and/or Purpose of Use $50.00

Application for Extension of Time $50.00
Proof of Completion of Well $25.00
Proof of Beneficial Use $25.00
Certificate of Construction $25.00
License to Appropriate $25.00
Application to Change Point of Diversion and Place and/or Purpose of Use
from Ground to Surface Water $50.00

Miscellaneous Applications

Change of Ownership $2.00
Application for Well Driller's License $50.00
Application for Renewal of Well Driller's License $20.00
Application to Amend Well Driller's License $5.00
Hearing Fee $25.00

Appendix 2: Cases When the State Engineer is Required to Issue a Permit within an Underground Water Basin

1. An appropriation of up to three acre-feet per year for a well for water live-stock, domestic use, or non-commercial irrigation (trees, lawn, garden, etc.) smaller than one acre.
2. An appropriation of up to three acre-feet for a maximum of one year for water use in prospecting, mining, or constructing public works, if the State Engineer finds that the proposed use will not impair existing water rights.

Available at http://www.blm.gov/nstc/WaterLaws/newmexico.html, September 2007.

OREGON

Water Rights System

Although Oregon's water rights system is based primarily on the doctrine of prior appropriation, remnants of riparian water rights still exist. Oregon can therefore be said to have a dual system of water rights. Riparian rights exist because until the enactment of Oregon's water code in 1909, the state recognized riparian water rights, and a few vestiges of these rights still remain. Water uses that were established prior to 1909, have not been abandoned or forfeited, and are verified and quantified through an adjudication process in the circuit court, are said to be vested rights.

The dominant system in Oregon, however, is prior appropriation or "first in time—first in right." Under Oregon law, all water is publicly owned and users must obtain a permit from the Water Resources Department to use water from any source. The four fundamental provisions of Oregon's water code are:

Beneficial purpose without waste—Surface or groundwater may be legally diverted for use only if it is for a beneficial purpose without waste.

Priority—The water right priority date determines who gets water in a time of shortage. The more senior the water right, the longer water is available in a time or shortage.

Appurtenancy—A water right is attached to the land where use was established. If the land is sold, the water right goes with the land to the new owner.

Must be used—Once established, a water right must be used as provided in the water right at least once every five years. With some exceptions established in the law, after five years of nonuse, the right is considered forfeited and is subject to cancellation.

Oregon has established basin programs in which all the land area, surface water bodies, aquifers, and tributaries that drain into a major river are managed together. This basin program includes water use "classifications" that describe the types of new water right applications that may be considered by the Water Resources Department. The Water Resources Commission (Commision) has adopted basin programs for all but two of the state's 20 major river basins. Within a basin, action by the state legislature or administrative procedures by the Commission can close an area to new appropriations. These restrictions on new uses from streams and aquifers are adopted to assure sustained supplies for existing water users and to protect important natural resources. Except in the case of "Critical Groundwater Areas" (see below), these restrictions do not affect existing water uses.

Oregon's water laws are contained in Oregon Revised Statutes, Chapters 536 through 541 and can be found at: http://www.leg.state.or.us/ors/home.html.

Responsible Agency

Water use in Oregon (both surface and groundwater) is administered by the Water Resources Department which is responsible for implementing Oregon's water policy. This general water policy is set by the seven-member Water Resources Commission, which is appointed by the Governor. The Commission also acts as the board of directors for the department.

Application Process

The development of a new surface or groundwater right in Oregon requires the submission of a permit application to the Water Resources Department (certain uses are "exempt uses"; see Appendix 1). Upon receipt of an application, the Water Resources Department reviews the application and verifies its completeness. Once completion is verified, the application is given a tentative priority date and then reviewed according to statutory criteria. Public notice of the application is then given and a proposed final order is prepared and distributed for public comment. A protest period is then open for the next forty-five days. If a protest is filed, it must be accompanied by a $200 protest fee. If a protest is filed against the application, the director of the Water Resources Department may or may not hold a hearing before making a final determination. The applicant has the right to protest the proposed final order, in which case the director is required to a hold a hearing. Following a hearing, a proposed order is issued by the hearings officer. Final orders are issued by the director of the Water Resources Department. An appeal of the final order goes to the Water Resources Commission.

When a Final Order is issued, development of the water must be initiated within one year. Deadlines for completion of the development are further specified on the permit, but generally must be completed within five years. Upon completion of the development (or "proving up" the water use), a final proof survey is submitted to the Water Resources Department. This involves having a certified water rights examiner (CWRE) conduct a survey and prepare a map and claim of beneficial use. Assuming all conditions of the permit have been met, a certificate of water right is granted. The types of applications that can be filed in Oregon can be seen in Appendix 2.

Assuming there are no complications with a water right application it takes a minimum of 190 days to obtain a final order or a water right permit. The time frame to obtain a certificate of water right depends upon the work involved in "proving up" the water use.

POINT OF DIVERSION AND CHANGE OF USE PROCEDURES

A point of diversion is required for consumptive uses of water, but not when establishing instream flows. Changes in the point of diversion, point of appropriation, place of use, and nature of use can be done, but must have approval from the Water Resources Department. If an applicant wants to change one of these specifications on the permit, a transfer application must be filed with the department. Both temporary and permanent transfers are allowed.

In order to approve a permanent transfer application, the department must determine that the proposed change will not injure other water rights. The public is offered a chance to comment and protest if an existing water right would be injured. Only protests that claim injury to another water right can be accepted. The department may attach conditions to an approval order to eliminate potential injury to other water rights. Once the transfer application is approved, the permittee must submit proof of completion of the change, at which time a new certificate is issued which confirms the modified water right.

Temporary transfers are allowed for a change in the place of use and may not exceed a period of five years. The application for a temporary transfer is the same as for a permanent transfer except that the proof of completion is not as rigorous.

STATE RECOGNIZED BENEFICIAL USES

Recognized beneficial uses of water in Oregon include:

Aquatic Life
Commercial
Domestic

Fire Protection
Fish
Groundwater Recharge
Industrial
Instream Flow
Irrigation
Mining
Municipal
Pollution Abatement
Power
Recreation Uses
Wildlife

GROUNDWATER

Groundwater in Oregon is declared to be part of the public waters of the state and must be appropriated by the application/permit/certificate process described above. Due to the basin program, groundwater and surface water are managed conjunctively within basins. Applications for groundwater use are examined for their interference with existing wells as well as surface water claims. The permit process is not required for certain uses of groundwater (see Appendix 1).

In order to regulate the use of groundwater (besides the regulation on new appropriations which results from the basin program), the Water Resources Commission may declare certain areas as "Critical Groundwater Areas." The law in Oregon requires that when pumping of groundwater exceeds the long-term natural replenishment of the aquifer, the Water Resources Commission must declare the source a Critical Groundwater Area and restrict water use. The purpose of this designation is to prevent excessive decline in groundwater levels and to stop quality degradation. Within Critical Groundwater Areas, certain users of water have preference over other users, regardless of established water right priority dates. There are currently six Critical Groundwater Areas in Oregon: The Dalles in Wasco County, Cooper Mountain—Bull Mountain southwest of Beaverton and Tigard, Butter Creek, Ordnance, and Stage Gulch.

WATER RIGHTS

A water right in Oregon can be held by any legal entity. In other words, a water right can be held by an individual, group of individuals, organization, corporation, government agency, etc. Although the name on a water right can be any entity, a water right in Oregon is specific to the place of use. The owner of the land to which the water right is attached has the authority to make decisions and modifications concerning the water right.

Water rights in Oregon can be transferred from one owner to another. There are two types of transfers allowed in Oregon: permanent and temporary. The approval process for a transfer application is that same as described above for a change in use or change in point of diversion. The transfer of water rights can occur through the buying and selling of rights. Oregon law states, however, that a profit cannot be made from the sale of a water right; the sale can only recover the costs incurred regarding the operation and sale of the water right. This provision, however, is not strictly enforced.

Water rights in Oregon can be lost through abandonment or forfeiture. Abandonment is voluntary by the owner, where as forfeiture occurs through five consecutive years of nonuse. Once a water right has been unused for five or more years, it is subject to cancellation. Cancellation requires a legal proceeding to determine whether or not the period of nonuse has occurred. A water right is subject to cancellation even if the property owner begins to use the water again after a period of nonuse. This is true even if the current owner did not own the property when use was discontinued. However, if more than 15 years have passed since the period of nonuse, the water right is not subject to cancellation under the law.

ADJUDICATIONS

General adjudications in Oregon are used to determine all pre-1909 and federal reserved water rights. The general adjudication of a river basin is initiated by the local circuit court or the director of the Water Resources Department. In order to claim a right during adjudication, a "proof of claim" must be filed with the department. Claims are reviewed and may be contested. The department issues an order of determination, and the circuit court reviews the order and affirms or modifies it. The final judgment by the circuit court is called a decree. The decree is the final determination of all pre-1909 and federal reserved water rights in that river basin. Individual certificates are then issued to water claimants according to the terms of the decree.

ONGOING ADJUDICATIONS

Pre-1909 rights have been adjudicated in approximately two-thirds of Oregon. Adjudication proceedings have been completed for most of the major stream systems in eastern and southern Oregon and a few of the larger tributaries to the Willamette River. A major adjudication proceeding involving federal agencies and the Klamath Tribe is underway in the Klamath Basin.

INSTREAM FLOWS

Instream flows in Oregon can be acquired through new appropriation or through transfers. Oregon was one of the first western states to recognize instream flow as a beneficial use. In 1915, the legislature prohibited the appropriation of creeks that form waterfalls in the Columbia River Gorge. In 1955, they expanded their instream flow program by adopting minimum stream flows to support aquatic life, minimize pollution, and maintain recreational opportunities. These minimum flows were administrative rules and were not full water rights. In 1983, amendments were adopted that authorized the Department of Fish and Wildlife, the Department of Environmental Quality, and the Department of Parks and Recreation to apply for minimum instream flow rights. In 1987, and again in 1993, further amendments were made to the water code strengthening instream flow rights, allowing for transfers, and allowing for the use of water markets to acquire instream flow rights.

Currently, only the Departments of Fish and Wildlife, Environmental Quality, and Parks and Recreation may apply for new appropriations for water for instream flow. Although these Departments apply for instream flow rights, the rights are not issued to the agencies, but are held in trust by the Water Resources Department. Instream flow rights can also be established through water right transfers (either permanent or temporary). Oregon water law allows any entity (public or private) to purchase, lease, or receive as a gift any water right for instream use. The converted rights, however, must be held in trust by the Water Resources Department.

RECOGNIZED BENEFICIAL USES FOR INSTREAM FLOW

Instream flow rights must be held in trust by the Water Resources Department for "public use." Public uses include recreation, conservation, fish and wildlife maintenance and habitat, other ecological values, pollution abatement, and navigation.

HOLDERSHIP OF INSTREAM FLOW WATER RIGHTS

The Oregon Water Resources Department is the only entity that may hold instream flow rights. The Departments of Fish and Wildlife, Environmental Quality, and Parks and Recreation can request new appropriations of instream flow rights. Individuals and other entities may acquire existing rights and take responsibility for changing the use to instream flow, but then they must turn the right over to the department to be held in trust.

QUANTIFICATION REQUIREMENTS AND PROCEDURES

In Oregon, the amount of water reserved for an instream water right cannot exceed the amount needed to provide increased public benefits. When natural stream flows are the source for meeting instream water rights, the amount allowed for the water right cannot exceed the estimated average natural flow. Applications to establish instream water rights must include the requested amount by month and year in cubic feet per second or acre-feet, a description of the technical data, and methods used to determine the requested amounts.

All the required procedures for establishing an instream flow right can be found in Oregon Administrative Rules, Chapter 690, Division 77.

FEDERAL RESERVED WATER RIGHTS

Adjudication proceeding are used to determine the water rights for federal reservations of land including Indian reservations. Legislation passed in 1987 and amended in 1993, allows the director of the department to act on behalf of the State of Oregon to negotiate settlements for Federal reserved water rights. These negotiations allow the director to include claimants, state and federal agencies, other water users, and public interest groups in discussions to resolve and quantify the use of water on federal and Indian reservations.

BLM-Specific Information

The requirement under Oregon state law that a certified water rights examiner (CWRE) conduct a survey and prepare a map and claim of beneficial use has been somewhat problematic for the BLM. The BLM has a large number of water rights claims and therefore has a backlog of water rights filings. It has been a priority to have districts eliminate their backlog, but funding has often hindered this. There are several employees in the Oregon State Office Cadastral Survey branch who are CWREs. In the past, the State Office Soil, Water and Air program provided funding to the Cadastral Survey branch. This funding covered the expenses of having these BLM CWREs travel to the districts to assist them with their water rights workload. Unfortunately, budget restrictions have reduced the amount of funding available. The need for CWREs now exceeds what the State Office can provide and thus the remaining CWRE need is passed to the district. Some districts have invested in training their own CWREs. Others have not, and these districts must contract with a local engineering firm to obtain CWRE services. In the latter case, the expense can be high which means that fewer water rights applications get filed.

Water applicants in Oregon must have the necessary right-of-way approval from the BLM prior to approval by the State. The applicant must provide proof that an easement or other authorization exists for a water right application on land that is not owned by the applicant.

The BLM is required to pay filing fees, the amounts for which depend on the type of application.

When the BLM seeks to obtain a water right for use on BLM land, the applicant must first determine if the proposed use qualifies as a federal reserved right. If it qualifies as a federal reserved right, the applicant should determine if the purpose of the reservation would be best served through the assertion of the federal reserved right. If the assertion is the best way to obtain a right, the applicant must follow a certain set of requirements depending upon whether or not the area has been adjudicated. If a federal reserved right cannot be asserted, or if the purpose would be more effectively served through the state application process, the applicant must follow a different set of requirements. The decision criteria and list of requirements for obtaining water for BLM purposes in Oregon are outlined in Appendix 5.

Adjudications in Oregon have not given the the BLM the opportunity to assert federal reserved water rights for wilderness areas. Another situation facing the BLM is that many of the basins in Oregon have already been adjudicated for pre-1909 water rights (back in the 1960s–1970s). In those adjudications, federal reserved water rights were not addressed. Therefore, federal reserved water rights exist in these basins that have not been quantified and asserted and they will not be unless the state initiates a new adjudication or the federal government brings suit against the state to have its claims quantified. The BLM is currently participating in the Klamath adjudication in Oregon.

To date, the BLM has worked well with the state of Oregon. However, recent tensions in the Upper Klamath Basin over water issues (a partial denial of the BLM's instream flow claim for the Upper Klamath River in the Klamath Basin Adjudication) and a recent bill in the Oregon Legislature (HB 3343) could signal change. House Bill (HB) 3343 attempted to prohibit the Oregon Water Resources Department from granting a water right or other control over waters of the state to the federal government, United Nations or other entity acting on behalf of the federal government or the United Nations. This prohibition would retroactively affect applications submitted by the BLM and other Federal agencies that have not received a permit from the state before the effective date of the measure. The Vale District would be particularly affected, as it has hundreds of such applications pending permit. The bill was referred to the House Water and Environment Committee on March 13, 2001. Public hearings were held March 16, 2001 and May 18, 2001. The bill, which was introduced at the request of the Oregon Cattlemen's Association, died in committee. Another potential issue surrounds the measurement of water use and water use reporting. HB 3623, recently introduced, would have required all water users to measure the amount of water withdrawn or stored. Oregon water law currently requires that annual water use reports be submitted for all reservoirs and large dams (those over 9.2 acre feet or over 10 feet in height). These are to be monitored monthly, and the use is to be reported by month for the year. The current approach being used by the BLM is to submit this type of information to the Water Resources Department as a matter of comity. Districts vary in their adherence to this requirement; some only submit a blank water use form. The BLM has thousands of water developments on the public lands in Oregon; therefore, the cost of installing measuring devices and recording measurements would be prohibitive. This bill did not pass this legislative session. If it had, the Oregon State Office BLM would likely have sought to obtain a waiver of this requirement or would have consulted with legal counsel about having to submit this information.

APPENDIX 1: PERMIT APPLICATION "EXEMPT USES"

Uses of Surface Water That Do Not Require a Permit

Natural Springs—A landowner's use of a spring, which, under natural conditions, does not form a natural channel and flow off the property where it originates at any time of the year.

Stockwatering—Where stock drink directly from a surface water source and there is no diversion or other modification to the source. Also, use of water for stockwatering from a permitted reservoir to a tank or trough, and under certain conditions, use of water piped from a surface source to an off-stream livestock watering tank or trough.

Salmon—Egg incubation projects under the Salmon and Trout Enhancement Program (STEP) are exempt. Also, water used for fish screens, fishways, and bypass structures.

Fire Control—The withdraw of water for use in, or training of, emergency fire fighting.

Forest Management—Certain activities such as slash burning and mixing pesticides. To be eligible, a user must notify the department and the Oregon Department of Fish and Wildlife and must comply with any restrictions imposed by the department relating to the source of water that may be used.

Land Management Practices—Where water use is not the primary intended activity.

Rainwater—Collection and use of rainwater from an impervious surface (like parking lot or a building's roof).

Uses of Groundwater That Do Not Require a Permit

Stockwatering

Lawn watering or noncommercial gardening of less than one-half acre

Limited school ground uses

Single or group domestic uses not exceeding 15,000 gallons

Down-hole heat exchanges

Single industrial or commercial uses not exceeding 5,000 gallons per day

APPENDIX 2: TYPES OF APPLICATIONS

Water Right Application Forms

Application for Surface Water Permits

Application for Groundwater Permits

Application to Store Water

Supplemental Application Forms

Land Use Form

Irrigation (Form I)

Commercial/Industrial (Form Q)

Mining Use (Form R)

Municipal/Quasi-Municipal Water Use (Form M)

Reclaimed Water Use Registration Form

Water Right Transfer Application

Other Forms

Application for Allocation of Conserved Water

Instream Lease Agreement

Application for Limited Water Use License

Available at http://www.blm.gov/nstc/WaterLaws/oregon.html, September 2007.

UTAH

WATER RIGHTS SYSTEM

The prior appropriation doctrine is the basis of water appropriation in Utah. State statutes provide that all water is the property of the public, and a water right is the right to the use of water based upon quantity, source, priority date, nature of use, point of diversion, and physically putting water to beneficial use. The basis of all water rights in Utah is beneficial use, and a water right is defined by the point of diversion, place of use, amount diverted, purpose of use, and period of use. A complete "water code" was enacted in 1903 and was revised and reenacted in 1919. This law, as amended, is presently in force as Utah Code, Title 73, which can be seen at: http://www.le.state.ut.us/~code/TITLE73/TITLE73.htm. Today, much of the State of Utah is closed to new appropriations of water, so new projects and allocations will require obtaining existing rights and amending them for new purposes.

RESPONSIBLE AGENCY

The State Engineer, through the Division of Water Rights, is responsible for the administration of water rights, including the appropriation, distribution, and management of the state's surface and groundwater. This office has broad discretionary powers to implement the duties required by the office. The Utah State Engineer's Office was created in 1897, and the State Engineer is the chief water rights administrative officer.

APPLICATION PROCESS

The establishment of a new water right or changing an existing right requires the filing of an application with the State Engineer. The types of applications that can be filed in Utah can be seen in Appendix 1. To initiate an application, the applicant must describe the proposed development. The application is reviewed, and upon verification of its completeness and adherence to existing policies, a legal notice is prepared and advertised for two consecutive weeks in a local newspaper. Applications can either be processed under formal or informal administrative procedures (this determination must be made by the applicant prior to advertising). The predominant difference between the two procedures relates to the appeal process. Under the formal procedures, an appeal is reviewed based upon the existing record, where as under the informal proceedings, the appeal is handled as a new trial. Following the legal advertisement, there is a 20-day protest period during which time protests can be filed against the application. Protests are not limited to water right holders; anyone who has an interest can file a protest. If an application is protested, a hearing is held to allow the applicant and protestant to present information to the State Engineer (these considerations can be seen in Appendix 3). The status of an application which has not been acted upon is referred to as "unapproved".

In approving or rejecting an application, the State Engineer considers items outlined in UCA section 73-3-8, as well as water quality issues (see Appendix 2—Assessing an Application). In approving the application, the State Engineer can impose conditions to protect prior water rights, better define the extent of the application, or address other issues such as required permits by other regulatory agencies or requiring minimum stream flow bypasses. The status of an application which has been approved is referred to as "approved," and the approval of an application takes a minimum of three months.

When applications are approved, they are granted for a specific time period (usually three years) in which to develop the project. Once the project is complete and the water has been put to beneficial use, the applicant is required to file proof of appropriation with the State Engineer. This file of proof affirms the quantity of water that has been developed, the extent of use, exact location of the point of diversion, and other related information. Upon filing of proof, the State Engineer then issues the "certificate of appropriation" and the status of the application is referred to as "perfected."

POINT OF DIVERSION AND CHANGE OF USE PROCEDURES

In most cases, a point of diversion is required in order to obtain a water right. Certain beneficial uses (such as instream flow), however, do not require diversion. Both the point of diversion and the purpose and place of use can be changed. To change the point of diversion, purpose of use, or place of use, a change application describing the proposed change must be filed with the State Engineer. The change application is processed in the same manner as an application to appropriate water and is evaluated using the same criteria. In addition to the criteria used to evaluate an application, the State Engineer also considers if the proposed change will exceed historical levels, and if intervening rights will be impaired due to the proposed change.

STATE-RECOGNIZED BENEFICIAL USES

Utah recognizes the following beneficial uses:

Agriculture
Culinary
Domestic
Industrial
Irrigation
Manufacturing
Milling
Mining
Municipal
Power
Stock watering
Instream flow—fish, recreation and the reasonable preservation or enhancement of the natural stream environment
Storage—irrigation, power generation, water supply, aquatic culture, and recreation

See Appendix 4 for Beneficial Use Quantification.

GROUNDWATER

The State Engineer, through the Division of Water Rights, is responsible for administering both surface and groundwater. The process for obtaining a groundwater permit (either a new application or a change application) requires the same forms and process as a surface water permit. Groundwater policy, however, is different than surface water, therefore the criteria used to evaluate the groundwater application may be different. Utah is divided into groundwater areas and policy is determined by area.

Utah also regulates the drilling of wells. Any well drilled to a depth of thirty feet or greater must be constructed by a licensed Utah water well driller. The State Engineer, through the Division of Water Rights, is responsible for licensing requirements and well construction criteria, and the development and publication of the Administrative Rules for Water Well Drillers.

Water Rights:

Water rights in Utah can be held by any legal entity. In other words, they can be held solely, jointly, collectively, or in the name of a corporation, organization, or government agency. Regardless of how the right is held, any change application must be titled in that entity's name. Water rights can be transferred from one entity to another, but a change application must be filed and approved by the State Engineer. Water rights can be bought and sold as means for transfer, but approval by the State Engineer is still required. An unapproved or approved application is considered personal property, where as a certificated application or "perfected" water right is considered real property. Since applications for a new water right are considered personal property, they may be bought and sold using a conveyance or assignment. When water rights are perfected, they are considered real property; therefore they must be conveyed by deed to the new owner.

A water right in Utah can be lost by either abandonment or forfeiture. Abandonment is determined by the intent of the water user and does not require a statutory time period. A water right is lost by forfeiture if the right is not used for five year. Water lost through abandonment or forfeiture reverts back to the public and is subject to future appropriation.

Adjudications

An adjudication of water rights is a State action addressed in district court to determine the water rights on the source or in the area involved in the action. The State Engineer is a party to the action with the statutory responsibility to prepare a "proposed determination of water rights" (PDET), which serves as the basis for the court's decree on the water rights in the area.

A PDE requires a thorough search of the division's records, files, and databases, which relate to the adjudication area. Further research is required at the County Recorder and Clerk Offices to identify land ownership where necessary and to obtain information or legal documents that help establish water rights that are not on the division's records or that help clarify or define water rights that are part of the division's records.

Maps of the area are created using digital aerial imagery and location coordinates gathered by GPS methods. A hydrographic survey of the area is conducted and field investigations are made with the water user or water provider to verify his sources of water, points of diversion, and specific places and nature of use. An evaluation is made of the water right based on the current use of water or the use of water within the recent past (five years).

When the various aspects of the water rights are gathered and evaluated, the water user prepares a Statement of Water Users Claim for each perfected water right, or group of water rights, and requests the water user to review and sign the claim form. When all of the perfected water rights in the adjudication area have been defined by a Statement of Water Users Claim, a PDET book is compiled and published. A copy of the book is distributed to each water user that is listed in the book. The PDET is the State Engineer's recommendation to the court regarding the status and quantification of the water rights. A copy of the PDET, the hydrographic survey maps, the original Statements of Water Users Claims, and other required supporting documentation are filed with the district court.

After the PDET book has been distributed, the statute provides for a 90-day protest period within which protests may be filed objecting to a particular water right listed in the PDET, an attribute of a water right, or the omission of a water right. Objections are filed with the appropriate district court.

Following the protest period, the division staff works with the Attorney General's office to resolve the protests that were filed. This effort often involves additional field work and discussions with the protestant and the water user (if the protestant is not the water user). Once this effort is completed, a pre-trial order is prepared for the court's signature. The pre-trial order essentially decrees those rights listed in the PDET that were not protested and those which were protested but resolved. The pre-trial order sets forth those protests which could not be resolved and which must be determined by the court. Once the remaining protests have been settled or determined by the court, an interlocutory decree is prepared and signed by the court. This decree supersedes all prior findings or decrees.

ONGOING ADJUDICATIONS

All of the hydrologic areas of the state are currently involved in a court ordered adjudication of water rights, except the Weber River and Sevier River drainages. The water rights on the Sevier and Weber Rivers were adjudicated and decreed in the 1920s and 1930s. The adjudications in most of the other areas of the state were started between 1950 and the early 1970s.

INSTREAM FLOWS

In 1986, Utah enacted an amendment to its water code recognizing instream flows as a beneficial use not subject to diversion requirements (UC 73-3-3-11). Utah's instream flow laws allow the Utah Division of Wildlife Resources or the Division of Parks and Recreation to file for temporary or permanent changes for instream flow rights. The law specifically states that unappropriated water cannot be appropriated for instream purposes. Change application can be filed on rights presently owned by either division; on perfected water rights purchased by either division through funding provided for that purpose, or acquired by lease, agreement, gift, exchange, or contribution; or on water rights acquired by either division with the acquisition of real property. Legislative approval, however, is required before either division can purchase water rights specifically for instream flow purposes. Instream flow rights held by either division retain the priority date of the original right.

Change applications for instream flow must go through the normal application process through the State Engineer, and are subject to the same assessment criteria. Change applications must identify the points on the stream between which the instream flow will be provided and must document the public benefits derived from the instream flow. The State Engineer retains the right to request additional information for the purpose of evaluating the application. There are few restrictions on the change of use of a water right, apart from the criteria used to assess the change application (see Appendix 3).

Although the above mentioned divisions are the only entities allowed to hold instream flow rights, the State Engineer has the legal power, through the application approval process, to preserve water for natural flows. Utah water law empowers the State Engineer to withhold approval or reject applications that would unreasonably affect public recreation or the natural stream environment.

Recognized Beneficial Uses for Instream Flow

Either division may file applications for permanent or temporary changes for the purpose of providing water for instream flows within a designated section of a natural stream channel or altered natural stream channel for the following purposes: the propagation of fish, public recreation, or the reasonable preservation or enhancement of the natural stream environment.

Holdership of Instream Flow Water Rights

Although the Division of Wildlife Resources and the Division of Parks and Recreation are the only two entities that may hold instream flow rights, individuals may acquire an existing right and transfer it to these agencies to hold as an instream flow right.

Quantification Requirements and Procedures

There are no specific quantification requirements for an instream flow right in Utah. Since instream flow rights can only be obtained through transfer, the quantification requirements depend upon the underlying right and how it was originally established.

BLM-Specific Information

The only Federal reserved water rights that the BLM holds in Utah result from Public Water Reserve 107 (PWR 107s).

The Utah BLM has a very good working relationship with the State Engineer's office. BLM receives careful consideration of water right requests. One of the keys to maintaining this relationship is to work through the process set up in the Utah Law of Water Rights.

APPENDIX 1: TYPES OF APPLICATIONS

Application to Appropriate Water—Used to acquire a new water right. These applications can be permanent, temporary, or fixed time.

Diligence Claim—Filed when it can be shown that a surface water source has been in continuous use since before 1903 or an underground water source has been in continuous since before 1935.

Application to Segregate a Water Right—Used to divide an unperfected water right into two or more separate and distinct water rights.

Application for Change of Water—Used to change the point of diversion, place, or nature of use of an existing water right. These applications can be permanent or temporary (less than one year).

Application for Temporary Appropriation of Water—Used to appropriate water for a period of time less than one year.

Application for Appropriation for Fixed Time—Used to appropriate water for a specific amount of time when the State Engineer feels that water is available for a limited period.

Application for Exchange of Water—Used to exchange points of diversion.

Application for Extension to Resume Use

Application for Groundwater Recovery

Application for Groundwater Recharge

APPENDIX 2: IF AN APPLICATION IS PROTESTED

Applicant will receive a copy of any protest and will have the opportunity to submit a response. An application may be protested because of concern for water supply, environment, etc.

An informal hearing may be held on both protested and unprotested applications. If a hearing is to be held, a date and place will be set. Hearings are held twice a year in each county throughout the state. The elapsed time before a hearing may depend on the schedule.

Hearings are conducted by division representatives. Both applicant and protestants may state their positions. Each has the opportunity for rebuttal. They may represent themselves or obtain legal counsel.

After the hearing, the State Engineer will review the evidence. He then will approve, reject, or hold the application for further study.

Applicants and protestants will be notified in writing of the State Engineer's decision.

An aggrieved party may file a Request for Reconsideration with the State Engineer within 20 days, and/or appeal to the district court within 30 days of the decision.

APPENDIX 3: POINTS THE STATE ENGINEER CONSIDERS IN ASSESSING A NEW APPLICATION (OR CHANGE IN USE OR DIVERSION POINT)

Is there unappropriated water in a proposed source?

Will the proposed use impair existing rights or interfere with more beneficial uses of the water?

Is the proposed plan physically and economically feasible?

Does the applicant have the financial ability to complete the proposed works?

Was the application filed in good faith and not for purposes of speculation or monopoly?

Will it unreasonably affect public recreation or the natural stream environment?

Will it be detrimental to the public welfare? Public welfare is not defined specifically by state law.

APPENDIX 4: BENEFICIAL USE QUANTIFICATION

The quantity of water appropriated for beneficial use is expressed as a flow rate in cfs (cubic feet per second) and/or as a volume in acre-feet to be taken from a well, river, spring, etc. for the required purpose. The depletion figure is the quantity of water consumed which will be lost to the hydrologic system through the said use. Depleted water does not return to the surface water sources or underground aquifers via seepage, drainage, etc. but is consumed in the growth of plants and animals, evaporation, and transmission away from the area. The following figures are used for general quantification.

DOMESTIC (inside use only): Water diversion for a full-time (permanent residence) use is evaluated at 0.45 acre-foot per family. Part-time (seasonal or recreational) use is equated at 0.25 acre-foot per family. Depletion is generally 20 percent if using a septic tank or drain field system but varies if the residence is connected to a community sewage system depending on the treatment method used and its distance away from the diverted source.

IRRIGATION (any outside watering): This purpose includes watering of crops, lawns, gardens, orchards, and landscaping. The diversion amount (irrigation duty) ranges from 2 acre-feet per acre in cool, mountain meadow areas to 6 acre-feet per acre in low, hot southern areas of the state. Higher, cooler valleys are generally 3 acre-feet per acre, and lower moderate areas 4 or 5 acre-feet per acre. If land is subirrigated or supplemented by other rights or supplies, the diversion rate may be less than average for the area. Generally, the irrigation season is described as April 1 to October 31 and/or the general frost-free period in the area. Some court decrees and early rights authorize differing periods. Depletion varies considerably due to differing soils, temperatures, wind factors, etc. and can range from about 40 percent to about 70 percent. Figures are taken from available studies (particularly "Consumptive Use of Irrigated Crops in Utah," Research Report 145.

STOCKWATERING: The diversion figures for this purpose are based on year-round watering. Stock operations for lesser or intermittent periods would need adjustment accordingly. Water diverted for this use is generally considered to be 100 percent depleted by the animal, evaporation, phreatophytes, and/or wastewater collection.

cow or horse: 0.028 acre-foot
sheep, goat, swine, moose, or elk: 0.0056 acre-foot
ostrich or emu: 0.0036 acre-foot

llama: 0.0022 acre-foot

deer, antelope, bighorn sheep, or mt. goat: 0.0014 acre-foot

chicken, turkey, chukar, sage hen, or pheasant: 0.00084 acre-foot

mink or fox (caged): 0.00005 acre-foot

INDUSTRIAL, COMMERCIAL, RECREATIONAL, COMMUNITY, AND MINING: Projects are evaluated on an individual basis. Parameters include method of processing or manufacturing, number of employees, length of workshift and period of operation, type of waste processing and/or discharge, and types of employee and/or public facilities (showers, food preparation, etc.). The Utah State Administrative Rules for Public Drinking Water Systems (particularly R309-203) are guidelines for such estimates.

Available at http://www.blm.gov/nstc/WaterLaws/utah.html September 2007.

WASHINGTON*

WATER RIGHTS SYSTEM

Washington's water law system is largely based on the doctrine of prior appropriation, recognizing prior existing riparian rights. The state is listed as a hybrid state, though it does not recognize new riparian rights. The official water code of Washington is found in R.C.W.A. 90.03. Water rights are subject to first in time use and priority of uses is determined accordingly. Washington participates in a water rights acquisition program for the protection of vulnerable salmon. The program began in 2003 and is voluntary, meaning interested water rights holders in affected areas may grant their rights to the state to help preserve the salmon.

RESPONSIBLE AGENCY

The agency in charge of Washington's water law is the Department of Ecology (see R.C.W.A. 90.03.005 & 90.03.015 (1)). The department is charged with averting wasteful practices in the exercise of water rights to "the maximum extent practicable, taking into account improved water use efficiency, and the most effective use of public and private funds." Agency functions are divided up among four regional offices: the northwest (Bellevue), southwest (Lacey), central (Yakima) and eastern (Spokane). There are field offices in Bellingham and Vancouver.

APPLICATION PROCESS

All new water rights in Washington State, since 1917 for surface water and 1945 for ground water, require a permit. Applications are available online through the Department of Ecology and must be submitted to the department at the address printed on the application. There are four types of ground water exemptions to the

---

* Note that BLM did not create a fact sheet for Washington State. This fact sheet was created by the author for this text in the general format of the BLM sheets.

permit requirement in Washington. These are livestock watering, watering non commercial lawn of ½ acre or less, domestic water for single family home or groups of homes not more than 5000 gallons a day and industrial and irrigation uses not more than 5000 gallons per day.

The Department of Ecology takes into account the costs and benefits of the impoundment, including environmental effects and consideration is given to increased supply from a project including offsets. The department does not make provisions for certain techniques as a condition for approving an application though such techniques are undoubtedly considered in the approval process. See R.C.W.A. 90.03.255. Applications should include maps and other information that might be needed by the department in making its decision. Upon receipt of an application the department makes a record of its receipt, including the date. The application may be returned afterwards if incomplete or for correction of errors but retains its priority date as of the date first received. R.C.W.A. 90.03.270. After the application is properly completed the department instructs the applicant to publish notice in a newspaper of general circulation. R.C.W.A. 90.03.280.

The department then investigates the application to determine what if any water is available and to what uses it may be applied. The investigation includes the public interest and uses detrimental to that interest will be denied. Preliminary permits may be issued for not more than three years, in the event insufficient information is provided to complete the investigation. An extension of the preliminary permit for an additional two years is possible after a showing of good faith. The department will then issue written findings of fact based on the investigation. The department will reject applications where no unappropriated water exists, the proposed use conflicts with existing rights or is detrimental to public interest. R.C.W.A. 90.03.290.

A permit, once approved, may be assigned subject to the conditions of the permit. R.C.W.A. 90.03.310. When an appropriation is perfected and this is demonstrated to the satisfaction of the department, the department will issue a certificate of rights. R.C.W.A. 90.03.330. Water rights relate back to the date of filing of the original application. R.C.W.A. 90.03.340. Water rights are appurtenant to the land but may be transferred in such a way as to become appurtenant to another parcel of land. R.C.W.A. 90.03.380.

CRIMINAL USES

It is a misdemeanor in Washington to take the water of another or to waste water to the detriment of another. R.C.W.A. 90.03.400. The willful interference with the rights of another also is a misdemeanor but may be a felony if damage meets the criteria of R.C.W.A. 9.61.070. R.C.W.A. 90.03.410. Obstruction of a right of way is also unlawful. R.C.W.A. 90.03.420.

FORMS

Washington uses the same form for both surface and groundwater rights. The application requires a nonrefundable $50.00 fee. The application includes a section for storage if necessary. The first section is biographical about the applicant. Section 2 is a statement of intent. Section 3 is the point of diversion. Section 4 is the place of use. Section 5 is a description of the water system. Section 6 is for domestic or municipal users. Section 7 is for irrigation and stock watering. Section 8 is for other water uses. Section 9 is for storage. Section 10 is for driving directions. Section 11 is for signatures.

Washington also uses a separate form for changes in use and transfer of a water right, which also requires a non-refundable $50.00 fee. This form covers changes of use, diversion, ownership, additional diversions, or additional purposes. The form includes detailed instructions. Washington does have an application for a permit to construct a reservoir and a separate application for drilling a well. The State of Washington has forms for notice of beginning and completion of construction and proof of appropriation. Additional forms are used for assignment of application or permit to appropriate or store water, requests for information, documents, or showing compliance, voluntary abandonment, administrative confirmation, and trust water rights.

TRACKING APPLICATIONS

Washington is divided into 62 water-resource inventory areas based on drainage systems. The state uses a water right tracking system (WRTS) where applications are divided by county. The system has an extensive key available at http://www.ecy. wa.gov/programs/wr/rights/tracking-apps.html.

WATER RIGHTS

A water right requires beneficial use in Washington. Such uses include dairy, multiple or single domestic uses, dust control, fish propagation, frost protection, heat exchange, hydropower, industrial, manufacturing and commercial, irrigation, mining, municipal, stock water, etc. A right may be relinquished after a period of five or more years of nonuse. Washington law specifies several reasons why a right might be exempt from relinquishment. See R.C.W.A. 90.14.140. Washington allows individuals to voluntarily relinquish a water right or the Department of Ecology must notify the holder that evidence shows the right has been relinquished in whole or in part (called an administrative order). The holder will be given opportunity to show cause and if they fail to do so the right will be relinquished.

RIGHTS TO INSTREAM FLOWS

Instream beneficial uses include preservation of uses such as fish, wildlife, and recreation. Instream uses may be considered beneficial uses under R.C.W.A. 90.03.345.

ADJUDICATION

When rights are disputed, adjudications can settle the disputes. When the area of the dispute is defined, the case is filed and rights holders in the defined area are notified. An evidentiary hearing is then held and evidence presented to a department referee in support of claims. The referee files a report of findings and recommendations to the court. If a claimant disagrees with the report he may file an exception and once the exceptions are all decided, the court issues an official decree. Finally, a certificate of adjudicated water rights is issued including, priority date, purpose of use, quantity, point of diversion, place of use and any limitations.

For help with claims and evidentiary support, see http://www.ecy.wa.gov/programs/wr/rights/adjhome.html.

BENEFICIAL USES (NOT EXCLUSIVE)

Dairy
Multiple or single domestic uses
Dust control
Fish propagation
Frost protection
Heat exchange
Hydropower
Industrial, manufacturing, and commercial
Irrigation
Mining
Municipal
Stock water
Recreation
Wildlife preservation

APPLICATIONS

Most applications in Washington require a $50.00 fee.
Application for appropriation of surface or groundwater, including storage, is $50.00.
Application for change or use or diversion is $50.00.
Application for permit to drill well (fees vary depending on the type $200–300 for wells, $40 for dewatering system and $50 for decommissioning a water well)
Application for construction of reservoir
Assignment of application or permit
Application for Trust Rights
Notice of beginning or completion of construction
Notice of proof of appropriation

WYOMING

WATER RIGHTS SYSTEM

Wyoming water law is founded on the doctrine of prior appropriation, or "first in time, first in right." The Wyoming constitution states that all natural waters within the boundaries of the state are property of the State. The State Engineer is charged with the regulation and administration of the state's water resources. Wyoming's water law is contained in Title 41, Wyoming Statutes Annotated, 1977, and can be found at: http://legisweb.state.wy.us/statutes/sub41.htm.

RESPONSIBLE AGENCY

The State Engineer's Office is the water rights administrator and is responsible for the appropriation, distribution, and management of the surface and groundwater throughout the state. Wyoming is divided into four water divisions for administration purposes. Each of these divisions is headed by a superintendent who administers the waters of each water division. These four superintendents and the State Engineer comprise the Wyoming Board of Control. The Board of Control meets quarterly to adjudicate water rights and to consider other matters pertaining to water rights and water appropriation. The Board of Control is also responsible for any requests for changes in point of diversion, change in use, change in the area of use, or abandonment of a water right.

APPLICATION PROCESS

Prior to statehood in 1890, a water right could be established by the use of water and the filing of a claim with the territorial officials. Water rights with priority dates before 1890 are termed "territorial" rights. Since statehood, however, the only way to obtain a surface or a groundwater right, is by filing an application with the State Engineer. The types of applications that can be filed in Wyoming can be seen in Appendix 1. The date the application is filed establishes the water right's priority date. The application is then reviewed and evaluated to ensure that the proposed use does not interfere with any existing rights or harm the public welfare. In addition to this review, groundwater applications (for projects over 25 gallons per minute) within a groundwater control area must be approved by the control area's advisory board. In these control areas, an application also must be advertised in a local newspaper.

For both surface and groundwater, the State Engineer has the authority to approve or reject the application. In approving an application, the State Engineer can impose conditions or limitations on the application to protect existing water rights, further define the extent of the application, and address any other issue deemed necessary. The applicant may appeal the State Engineer's final decision to the Board of Control if the applicant disputes the findings.

If an application is approved by the State Engineer, the application achieves the status of "permit." The permittee is then given a specified time period (usually one year) within which to commence any necessary construction, and an additional time period (usually five years) within which to complete the project and put the water to beneficial use. The permittee is required to submit a notice of commencement and a notice of completion with the State Engineer's office. When the notice of completion is received, a proof of completion is prepared. The proof is sent to the appropriate water division superintendent for field inspection and advertised for public comment. For groundwater rights, the State Engineer, not the superintendent, verifies the information through field inspections. Protests can be brought against the permit, and can lead to public hearings. Once proof of beneficial use is verified and any disputes are settled, the Board of Control is notified and they issue a "Certificate of Appropriation" (or a "Certificate of Construction" for reservoirs). It normally takes about three months to get an approved water right application back from the State Engineer's office. Therefore, the typical time frame for a permit is three months, but the final approval of the water right does not occur until the project has been constructed.

Once a certificate is issued, the water right is referred to as having "adjudicated status," and the right is listed in the tabulation of adjudicated rights. A water right that is not adjudicated (a water right that is going through the application process) is often referred to as an "inchoate right." Once adjudicated, the water right is permanently attached to the specific land or place of use described on the certificate, and it cannot be removed except by action of the Board of Control. Any disputes with the Board of Control can be appealed to district court.

POINT OF DIVERSION AND CHANGE OF USE PROCEDURES

A point of diversion is required for all water rights (except for instream flow rights which require the identification of the appropriate stream segment). Changes in the point of diversion require the filing of a petition with the State Engineer's office for unadjudicated rights and with the Board of Control for adjudicated rights. Although a point of diversion is required for all water rights, the water right is attached to and defined by the place of use, not the point of diversion.

Any changes in point of diversion, conveyance, or use is done through a petition. The petition goes to the Board of Control for adjudicated rights or to the State Engineer if the water right is inchoate. Changes of use are only granted if the quantity of transferred water does not exceed historic consumptive use or diversion rates, does not decrease the amount of historic runoff, and does not impair other existing rights.

STATE RECOGNIZED BENEFICIAL USES

Wyoming recognizes the following beneficial uses. Although these categories apply to both surface and ground water, the definition may be different when pertaining to surface as opposed to groundwater. In addition, water rights holders are limited to withdrawals necessary for the beneficial purpose, and these limits are established for each use (for example, irrigators are allowed to divert up to 1 cfs for each 70 acres under irrigation).

Irrigation
Municipal
Industrial
Power generation
Recreational
Stock
Domestic
Pollution control
Instream flows
Miscellaneous

GROUNDWATER

The application process for groundwater is quite similar to that for surface water (see above). In Wyoming, however, surface and groundwater are treated as hydrologically separate. If, however, a user protests that ground and surface water appear to be part of the same source, the state will investigate (using monitoring wells). If a hydrologic connection is found between the two sources, the water use is treated as one source. Until this hydrologic connection is established, groundwater and surface water are assumed to be separate. In addition, springs producing more that 25 gallons per minute are treated as surface water, and those producing less than 25 gallons per minute are treated as groundwater.

Prior to 1947, the groundwater division was responsible for maintaining a registration of groundwater rights for all uses except stock and domestic. In 1955, legislation was passed requiring that a permit be obtained from the State Engineer's office prior to the drilling of all wells, except stock and domestic wells. In 1969, the law was amended requiring a permit for the drilling of any water well. As a result, groundwater rights can only be obtained through the State Engineer. Groundwater rights are issued for the same beneficial uses as surface water rights.

Due to the large-scale development of groundwater for irrigation use in some areas of Wyoming, three groundwater management districts called Control Areas have been established. An advisory group is elected in each of the Control Areas to review new permit applications, review requests for water right changes, and advise the State Engineer's office regarding such items.

WATER RIGHTS

There are no restrictions in Wyoming as to who can hold a water right (with the exception of instream flows which can only be held by the state). Any entity including a federal agency, state board, corporation, district, or individual may hold a water right. In addition, water rights can be held jointly by a group of individuals where each individual is listed as a co-owner.

A water right in Wyoming is considered a property right, but it is a right which is attached to the lands or to the place of use specified in the permit. Water rights can be transferred to a new place of use through a petition to the State Engineer (or to the Board of Controls for adjudicated rights). Wyoming water law, however, expressly prohibits the sale of water rights. Since water rights are attached to the land, they cannot be sold separately from that land, but can be included in the sale of land.

A water right in Wyoming can be lost by abandonment. There are three ways in which abandonment can be initiated. The first is voluntary abandonment by the water right holder. In second another water user can claim that the reactivation of an allegedly abandoned water right would injure their right. This occurs if a right has not been used for a period of five consecutive years, and a junior (in some case a senior) appropriator brings a declaration of abandonment to the Board of Control. The third way abandonment can occur is the State Engineer can initiate it if it is felt water has not been put to beneficial use for five consecutive years and a reallocation would be in the public interest. Water lost through abandonment reverts back to the public and is subject to future appropriation.

ADJUDICATIONS

Adjudications are conducted for both surface and groundwater in Wyoming, and adjudicated rights can be obtain both through the administrative process and through court order. The application process discussed above results in an adjudicated right through the administrative procedures. Once a certificate is issued by the Board of Control, the water right is adjudicated and listed in the tabulation of adjudicated rights. General adjudications through the courts also result in adjudicated rights. The primary reason for general adjudications in Wyoming is the determination and integration of tribal and federal water rights.

Once a water right is adjudicated, any action on that right (change of use, place of diversion, etc.) must go through the Board of Control. When an entity holds an adjudicated water right, no further inspection is required and the owner is not required to continually submit proof of beneficial use. An adjudicated right exists in perpetuity and can only be lost through abandonment (see above).

ONGOING ADJUDICATIONS

The adjudication of water rights in Wyoming is part of the ongoing application process. In addition to these administrative adjudications, general adjudications can take place through the courts. The only general adjudication taking place is in division three (the Big Horn Basin). This adjudication is almost complete and no other general adjudications are currently occurring.

INSTREAM FLOWS

Instream flow legislation was enacted in Wyoming in 1986. Only the State of Wyoming may hold a right for instream flow, but no single agency has sole responsibility for the instream flow program. The Game and Fish Department identifies priority streams, prepares biological assessments, and makes instream flow recommendations to the Water Development Commission. The commission prepares hydrologic analysis and then applies to the State Engineer for an instream flow water right. The State Engineer studies the feasibility of the instream flow segment and has the authority to approve the application. A public hearing is required, at which, information is presented and there is an opportunity for public comment. If approved by the State Engineer, an instream flow right is established. Water for instream flow can come from new appropriation or through the transfer of existing rights. The transfer of existing water rights, however, can only be done by voluntary transfer or gift. In order to ensure "voluntary" transfers, Wyoming law expressly denies any power of condemnation or the purchase of existing rights for instream flow.

Recognized Beneficial Uses for Instream Flow

Instream flow rights in Wyoming may only be used to establish or maintain new or existing fisheries. Other uses commonly associated with instream flow (recreation, aesthetics, water quality, etc.) are not defined as beneficial uses under Wyoming water law.

Holdership of Instream Flow Water Rights

Only the State of Wyoming may apply for and hold an instream flow right. Other entities, however, may request application for an instream flow right. In addition, the State of Wyoming can accept water rights as a gift and convert them to instream flow (as long as the purpose is to support fisheries).

Quantification Requirements and Procedures

Wyoming requires an assessment of the entire reach of the stream covered by the proposed instream flow right. The Game and Fish Department must analyze the stream and determine that the proposed flows are adequate to support fisheries.

BLM-Specific Information

The State of Wyoming does not require a right-of-way approval by the BLM prior to approving an application. There is a statement on the water-right permit form which states that the granting of a water right does not grant an easement and that the applicant is responsible for obtaining any rights-of-way needed to perfect the permit.

The BLM pays filing fees for water-rights applications. The fee for stock reservoirs, wells, and springs are $25, and the fee for any dam over twenty feet high or impounding more than twenty acre-fee is $100.

Appendix 1: Types of Applications

Permits are issued in Wyoming for:

Transporting water through ditches or pipelines
Storage in reservoirs
Storage in smaller (under 20 acre-feet of capacity and a dam height less than
   20 feet) reservoir facilities for stockwater or wildlife purposes
Enlargements to existing ditch or storage facilities
Instream flow purposes

Available at http://www.blm.gov/nstc/WaterLaws/wyoming.htm, September 2007.

Appendix 5

Forms and Illustrations

Form 1 (United States Dept. of the Army)

ENG Form 4345 Application for Dept. of the Army permit

APPLICATION FOR DEPARTMENT OF THE ARMY PERMIT ENG FORM 4345 (33 CFR 325)	OMB APPROVAL NO.0710-0003 Expires December 31, 2004

Public reporting burden for this collection of information is es timated to average 5 hours per response, including the time for reviewing instructions, searching existing data sources, gathering and maintaining the data needed, and completi ng and reviewing the collection of information. Send comments regarding this burden estimate or any other aspect of this collection of information, including suggestions for reducing this burden, to Department of Defense, Washington Headquarters Service Directorate of InformationOperations and Reports, 1215 Jefferson Davis Highway, Suite 1204, Arlington, VA 22202-4302; and to the Office of Man agement and Budget, Paperwork Reduction Project (0710-00031, Washington, DC 20503. Please DO NO RETU RN your form to either of those addresses. Completed ap plications must be submitted to the District Engineer having jurisdiction over the location of the proposed activity.

PRIVACY ACT STATEMENT

Authority: 33 USC 401, Section 10; 1413, Section 404. Principal Purpose: These laws require permits authorizing activities in, or affecting, navigable waters of the United States, the discharge of dredged or fill ma terial into waters of the United States, and the transportation of dredged material for the purpose of dumping it into ocean waters. Routine Uses. Information provided on this form will be used in evaluating the application for a permit. Disclosure: Disclosure of requested information is voluntary. If information is not provided, however, the permit applic ation cannot be processed nor can a permit be issued.

One set of original drawings or good reproducible copies which show the location and character of the proposed activity must be attached to this application (see sample drawings and instructions) and be submitted to the District Engineer having jurisdiction over the location of the proposed acti vity. An application that is not completed in full will be returned.

(ITEMS 1 THRU 4 TO BE FILLED BY THE CORPS)			
1. APPLICATION NO.	2. FIELD OFFICE CODE	3. DATE RECEIVED	4. DATE APPLICATION COMPLETED

(ITEMS BELOW TO BE FILLED BY APPLICANT)	
5. APPLICANT'S NAME	8. AUTHORIZED AGENT'S NAME AND TITLE (an agent is not required)
6. APPLICANT'S ADDRESS	9. AGENT'S ADDRESS
7. APPLICANT'S PHONE NUMBERS W/AREA CODE	10. AGENT'S PHONE NUMBER W /AREA CODE
a. Residence	a. Residence
b. Business	b. Business
11. STATEMENT OF AUTHORIZATION	

I hereby authorize _____ to act in behalf as my agent in the processing of this application and to furnish, upon request, supplemental information in support of this permit application.

APPLICANT'S SIGNATURE	DATE

NAME, LOCATION, AND DESCRIPTION OF PROJECT OR ACT IVITY	
12. PROJECT NAME OR TITLE (see instructions)	
13. NAME OF WATERBODY, IF KNOWN (if applicable)	14. PROJECT STREET ADDRESS (if applicable)

15. LOCATION OF PROJECT

_____   _____
        COUNTY   STATE

16. OTHER LOCATION DESCRIPTIONS, IF KNOWN, (see instructions)

17. DIRECTIONS TO THE SITE

18. Nature of Activity (Description of Project, include all features)

19. PROJECT PURPOSE (Describe the reason or purpose of the project, see instructions)

USE BLOCKS 20-22 IF DREDGED AND/OR FILL MATERIAL IS TO BE DISCHARGED

20. REASON(S) FOR DISCHARGE

21. TYPE(S) OF MATERIAL BEING DISCHARGED AND THE AMOUNT OF EACH TYPE IN CUBIC YARDS

22. SURFACE AREA IN ACRES OF WETLANDS OR OTHER WATERS FILLED (see instruction)

23. IS ANY PORTION OF THE WORK ALREADY COMPLETE?   YES   NO   IF YES DESCRIBE THE COMPLETED WORK

24. ADDRESSES OF ADJOINING PROPERTY OWNERS, LESSEES, ETC., WHOSE PROPERTY ADJOINS THE WATERBODY
(if you have more that can be here,
please attach a supplemental list).

25. LIST OF OTHER CERTIFICATIONS OR APPROVAL/DENIALS RECEIVED FROM OTHER FEDERAL, STATE, OR LOCAL AGENCIES FOR WORK DESCRIBED IN THIS APPLICATION.

AGENCY   TYPE APPROVAL*   IDENTIFICATION NUMBER   DATE APPLIED   DATE APPROVED   DATE DENIED

* Would include but is not restricted to zoning, building and floodplain permits

26. Application is hereby made for a permit or permits to authorize the work described in this application. I certify that the information in this application is complete and accurate. I further certify that I possess the authority to undertake the work described herein or am acting as the duly authorized agent of the applicant.

_____   _____   _____   _____
  Signature of Applicant   Date   Signature of Agent   Date

The application must be signed by the pers
11 has been filled out and signed.

  18 U.S.C. Section 1001 provides that: whoever, in any manner within the jurisdiction of any department or agency of the United States knowingly and willfully falsifies, conceals, or covers up any trick, scheme, or disguises a material fact or makes any false, fictitious or fraudulent statements or representations or makes or uses any false writing or document knowing same to contain any false, fictitious or fraudulent statements or entry, shall be fined not more than $10,000 or imprisoned not more than five years or both.

Form available at 13 Fed Proc Forms § 51:31, see § 51:32 for checklist; Procedures found in 25 Fed Proc L Ed §§ 57:229 – 57:247. Also available online
http://www.nwk.usace.army.mil/regulatory/permit_application.pdf March 2008.

Form 2 (Oregon)
Sample Application for Water Right Transfer

Oregon Water Resources Department
725 Summer Street NE, Suite A
Salem Oregon 97301-1266
(503) 986-0900
www.wrd.state.or.us

Application for Water Right

Transfer

Please type or print legibly in dark ink. If your application is incomplete or inaccurate, we will return it to you. If any requested information does not apply to your application, insert "N/A" to indicate "Not Applicable." As you complete this form, please refer to notes and guidance included on the application. A summary of review criteria and procedures that are generally applicable to these applications is available at www.wrd.state.or.us/OWRD/PUBS/forms.shtml.

1. TYPE OF TRANSFER APPLICATION

Please check one

☐ Permanent Transfer	☐ Instream Transfer
☐ Temporary Transfer (1 to 5 yrs.)	☐ Permanent
● total number of years: ____	☐ Time-Limited
(begin year:)	☐ Drought Transfer
(end year: _____)	☐ Other

2. APPLICANT INFORMATION

Name: _____
           First   Last

Address: _____

_____
         City   State   Zip

Phone: _____
         Home   Work   Other

Fax: _____ E-Mail address: _____

3. AGENT INFORMATION

(The agent listed is authorized to represent the applicant in all matters relating to this transfer application.)

Name: _____
           First   Last

Address: _____

_____
         City   State   Zip

Phone: _____
         Home   Work   Other

Fax: _____ E-Mail address:_____

- If an agent is listed above, please check **_one_** of the following:
  - ☐ Please send all correspondence to Agent. Send *copies* of correspondence to Applicant; **_or_**
  - ☐ Please send all correspondence to Applicant. Send *copies* of correspondence to Agent.

4. PROPOSED CHANGE(S) TO WATER RIGHT(S)

- List **all** water rights to be affected by this transfer. Indicate the certificate, permit, decree or other identifying number(s) in the table below: *(Attach additional pages as necessary.)*

	Application / Decree	Permit / Previous Transfer	Certificate
1.			
2.			
3.			
4.			
5.			
6.			

- Attach a **separate** *Supplemental Form A (Description of Proposed Change(s) to a Water Right)* for **each** water right listed above.

- Check **all** proposed change(s) included in this transfer application:

  ☐ Place of Use   ☐ Point of Diversion (POD)   ☐ Additional Point of Diversion

  ☐ Character of Use   ☐ Point of Appropriation (POA, or well)   ☐ Additional POA

  ☐ Instream Transfer   ☐ Surface Water POD to Ground Water POA

  ☐ Substitution of Supplemental Groundwater right for Primary Surface water right

  ☐ Historic POD change   ☐ Other

- Reason(s) for change(s): _____

  _____

5. WATER DELIVERY SYSTEM

- Describe the *current* water delivery system <u>or</u> the system that *was in place* at some time <u>within the last 5 years</u>. Include information on the pumps, canals, pipelines and sprinklers used to divert, convey and apply the water at the authorized place of use. If the transfer involves multiple rights that have independent systems, describe each system separately.

  The description must be sufficient to demonstrate that the full quantity of water to be transferred can be conveyed from the authorized source and applied at the authorized location and that the applicant is ready, willing, and able to exercise the right.

  _____

  _____

  _____

- System capacity: _____ cubic feet per second (cfs). If the transfer involves multiple rights that have independent systems, describe the capacity for each system separately.

  _____

  _____

6. EVIDENCE OF BENEFICIAL WATER USE

- Attach one or more **Evidence of Use Affidavits** (Supplemental Form B) demonstrating that each of the right(s) involved in the transfer have been exercised in the last five years in accordance with the terms and conditions of the right or that a presumption of forfeiture for non-use could be rebutted. The Evidence of Use Affidavit(s) **must include supporting documentation** such as the following:

  ► Copies of receipts from sales of irrigated crops or for expenditures relating to use of water;

  ► Records such as Farm Service Agency crop reports, irrigation district records, an NRCS farm management plan, or records of other water suppliers;

  ► Dated aerial photographs of the lands or other photographs containing sufficient detail to establish location and date of the photograph; *or*

  ► If the right has **not** been used during the past five years, documentation that the presumption of forfeiture would be rebutted under ORS 540.610(2).

7. AFFECTED DISTRICTS

- Are any of the water rights proposed for transfer located within or served by an irrigation or other water district? ☐ Yes ☐ No

- Will any of the water rights be located within or served by an irrigation or other water district after the proposed transfer? ☐ Yes ☐ No

- Is water for any of the rights supplied under a water service agreement or other contract for stored water with a federal agency or other entity? ☐ Yes ☐ No

  If "Yes", for any of the above, list the name and mailing address of the district, agency and/or entity:

  _____

  _____

  _____

8. LOCAL GOVERNMENTS

- List the name and mailing address of all local governments (i.e., each county, city, municipal corporation, or tribal government within whose jurisdiction water will be diverted, conveyed or used).

  _____

  _____

  _____

9. LAND OWNERSHIP

- Does the applicant own the lands **FROM** which the right is being transferred? ☐ Yes ☐ No

  *If "No", provide the following information. **For Temporary Transfers**, also include a **notarized statement granting consent** to the transfer from **each** of the landowners (for Permanent Transfers see Section 12(c)):*

  Names of Current Landowner(s): _____
                                                First   Last

  Address: _____

  _____
            City   State   Zip

- Does the applicant own the lands **TO** which the right is being transferred?

      ☐ Yes   ☐ No   ☐ N/A - *NOT APPLICABLE TO INSTREAM TRANSFERS*

  If "No", provide the following information:

  Names of Receiving Landowner(s): _____
                                                First   Last

  Address: _____

  _____
            City   State   Zip

- Check **one** of the following:

  ☐ The receiving landowner will be responsible for completion of the proposed changes after the final order is issued. All notices and correspondence should be sent to this landowner.

  ☐ The applicant will remain responsible for completion of changes. Notices and correspondence should continue to be sent to the applicant and applicant's agent.

  ☐ N/A. *(Not applicable. Application is for an Instream Water Right Transfer.)*

10. Other Remarks (optional)

_____

_____

_____

_____

_____

_____

11. ATTACHMENTS

Check each of the following attachments included with this application.
*The application will be returned if all required attachments are **not** included.*

Supplemental Form A –
Description of Proposed Change(s) to a
Water Right

☐ A **separate** Supplemental Form A is enclosed
for **each** water right to be affected by this
transfer.

Supplemental Form B –
Evidence of Use Affidavit(s)

☐ At least one Evidence of Use Affidavit
documenting that the right has been used
during the last five years or that the right is not
subject to forfeiture under ORS 540.610 is
attached. The affidavit provided must be the
original (not a copy), **and**

☐ The Evidence of Use Affidavit **must be**
accompanied by **supporting documentation.**

Map

☐ *Water Right Transfer*
The map must be prepared by a Certified Water
Right Examiner and meet the requirements of
OAR 690-380-3100 unless a waiver has been
granted. The map provided must be the
original, not a copy.

☐ *Temporary Transfer or Historical POD*
Change
A map meeting the requirements of OAR 690-
380-3100 must be included but need **not** be
prepared by a Certified Water Right Examiner.

Water Well Report(s)/Well Log(s):

☐ The application is for a change in point of
appropriation or change from surface water to
ground water and copies of all water well
reports are attached.

☐ Water well reports are not available and a
description of construction details including
well depth, static water level, and information
necessary to establish the ground water body
developed or proposed to be developed is
attached.

☐ N/A. The application does **not** involve a
change in point of appropriation or a change
from surface water to ground water, so water
well reports are **not** required.)

Land Use Information For Proposed Changes:

For Instream Transfers:

☐ Notice of the intent to file an instream transfer
application has been provided to each affected
local government along the proposed reach, and
copies of the notices are enclosed. (*For instream*
transfers a Land Use Information Form is not
required.)

For All Other Transfers:

☐ Land Use Information Form is enclosed; *or*

☐ **All** of the following criteria are met, therefore a
Land Use Information Form is not required:

❶ In EFU zone or irrigation district,
❷ Change in place of use only,
❸ No structural changes needed, including
diversion works, delivery facilities, other
structures, *and*
❹ Irrigation only.

Fees:

☐ Amount enclosed: $_____
See the Department's Fee Schedule at
www.wrd.state.or.us or call (503) 986-0900.

Instream Water Right Transfers, also include:

Supplemental Form C –
Instream Water Right Transfer

☐ Complete this form to describe the desired nature
and attributes for the proposed instream water
right.

Temporary Transfers, also include:

Recorded Deed:

☐ The applicant must submit a copy of the current
deed of record for the land **from** which the
authorized place of use or point of
diversion/appropriation is being moved.

Affidavit of Consent:

☐ If the applicant is **NOT** the owner of record for the
land **from** which the authorized place of use or
point of diversion/appropriation is being moved, a
notarized statement from the actual owner of
record consenting to the proposed transfer must be
submitted.

Before submitting your application to the Department, be sure you have:
- Answered each question completely.
- Included all the required attachments.
- Provided original signatures for **all** named deed holders, or other parties, with an interest in the water right.
- Included a check payable to the Oregon Water Resources Department for the appropriate amount.

---

12. SIGNATURES

▪ **Check *one* of the following, as appropriate, and sign the application in the signature box below:**

☐ In accordance with OAR 690-380-3000(13)(a), **I (we) understand that prior to Department approval of a permanent transfer and upon my receipt of a draft Preliminary Determination for the proposed transfer, I (we) will be required** *[pursuant to OAR 690-380-4010(5)]* **to provide the following landownership information and evidence demonstrating that I (we) are authorized to pursue the transfer:**

    (a)   A **report of ownership and lien information** that has been prepared by a title company *within the last three months;*

    (b)   A **copy of written notification** of the proposed transfer provided by the applicant to **all lien holders** on the subject lands unless the report of ownership and lien information shows that a water right conveyance agreement has been recorded for the subject lands. *If a water right conveyance agreement has been recorded for the subject lands, a copy of the agreement and identification of the owner of the lands at the time the agreement was recorded must be submitted;* **and**

    (c)   If the landowner identified in the report of ownership and lien information is **not** the applicant, a **notarized statement consenting to the transfer** *(attached)* signed by the landowner identified in the report or an authorized representative of the entity to whom the interest in the water right has been conveyed as identified in a water right conveyance agreement or other documentation demonstrating that the applicant is authorized to pursue the transfer in the absence of the consent of the landowner.

☐ **I (we) affirm that interest in the water right has been conveyed to someone other than the landowner**, and documents, **including a water right conveyance agreement**, report of ownership and lien information, and any required lienholder notification **are enclosed**, demonstrating that I am (we are) authorized to pursue the transfer. **I (we) understand that we may be required to provide additional information upon receipt of the draft Preliminary Determination.**

☐ **I (we) affirm that the applicant is a municipality, as defined in ORS 540.510(3)(b), and that the right is in the name of the municipality or a predecessor.** Therefore, pursuant to OAR 690-380-3000(13)(b), the applicant is **NOT** required to provide the above described report of ownership and lien information.

☐ **I (we) affirm that the applicant is an entity with the authority to condemn property and is acquiring the property to which the water right proposed for transfer is appurtenant by condemnation.** Documentation is provided with this application supporting this statement. Therefore, pursuant to OAR 690-380-3000(13)(c), the applicant is **NOT** required to provide the above described report of ownership and lien information. *(NOTE: Such an entity may only apply for a transfer under this subsection if it has filed a condemnation action to acquire the property.*

☐ **I (we) affirm that this is a temporary transfer and a copy of the deed** for the "from" land (and affidavits of consent from any other landowners, if applicable) **is enclosed.**

▪ **I (we) affirm that the information contained in this application is true and accurate.**

_____	_____	_____
applicant signature	name *and title if applicable* (print)	date
_____	_____	_____
applicant signature	name *and title if applicable* (print)	date

Form 3 (Washington)
Sample Water Right Application

State of Washington
Application for a Water Right Permit
☐ SURFACE WATER ☐ GROUND WATER
☐ Permanent ☐ Temporary ☐ Short Term

Follow the attached instructions. Attach additional sheets as necessary.

For Ecology Use
(Date Stamp)

A NON-REFUNDABLE **MINIMUM** FEE OF $50.00 PAYABLE TO
THE DEPARTMENT OF ECOLOGY MUST ACCOMPANY THIS APPLICATION.

Section 1. APPLICANT

Applicant/Business Name:	Phone No:	Other No:
Address:		
City:	State:	Zip:
Email Address (optional):		

Contact Name (if different from above):	Phone No:	Other No:
Relationship to Applicant:		
Address:		
City:	State:	Zip:
Email Address (optional):		

Section 2. STATEMENT OF INTENT

Briefly describe the purpose of your proposed project: _____
_____
_____

Anticipated length of time to complete your project:_____

Water Use List all purposes for which water will be applied to a beneficial use and list quantity required for each.

Purpose(s) of Use	Rate (check one box only) ☐ Cubic Feet per Second (CFS) ☐ Gallons per Minute (GPM)	Acre-Feet per Year (AF/YR) (If known)	Period of Use (Continuously or Seasonal)
TOTAL:			

Short Term/Temporary Water Use

Is this a request for a short term project (less than four months and non-recurring)? ☐ YES ☐ NO

Is this request for a temporary permit? ☐ YES ☐ NO

If yes to either question above, indicate the dates that the water will be needed:

FROM: ____/____/____ TO: ____/____/____

For Ecology Use	APPLICATION NO: _____ SEPA: Exempt/Not Exempt
	Fee Paid:_____ Check No:_____ ECY Coding: 001-001-WR1-0285-000011
Date Returned _____ By _____ Priority Date _____ By _____ WRIA:_____	

Section 3. POINT OF DIVERSION OR WITHDRAWAL

Complete A or B, and C below

A.) If Surface Water Source	B.) If Ground Water Source
☐ Spring ☐ Creek ☐ River ☐ Lake ☐ Other:_____	☐ Well(s) ☐ Other:_____
Source Name:_____	Well diameter & depth:_____
Tributary to:_____	Number of proposed points of withdrawal:____ Do you have an existing well? ☐ YES ☐ NO
Number of proposed diversion points:_____ Do you have an existing diversion? ☐ YES ☐ NO	If available, attach Water Well Report and pump test. Well Tag ID No._____

C.) Point of Diversion/Withdrawal – Legal Description

Parcel No.	¼	¼	Section	Township	Range	County

Lot(s)	Block(s)	Subdivision

If known, enter the distances in feet from the point of diversion or withdrawal to the nearest section corner:

_____ Feet (☐ North/☐ South) and _____ feet (☐ East/☐ West)

from the (☐NW ☐SW ☐NE ☐SE ☐ ____) corner of Section_____.

Parcel No.	¼	¼	Section	Township	Range	County

Lot(s)	Block(s)	Subdivision

If known, enter the distances in feet from the point of diversion or withdrawal to the nearest section corner:

_____ feet (☐ North/☐ South) and _____ feet (☐ East/☐ West)

from the (☐NW ☐SW ☐NE ☐SE ☐____) corner of Section_____

NOTE: *If more than two points of diversion/withdrawal attach additional information on a separate sheet of paper.*

Do you own the land on which the proposed point of diversion/withdrawal is located? ☐ YES ☐ NO
If no, do you have legal authority to make this application for use of another's land? ☐ YES ☐ NO
Provide the owner name(s), address, and phone number:_____

_____

Section 4. PLACE OF USE

Attach a copy of the legal description of the property (on which the water will be used) taken from a real estate contract, property deed or title insurance policy, or copy it carefully in the space below.

¼	¼	Section	Twp.	Range	County	Parcel No.

Do you own all the lands on which the proposed place of use is located? ☐ YES ☐ NO.

If no, do you have legal authority to make this application for use of another's land? ☐ YES ☐ NO
Provide owner name(s), address, and phone number:_____

_____

Are there any other water rights or claims associated with this property or water system? ☐ YES ☐ NO

If yes, provide the water right and/or claim numbers:_____

_____

Attach a map of your project showing the point of diversion/withdrawal and place of use. If platted property, be sure to include a complete copy of the plat map.

Section 5. WATER SYSTEM DESCRIPTION

Describe your proposed water system (include type and size of devices used to divert or withdraw water from source): _____

_____

_____

_____

_____

_____

Section 6. DOMESTIC WATER SUPPLY SYSTEM INFORMATION

Complete A or B, and C below

A.) Domestic Water Systems only	B.) Municipal Water Systems only (defined under RCW 90.03.015)
Projected number of connections to be served: _____	Present population to be served water: _____
Type of connections:_____ (e.g., home, recreational cabin)	Estimate future population to be served: _____ (20 year projection)

C.) Water System Planning

Do you have a Water System Plan approved by the Washington State Department of Health, Drinking Water Division? ☐ YES ☐ NO

If yes, date plan was approved ____/____/____ Water System Number:_____

Name of water system:_____

Are you within the service area of an existing water system? ☐ YES ☐ NO

If yes, explain why you are unable to connect to the system:_____

_____

_____

_____

_____

Section 7. IRRIGATION/STOCKWATER/OTHER FARM USES

Irrigation

Total number of acres requested to be irrigated under this application = _____ACRES
NOTE: Outline the area to be irrigated on your attached map.

Stockwater

List number and kind of stock:_____

_____

_____

Is the proposed project for a dairy farm? ☐ YES ☐ NO

Other Proposed Farm Uses
Describe all proposed uses: _____

_____

_____

_____

Family Farm Water Act (RCW 90.66):

Calculate the acreage in which you have a controlling interest, including only:
- Acreage irrigated under water rights acquired after December 8, 1977,
- Acreage proposed to be irrigated under this application, and
- Acreage proposed to be irrigated under other pending application(s).

Is the combined acreage under existing rights greater than 6000 acres? ☐ YES ☐ NO

Do you have a controlling interest in a Family Farm Development Permit? ☐ YES ☐ NO

If yes, enter Permit No: _____

Section 8. OTHER WATER USES

Hydropower

Indicate total feet of head _____ and proposed capacity in kilowatts:_____

Describe works:_____

_____

_____

Indicate all uses to which power is to be applied:_____

FERC License No: _____

Mining/Industrial Use

Describe use, method of supplying and utilizing water:_____

_____

_____

Other Use

_____

_____

_____

_____

Section 9. WATER STORAGE

Will you be using a dam, dike, or other structure to retain or store water? ☐ YES ☐ NO

Are you proposing to store more than 10 acre-feet of water? ☐ YES ☐ NO

Will the water depth be 10 feet or more? ☐ YES ☐ NO

If you answered yes to any of the above questions, please describe:_____

_____

NOTE: If you will be storing 10 acre-feet or more of water and/or if the water depth will be 10 feet or more at the deepest point and some portion of the storage will be above grade, you must also complete an Application for Permit to Construct a Reservoir and a Dam Construction Permit and Application.

Section 10. DRIVING DIRECTIONS

Provide detailed driving directions to the project site:_____

_____

_____

_____

Site Address:_____

_____

Section 11. REQUIRED SIGNATURES

I certify that the information provided in this application is true and accurate to the best of my knowledge. I understand that in order to process my application, I grant staff from the Department of Ecology access to the site for inspection and monitoring purposes. Even though the employees of the Department of Ecology may have assisted me in the preparation of the above application, all responsibility for the accuracy of the information rests with me, the applicant.

Print Name (Applicant or authorized representative)	Signature	Date
Print Name (Landowner of Place of Use)	Signature	Date
Print Name (Landowner of Place of Use)	Signature	Date
Print Name (Landowner of Place of Use)	Signature	Date

Submit your application to: DEPARTMENT OF ECOLOGY
CASHIERING SECTION
PO BOX 5128
LACEY WA 98509-5128

Please check the region in which your proposed project is located.
☐ Southwest ☐ Northwest ☐ Central ☐ Eastern

Below is a map of the State of Washington, with outlines of the four Ecology regional offices. If you have questions about your application, contact the Water Resources program at the regional office in which your project is located.

Southwest Regional Office: 360-407-6300

Northwest Regional Office: 425-649-7000

Central Regional Office: 509-575-2490

Eastern Regional Office: 509-329-3400

If you need this document in an alternate format, please call the Water Resources Program at 360-407-6600. Persons with hearing loss can call 711 for Washington Relay Service. Persons with a speech disability can call 877-833-6341

State of Washington
INSTRUCTIONS for the
Application for a Water Right Permit

Please read these instructions carefully. Be accurate and complete in filling out your application, as the information you provide is very important in processing your application. Be sure to attach your fees, maps, and any additional information related to the water uses you are proposing.

If you need assistance, please contact the regional office in which your project will be located. A map of the Ecology regions is on the back page of the application. If your answers to any questions are longer than the space provided, you may attach additional sheets as necessary.

Check Boxes

Check the appropriate box for Surface or Ground Water.
Check the appropriate box for Permanent, Temporary, or Short Term use (duration of 4 months or less).

Application Fee

- A minimum fee of $50.00 is required for each new application for a water right permit.
- No fees are required for applications to be processed under a Cost Reimbursement contract.
- No fees are required for Emergency Drought Applications (only when a drought is declared).

If additional fees are required, Ecology will send you a letter requesting those fees. If you are unsure of the appropriate fee amount, contact your regional office for more information, or visit our website: <http://www.ecy.wa.gov/programs/wr/rights/wr_fees.html>.

**Please make checks or money orders payable to the "Department of Ecology." Cash cannot be accepted.
ALL FEES ARE NONREFUNDABLE.**

Section 1. APPLICANT

Enter the name of the person, organization, or water system for which the water right permit is requested. For instance, if the permit is required for a community water system, enter the name of the system (e.g. Green Acres Water Works). Enter a mailing address, including zip, daytime telephone, an alternate or cell phone number, and an Email address (if you have one).

Provide the name of a contact person (if different from above) to call in case we have questions about the application or proposed project. Describe the relationship of the contact person to the applicant, e.g. "consultant," "water systems engineer," "realtor," "chair of community well organization," etc.

Section 2. STATEMENT OF INTENT

Provide a brief description of the purpose of your proposed project and the anticipated length of time to complete the project.

Water Use
List the purpose(s) for which you are proposing to use the water (**see examples of purposes below**). Check the appropriate box to indicate if the rate you have provided is measured in cubic feet per second or gallons per minute. For each purpose provide the maximum rate at which water is proposed to be taken from the water source. If known, provide the maximum quantity to be used for the purposes in acre-feet per year. Provide period of use (months) in which the water will be used for each purpose. Total the water needs for each purpose of use and write the total within the space provided.

Short Term/Temporary Water Use
If this application is being submitted for a short term (less than four months – see Policy 1037) or temporary water use (see Policy 1035), check the appropriate box and indicate the dates the water will be needed.

For more information on Water Resources Program Policies, contact your regional office or visit our website: <http://www.ecy.wa.gov/programs/wr/rules/pol_pro.html#wradminpolicy>.

Examples of purpose(s)
Be sure that you include ALL uses that you propose, not just the major use of water. Some examples are:

- Dairy
- Domestic-Multiple
- Domestic-Single
- Dust Control
- Fish Propagation
- Frost Protection
- Heat Exchange

- Hydropower
- Industrial/Manufacturing/Commercial
- Irrigation
- Mining
- Municipal
- Stockwater
- Other (describe)

ECY 040-1-14 (Rev. 5/07)   [6]   APPLICATION FOR A WATER RIGHT PERMIT

Section 3. POINT OF DIVERSION OR WITHDRAWAL

Complete A or B, and C

A.) If Surface Water Source
Check the appropriate box if you plan to divert water from a spring, creek, river, lake, or other (describe). Enter the source name, e.g. "Wenatchee River." If the source feeds another body of surface water, give the name of the body of water to which the source is a tributary, e.g. "Columbia River." Enter the number of proposed diversion points. Check the appropriate box if you have an existing diversion.

B.) If Ground Water Source
Check the appropriate box if you plan to withdraw water from a well or other ground water system (describe). Enter the diameter, depth, and the number of proposed points of withdrawal (wells). Check the appropriate box if you have an existing well. If the well has been constructed, attach a Water Well Report. If you have already done a pump test, attach a copy of the pump test results. Provide the Well Tag ID number, if available.

C.) Point of Diversion/Withdrawal Location – Legal Description
Enter the parcel number, quarter-quarter (¼¼), section, township, range and county in which each point of diversion or withdrawal is located. If the location has been platted (subdivided), enter the lot, block, and subdivision name. You can generally obtain this information from a legal description or plat of the property, or from your county assessor's office. If there are more than two points of diversion or withdrawal, attach additional information on a separate sheet of paper.

If known, enter the distances in feet from the nearest section corner to each point of diversion or withdrawal (e.g. 420 feet south and 150 feet west from the Northeast Corner of Section 12). You can obtain this information by measuring the distance on a USGS map, other map drawn to scale, or by measurement on the ground.

Check if you own the land containing the proposed point of diversion/withdrawal. If you don't own the land, provide the owner's name(s), address, and phone number. Please check whether you have legal authority to make this application for use of another's land.

Section 4. PLACE OF USE

Attach a legal description of the lands where you propose to use the water or copy it carefully in the space provided. You can usually obtain a legal description from a survey, county assessor's office, real estate contract, title insurance policy, or property deed. Also include the tax parcel number(s) if available.

Check if you own all of the lands on which the proposed place of use is located. If you do not own the lands, provide the owner's name(s), address and phone number. If this is a community or municipal water system, please include a copy of your current and future service area map.
NOTE: Landowner's signature is required in Section 11.

Check if there are any other water rights or claims associated with this property or water system. If yes, provide the water right and/or claim numbers.

Attach a map of your project showing the point(s) of diversion/withdrawal and place of use. If platted property, be sure to include a complete copy of the plat map.

Section 5. WATER SYSTEM DESCRIPTION

Provide a description of your proposed project, explaining how you will divert, pump, distribute, and store the water, and any conservation measures you may be taking. Include proposed size, capacity, location, and motor horsepower of any pump.

Section 6. DOMESTIC WATER SUPPLY SYSTEM INFORMATION

Complete A or B, and C

A.) Domestic Water Systems
Enter the projected number of connections to be served and the type of connection (e.g. home, recreational cabin).

B.) Municipal Water Systems (as defined under RCW 90.03.015)
Enter the present population to be served water and estimate the future population to be served (20 year projection).

C.) Water System Planning
Check yes if you have a Water System Plan approved by the Washington State Department of Health, Drinking Water Division. Provide the date the plan was approved, as well as the water system number. Enter the name of the water system (e.g. Johnson Point Water Association).

Check yes, if you are within the service area of an existing water system and explain why you are unable to connect to the system.

Section 7. IRRIGATION/STOCKWATER/OTHER FARM USES

Irrigation

ECY 040-1-14 (Rev. 5/07)　　　　　[7]　　　　　APPLICATION FOR A WATER RIGHT PERMIT

Provide the total number of acres of land to be irrigated in the space provided. The number of acres to be irrigated should not include lands within the general irrigation area that may contain buildings, roads, etc. Outline the area to be irrigated on your attached map from Section 4.

Stockwater
Indicate total number of animals receiving stockwater and the type of animal (e.g. goats, chickens, llamas).

Check yes if the proposed project is for a dairy farm.

Other Proposed Farm Uses
Describe all other proposed farm uses (e.g. frost protection, heat control, or harvesting) listed in Section 2 and provide the proposed number of acres of land upon which each purpose would occur. Also note other uses of water on the farm (e.g. cleaning the milking parlor, washing cattle, or for a cooling system) and how much water is needed for each use.

Family Farm Water Act (RCW 90.66)
In order to comply with the Family Farm Water Act, indicate if you have a controlling interest in more than 6,000 acres of irrigation as defined in RCW 90.66.040(3). This includes the number of acres that are irrigated under water rights acquired after December 8, 1977, acreage that would be irrigated under this application, and acreage proposed to be irrigated under other pending applications on file with the Department of Ecology.

Check yes, if the proposed project is over 6,000 acres.

Enter the permit number(s) of any Family Farm Development Permit in which you hold controlling interest.

Section 8. OTHER WATER USES

Hydropower
For hydropower projects, indicate the total feet of head and proposed capacity in kilowatts. Describe the proposed diversion facility, including the bypass reach. Indicate all uses to which power is to be applied. Enter the FERC license number.

Mining/Industrial Use
Describe use, method of supplying and utilizing water.

Other Use
Describe any other use(s) of water.

Section 9. WATER STORAGE

Check the appropriate box for each question.

If you answered yes to any of the questions, your project may require a reservoir permit, or an approval from Ecology's Dam Safety Program. For criteria on reservoir permits contact the regional office in which your project is located.

Section 10. DRIVING DIRECTIONS

Provide detailed driving directions from the nearest town to the project site. If applicable, provide the site address.

Section 11. REQUIRED SIGNATURES

The applicant or authorized representative (e.g. the Public Works Director of a municipality, or the chair of a community water system) AND the landowner(s) of the place of use MUST sign the application.

If you require this document in an alternate format, please call the Water Resources Program at 360-407-6600. Persons with hearing loss can call 711 for Washington Relay Service. Persons with a speech disability can call 877-833-6341.

ECY 040-1-14 (Rev. 5/07)   [8]   APPLICATION FOR A WATER RIGHT PERMIT

Form 4 (Utah)
Sample Proof of Beneficial Use

PROOF OF BENEFICIAL USE OF WATER
STATE OF UTAH

1. **TYPE OF PROOF**   ☐ Appropriation   ☐ Change   ☐ Resumption of Use
   Water Right No._____ Application No. _____
2. **OWNER DATA** (Entity submitting proof MUST be the current owner of record.)
   Name _____ Telephone _____
   Mailing Address_____
3. **SOURCE OF WATER** _____
4. **POINT OF DIVERSION** (Must be based on a competent land survey; all ties must be given by rectangular coordinates with reference to a regularly established U.S. land corner.)
   Location _____

   Street Address _____
   Description of Diverting and Carrying Works_____

5. **NATURE, EXTENT AND PERIOD OF USE**
   Domestic: Number of Families _____ Part-Time/Recreational? _____ From ___ / ___ to ___ / ____
   Irrigation: Sole Supply Acres _____ Total Acres_____ From ___ / ___ to ___ / ____
   Stock: Number and Type_____ From ___ / ___ to ___ / ____
6. **QUANTITY OF WATER** _____ cfs and/or_____ ac-ft.
7. **PLACE OF USE** (ALL uses must be clearly shown on the map to establish appurtenance to the land. List the quarter-quarter sections for all uses.) _____

8. **SUPPLEMENTAL WATER RIGHTS** _____

9. **WATER MEASUREMENTS** (Report flow in units of cubic feet per second.)
   Name of Measurement Taker_____
   Date _____ Flow _____ cfs
   Measurement Method (Include equipment used and any other relevant information.) _____

10. **EXPLANATORY** (Extra space for above items and to provide additional information. Attach additional 8½" x 11" pages, if needed.)_____

11. **MAPS** (Must be submitted and must comply with rules and standards established by the State Engineer.)
12. **SIGNATURE PAGE** (Must be completed before filing with the State Engineer. Use additional signature pages if more than one applicant is signing the proof. Each signature must be notarized separately.)

CAUTION: File proof only if all desired development is done and the water is being fully put to beneficial use. Otherwise, consider filing an Extension of Time Request. The water right will be limited to the extent and nature of use in the accepted proof.

Proof

CERTIFICATE OF APPLICANT(S) (MUST be complete before filing with the State Engineer.)

STATE OF _____ COUNTY OF _____

Having been duly sworn, I hereby certify that_____ was employed to prepare
Proof of Beneficial Use for Water Right No._____ Application No._____, and that
to the best of my knowledge all information in the proof and all accompanying documents is accurate and complete
and is free of fraud, misrepresentation, and omission of material fact.

Name _____ For_____

Check One: □ Owner/Co-owner   □ Shareholder   □ Agent (Power of Attorney must be provided.)

          □ Appointed/Elected Representative (List title.) _____

_____
Applicant's Signature

Sworn to before me this _____ day of _____ , 20 _____

_____   Notary's Seal
Notary's Signature

Name _____ For_____

Check One: □ Owner/Co-owner   □ Shareholder   □ Agent (Power of Attorney must be provided.)

          □ Appointed/Elected Representative (List title.) _____

_____
Applicant's Signature

Sworn to before me this _____ day of _____ , 20 _____

_____   Notary's Seal
Notary's Signature

CERTIFICATE OF PROOF PROFESSIONAL (MUST be complete before filing with the State Engineer.)

STATE OF _____ COUNTY OF _____

Name _____ Phone No. _____

Address _____

_____

Having been duly sworn, I hereby certify that I was employed to prepare
Proof of Beneficial Use for Water Right No._____
Application No._____, and that to the best of my
knowledge all information in the proof and all accompanying documents is
accurate and complete and is free of fraud, misrepresentation, and omission
of material fact.

_____   Proof Professional's Seal
Proof Professional's Signature

Sworn to before me this _____ day of _____ , 20 _____

_____   Notary's Seal
Notary's Signature

This area is for Division of Water Rights use only

Maps and drawings filed: Herewith -or- Hanger_____ Page(s) _____

_____

_____

_____

_____

Form 5 (Idaho)

Sample Application for Extension of Time to Avoid Forfeiture

Form 2-222
12/99
FEE $100.00

Water Right No. _____

STATE OF IDAHO
DEPARTMENT OF WATER RESOURCES

APPLICATION FOR EXTENSION OF TIME TO AVOID
FORFEITURE OF A WATER RIGHT

Section 42-222, Idaho Code, provides that a water right shall be lost and forfeited and the water right revert to the state upon the failure of a water user to apply the water to beneficial use for a period of five (5) years. The Director may allow an additional five (5) year period of non-use without forfeiture upon a showing of reasonable cause.

1. Name of water right holder _____

2. Mailing address _____

3. The right was obtained or is evidenced by:

    a. Decree to _____ in Case of _____

    vs. _____, Dated _____ in _____
                                                                        (Name of Court)
    Court,   County   of   _____.   Fully   describe   decree

    _____

    b. License No. _____   c. Claim No. _____

    d. SRBA No. _____

4. Source of water _____ tributary to _____
                       (Name of stream or source)

5. Date of priority of right _____

6. Description of Right:

    amount _____ for _____ purposes from _____ to _____
             (cfs/ac-ft)

    amount _____ for _____ purposes from _____ to _____
             (cfs/ac-ft)

7. Point of Diversion _____ ¼ _____ ¼ of Sec. _____ Twp. _____ Rge. _____ County _____

8. Lands irrigated or Place(s) of Use:

TWP	RGE	SEC	NE				NW				SW				SE				Totals
			NE	NW	SW	SE	NE	NW	SW	SE	NE	NW	SW	SE	NE	NW	SW	SE	

Total Acres _____

9. Give the last date when water was put to beneficial use _____

   Length of extension of time requested _____ years, or until _____
   <small>(Date)</small>

10. Explain fully your reasons for requesting an extension of time within which to resume the use of your water right and

   reasons for non-use _____

   _____

   _____

   _____

   _____

   _____

   _____

The information contained in this application is true to the best of my knowledge.

_____
(Signature of Applicant)

Subscribed and sworn to before me this _____ day of _____, 20 _____

_____
(Notary Public)

My commission expires _____   Residing at _____

ACTION OF THE DIRECTOR, DEPARTMENT OF WATER RESOURCES

This is to certify that I have examined this application for extension of time within which to resume the use of a water right and I hereby _____ said application subject to the following conditions or reasons:

_____

_____

_____

_____

Signed this _____ day of _____, 20 _____

_____
Chief, Water Allocation Bureau

FOR DEPARTMENT USE ONLY

Date Received _____ FEE Receipted by _____ Date _____ Receipt # _____

Preliminary ✓ by _____ Publication prepared by _____ Pub. approved by _____ Date _____

Published in _____ Dates Published _____

Protests filed by _____

Copies of protests forwarded by _____ Hearing held by _____ Date _____

Recommended for - APPROVAL / DENIAL   By _____

Form 6 (California)

Sample Complaint Form

 State Water Resources Control Board

Division of Water Rights

Linda S. Adams
*Secretary for
Environmental Protection*

1001 I Street, 14th Floor ♦ Sacramento, California 95814 ♦ 916.341.5300
P.O. Box 2000 ♦ Sacramento, California ♦ 95812-2000
FAX: 916.341.5400 ♦ www.waterrights.ca.gov

Arnold Schwarzenegger
Governor

WATER RIGHT COMPLAINT

For information in filling out this form,
see pamphlet titled "Investigating Water Right Complaints"

CID# _____
File: __ __-__ __-__ __
(For staff use only)

Complaint

_____   _____
(Name)   (Phone No.)

_____   _____
(Address)   (Zip Code)

Party complained against (Respondent)

_____   _____
(Name)   (Phone No.)

_____   _____
(Address)   (Zip Code)

Location of Respondent's Diversion

The diversion is located on: _____
(Name of Spring, Stream, or Body of Water)

At a point within _____ ¼ of _____ ¼ of Section _____ T _____ R, _____ B&M

County of _____ Assessor's Parcel No. _____

The general location is as follows: _____
(Name of Road, Distance to Nearest Town, Etc.)

_____

Description of Complaint

The following situation or condition is occurring (attach additional sheets, photographs, maps, sketches,
reports, etc. as needed.)

_____

_____

_____

_____

_____

_____

COMP (2-05)

Injury to Complainant or Public Trust Resources

The situation is causing injury to me or public trust resources as follows (attach additional sheets if necessary):

_____

_____

_____

_____

Possible Resolution of Complaint

I offer the following possible solution to the situation (attach additional sheets if necessary):

_____

_____

_____

_____

Complainant's Diversion and Water Rights (Fill in if Injury Claimed)

My diversion is located on: _____
(Name of Spring, Stream, or Body of Water)

At a point within _____ ¼ of _____ ¼ of Section _____ T _____ R, _____ B&M

County of _____ Assessor's Parcel No. _____

I use water for (what and where): _____

_____

_____

The basis of my claim to divert water is:

❐ An appropriative right under License No. _____, Permit No. _____, Application No. _____
❐ A Riparian or pre-1914 claim supported by Statement of Water Diversion and Use No. _____
❐ Other (Describe): _____

A copy of this complaint has been sent to the Respondent by:

❐ Certified Mail   ❐ Regular Mail   ❐ Personal Delivery

I declare under penalty of perjury that the above is true and correct to the best of my knowledge and belief.

_____   _____
Signature   Date

NOTE: Send original Complaint to the Division of Water Rights and a copy to the Respondent.
Forms for submitting an Answer to Complaint will be sent to the Respondent by the
Division of Water Rights.

Form 7 (Arizona)

Sample Well Drilling/Opperation Permit

STATE OF ARIZONA
DEPARTMENT OF WATER RESOURCES
WATER MANAGEMENT DIVISION
MAIL TO: P.O. BOX 33589, PHOENIX, ARIZONA 85067-3589
3550 North Central Avenue, Phoenix, Arizona 85012
Phone (602) 771-8585 • Fax (602) 771-8688

APPLICATION FOR A PERMIT TO DRILL OR OPERATE A NON-EXEMPT WELL WITHIN AN ACTIVE MANAGEMENT AREA PURSUANT TO A.R.S. § 45-599

I. <u>INSTRUCTIONS:</u>
  1. This application should be used to obtain a permit to:
    (a) Drill a non-exempt well in conjunction with a new or existing General Industrial Use Permit Application, a Certificate of Grandfathered Right, a Service Area Right, or an Irrigation District Right.
    (b) Convert an existing well to a non-exempt well, or increase the annual permitted volume to be withdrawn from the well.
  2. Complete all appropriate items on this application, sign in the appropriate place and mail to P.O. Box 33589, Phoenix, Arizona 85067-3589 or hand deliver to 3550 North Central Avenue, Phoenix, Arizona 85012
  3. Pursuant to A.R.S. § 45-599, the application fee is $150.00. Pursuant to A.A.C. R12-15-151(B)(4), the permit fee is $30.00. You may submit both fees at the time of filing the application.

II. <u>GENERAL DATA:</u>

	FOR DEPARTMENT USE ONLY
1. Applicant _____	Application No. _____
Mailing Address _____	Registration No. _____
	File No. _____
City State Zip Code	Date Received _____
Contact Person _____	AMA _____
Telephone Number _____	W/S _____ S/B _____

  2. Name of Land Owner _____

    Mailing Address _____

    City   State   Zip   Telephone Number

  3. Applicant is: ☐ Owner   ☐ Lessee

  4. Proposed well is: ☐ New well   ☐ Conversion (enlargement) of existing well   ☐ Replacement well in a new location.

  5. Claim of entitlement to withdraw groundwater is based upon:

    ☐ Certificate of Grandfathered Right No:_____
    ☐ General Industrial Use Permit No. 59- _____
    ☐ Service Area Right No: _____
    ☐ Irrigation District Right No: _____

  6. The principal use(s) of groundwater will be (be **specific**)_____

  7. Well location: _____ ¼ _____ ¼ _____ ¼ Section _____ Township _____ N/S Range _____ E/W
               10 Acre   40 Acre   160 Acre

  8. Position location of the well: Latitude _____ ° _____ ' _____ " N   Longitude _____ ° _____ ' _____ " W

  9. Design Pump Capacity_____ gpm   Depth _____ feet

    Diameter_____inches   Type of casing_____

  10. Proposed annual volume of water_____ acre feet

DWR 55-0001   Revised 9/07

11. Well is located in the_____ subbasin of the_____ Active Management Area.

12. Approximate date construction will begin: MONTH _____ YEAR _____

      Estimated time to complete new well _____. (If longer than 1 year, attach explanation.)

13. Legal description of the land where the groundwater will be used:
      _____¼ _____¼ _____¼ Section _____Township _____ N/S Range _____ E/W. County _____
        10 Acre   40 Acre   160 Acre

14. Is the proposed well site within 100 feet of a septic tank system, sewage disposal area, landfill, hazardous waste facility or storage area of hazardous materials? ☐ Yes ☐ No (if yes, a request for a variance must accompany this application pursuant to R12-15-820.)

15. Driller's Name_____ DWR License No:_____ ROC License Category _____

     Mailing Address: _____
                  Street   City   State   Zip   Telephone Number

16. **Attach a Well Construction Supplement, DWR form 55-90, and include a detailed construction diagram as indicated on the form.**

III. FOR SERVICE AREA WELLS AND IRRIGATION DISTRICT WELLS ONLY:

17. Is the proposed well located in your service area? ☐ Yes ☐ No

18. Will groundwater withdrawn be used in your service area? ☐ Yes ☐ No **(If answer is no, attach explanation.)**

IV. FOR REPLACEMENT WELL IN NEW LOCATION ONLY:

19. Registration number of original well 55- _____.

20. Location of the original well: _____¼ _____¼ _____¼ Section _____ Township _____N/S Range _____E/W
                          10 Acre   40 Acre   160 Acre

21. Distance between original well and proposed replacement well_____ feet.

22. When determining impacts under the Department's well spacing rules, the director will take into account the collective efforts of reducing or terminating withdrawals from the well being replaced combined with the proposed withdrawals from the replacement well if the applicant submits a hydrological study demonstrating those collective effects to the satisfaction of the director. Will a hydrological study be submitted? ☐ Yes ☐ No

23. Will the original well be abandoned if applicant receives a permit to drill a replacement well? ☐ Yes ☐ No.
    (If yes, please submit a completed <u>Notice of Intent to Abandon a Well</u> along with this application.)
    If no, explain the planned use of the original well_____
    _____

V. FOR CONVERSION (ENLARGEMENT) OF EXISTING WELL ONLY:

24. Registration number of the existing well 55-_____ Present pump design capacity _____ gallons per minute. Present permitted volume _____acre-feet per year.

25. The new design pump capacity will be_____gallons per minute. New permitted volume will be _____ acre-feet per year.

26. The existing well has previously been used in conjunction with or for the following:_____
    _____

It is understood that the permit, if granted, will be in accordance with the Groundwater Management Act (Title 45, Chapter 2), and the rules adopted thereunder. The permittee will be bound by the provisions of such law and the provisions of the permit issued.

I (we), _____hereby affirm that all information provided in this application is true and correct to the best of my/our
        (print name)   knowledge and belief.

Signature of Applicant_____ Date_____

DWR 55-0001   Revised 9/07

Form 8 (New Mexico)

Sample Application to Change Point of Diversion

File Number: _____
(For OSE Use Only)

NEW MEXICO OFFICE OF THE STATE ENGINEER
APPLICATION FOR PERMIT TO CHANGE POINT OF DIVERSION
SURFACE WATERS

1. APPLICANT
   Name: _____   Work Phone: _____
   Contact: _____   Home Phone: _____
   Address: _____
   _____
   City: _____   State: __ Zip: _____

3. CHANGE FROM

A. POINT OF DIVERSION (a, b, c, or d required, e or f if known)

   a. ____1/4 ____1/4 ____1/4 Section:____ Township:____ Range:____N.M.P.M.
      in _____ County.

   b. X =_____ feet, Y = _____ feet, N.M. Coordinate System
      _____ Zone in the _____ Grant.
      U.S.G.S. Quad Map _____

   c. Latitude: _____d _____m _____s Longitude: _____d _____m _____s

   d. East _____ (m), North _____ (m), UTM Zone 13, NAD __ (27 or 83)

   e. Tract No. _____, Map No. _____ of the _____ Hydrographic Survey

   f. Lot No. _____, Block No. _____ of Unit/Tract _____ of the
      _____ Subdivision recorded in _____ County.

   g. Other: _____

   h. Give State Engineer File Number of existing well: _____

   i. On land owned by (required): _____

   j. Source of surface water supply:
      a. Name of ditch, acequia, or spring: _____
      b. Stream or water course: _____
      c. Tributary of: _____

B. QUANTITY

   Diversion Amount: _____ acre-feet per annum
   Consumptive Use: _____ acre-feet per annum

_____ Do Not Write Below This Line _____

File Number: _____   Trn Number: _____
   Form: wr-29   page 1 of 4

File Number: _____
(For OSE Use Only)

NEW MEXICO OFFICE OF THE STATE ENGINEER
APPLICATION FOR PERMIT TO CHANGE POINT OF DIVERSION
SURFACE WATERS

3. CHANGE TO

A. POINT OF DIVERSION (a, b, c, or d required, e or f if known)

a. _____1/4 _____1/4 _____1/4 Section:_____ Township:_____ Range:_____N.M.P.M.
   in _____ County.

b. X =_____ feet, Y = _____ feet, N.M. Coordinate System
   _____ Zone in the _____ Grant.
   U.S.G.S. Quad Map _____

c. Latitude: _____d _____m _____s Longitude: _____d _____m _____s

d. East _____ (m), North _____ (m), UTM Zone 13, NAD __ (27 or 83)

e. Tract No. _____, Map No. _____ of the _____ Hydrographic Survey

f. Lot No. _____, Block No. _____ of Unit/Tract _____ of the
   _____ Subdivision recorded in _____ County.

g. Other: _____

h. Give State Engineer File Number of existing well: _____

i. On land owned by (required): _____

j. Source of surface water supply:
   a. Name of ditch, acequia, or spring: _____
   b. Stream or water course: _____
   c. Tributary of: _____

B. QUANTITY

Diversion Amount: _____ acre-feet per annum
Consumptive Use: _____ acre-feet per annum

4. DIVERSION DAM (if applicable)

The diversion dam is constructed of _____;
Crest length _____ feet; Crest width _____ feet;
Height above stream bed _____ feet; Depth below stream bed _____ feet;
Side slopes of _____ horizontal to 1 (one) vertical on upstream face
And _____ horizontal to 1 (one) vertical on downstream face;
And contains about _____ cubic yards of material.

_____Do Not Write Below This Line_____

File Number: _____
(For OSE Use Only)

NEW MEXICO OFFICE OF THE STATE ENGINEER
APPLICATION FOR PERMIT TO CHANGE POINT OF DIVERSION
SURFACE WATERS

5. REASON FOR CHANGE

Application is made to change point of diversion for the following
reasons: _____
_____
_____

6. ADDITIONAL STATEMENTS OR EXPLANATIONS:

_____
_____
_____
_____
_____
_____
_____
_____
_____
_____
_____
_____
_____
_____
_____
_____
_____
_____
_____
_____
_____
_____
_____
_____
_____
_____
_____

ACKNOWLEDGEMENT

(I, We) _____ affirm that the
(Please Print)
foregoing statements are true to the best of (my, our) knowledge and belief.

_____   _____.
    Applicant Signature   Applicant Signature

_____ Do Not Write Below This Line _____

File Number: _____   Trn Number: _____
    Form: wr-29   page 3 of 4

NEW MEXICO OFFICE OF THE STATE ENGINEER
APPLICATION FOR PERMIT TO CHANGE POINT OF DIVERSION
SURFACE WATERS

ACTION OF STATE ENGINEER

This application is approved/denied/partially approved provided it is not exercised to the detriment of any others having existing rights, and is not contrary to the conservation of water in New Mexico nor detrimental to the public welfare; and further subject to the following conditions: _____
_____
_____
_____
_____
_____
_____
_____
_____

Witness my hand and seal this _____ day of _____, 20 _____

_____, State Engineer

By: _____

Form 9 (Nevada)

Sample Proof of Completion of Work

IN THE OFFICE OF THE STATE ENGINEER
OF THE STATE OF NEVADA
PROOF OF COMPLETION OF WORK

Permit No. ....

STATE OF ....
(When Sworn)

COUNTY OF ....
(When Sworn)

  Comes now ...., on behalf of

.... the Permittee, who after being first duly sworn,

deposes and says that at least .... dollars ($ ....) has been expended in work

performed or improvements made to develop water as set forth under the conditions in Permit No........................

  Said improvements consisted of ....

....

....

....

....

....

....

....

....

said work being essential to the actual diversion of the water applied for and required under said permit.

The works of diversion were completed prior to ....(date). The point of diversion is located

within the .... ¼ ....¼ of Section ...., Township.......N/S, Range................E, M. D. B. & M.

Well Drilling Contractor..

If this is a new well, please attach copy of the well log or provide a well log number ....

If possible, please provide the present static water level ....feet below land surface.

Subscribed and sworn to before me this ....day of

........................, 20................

....

   Signature of Notary Public

Notary Public in and for the County of ....

State of ....

My commission expires ...., 20................

Signed..
     Permittee or Agent

Address ....
    Street No. or P.O. Box No.

....
    City, State, Zip Code No

(Notary Seal)

A ten-dollar ($10.00) filing fee must accompany this proof (NRS 533.435)
Please see filing instructions of back of form and submit this form on blue paper

Revised 01/07

FILING THE PROOF OF COMPLETION OF WORK

1. Provide the water right permit number.
2. Indicate the State and County in which the proof is notarized.
3. Name of the person or persons signing the proof. If other than the permittee, give authority for signing.
4. The proof represents a sworn statement as the person signing the proof confirms the veracity of the information provided therein.
5. Indicate the approximate amount of money spent on the works to divert water.
6. Describe the work performed or improvements made to develop the water allowed under the conditions of the permit.
    (a) If this is an underground water right, describe the diameter and depth of casing, the size, name and type of pump, and the name and size (hp) of motor installed. Describe the flow meter that is installed as required by terms of the permit including the make, model, serial number and reading on the date of installation.
    (b) For a water right on a stream, spring, lake, or other water source, fully describe the completed works used to divert or store water, e.g., dams, ditches, pipelines, pumping stations, etc. Describe all water measuring devices that are installed as required by terms of the permit.
7. Indicate the date the well or other diversion works were completed.
8. Describe the point of diversion by public land survey. This description should match the legal description of the permit.
9. If applicable, provide the name of the well drilling contractor and please attach copy of the well log or provide the well log number.
10. If possible please provide the present static water level before pumping began.
11. Sign the form in the presence of the Notary Public.
12. Affix Notary Public stamp and seal.

The Proof of Completion of Work must be filled out entirely, signed, notarized and received in the Office of the State Engineer, Nevada Division of Water Resources, 901 S. Stewart Street, Suite 2002, Carson City, Nevada 89701, together with the $10.00 statutory filing fee on or before the due date on the permit and not later than 30 days from the date of any final notice received from this office. A separate Proof must be submitted with the $10.00 filing fee for each individual permit. If you have any questions please call 775-684-2800 or in Las Vegas 702-486-2770 or visit our web page at http://water.nv.gov

Revised 01/07

Form 10 (Texas)

Sample Storage Application

Aboveground Storage Tank Facility Plan Application
For Permanent Storage on The
Edwards Aquifer Recharge And Transition Zones
And Relating to 30 TAC §213.5(e), Effective June 1, 1999

REGULATED ENTITY NAME:_____
ABOVEGROUND STORAGE TANK (AST) FACILITY INFORMATION

1.   Tanks and substance stored:

AST Number	Size (Gallons)	Substance to be Stored	Tank Material
1			
2			
3			
4			
5			
Total	x 1.5 =	gallons	

2.   ___   The AST will be placed within a containment structure that is sized to capture one and one-half (1 1/2) times the storage capacity of the system. For facilities with more than one tank system, the containment structure is sized to capture one and one-half (1 1/2) times the cumulative storage capacity of all systems.

       ___   **ATTACHMENT A - Alternative Methods of Secondary Containment.** Alternative methods for providing secondary containment are proposed. Specifications that show equivalent protection for the Edwards Aquifer are found as **ATTACHMENT A** at the end of this form.

3.   Inside dimensions and capacity of containment structure(s):

Length (L) (Ft.)	Width (W) (Ft.)	Height (H) (Ft.)	L x W x H = (Ft³)	Gallons
Total				

4. __ All piping, hoses, and dispensers will be located inside the containment structure.
   __ Some of the piping to dispensers or equipment will extend outside the containment structure.
   - __ The piping will be aboveground
   - __ The piping will be underground

5. __ The containment area must be constructed of and in a material impervious to the substance(s) being stored. The proposed containment structure will be constructed of_____.

6. **ATTACHMENT B - Scaled Drawing(s) of Containment Structure.** A scaled drawing of the containment structure that shows the following is found as **ATTACHMENT B** at the end of this form:

   - __ Interior dimensions (length, width, depth and wall and floor thickness).
   - __ Internal drainage to a point convenient for the collection of any spillage.
   - __ Tanks clearly labeled
   - __ Piping clearly labeled
   - __ Dispenser clearly labeled

SITE PLAN

Items 7 through 17 must be included on the Site Plan.

7. The Site Plan must have a minimum scale of 1" = 400'.
   Site Plan Scale: 1" = ___'.

8. 100-year floodplain boundaries

   - __ Some part(s) of the project site is located within the 100-year floodplain. The floodplain is shown and labeled.
   - __ No part of the project site is located within the 100-year floodplain.

   The 100-year floodplain boundaries are based on the following specific (including date of material) sources(s):
   _____
   _____

9. __ The layout of the development is shown with existing and finished contours at appropriate, but not greater than ten-foot contour intervals. Show lots, recreation centers, buildings, roads, etc.
   __ The layout of the development is shown with existing contours. Finished topographic contours will not differ from the existing topographic configuration and are not shown.

10. All known wells (oil, water, unplugged, capped and/or abandoned, test holes, etc.):

    - __ There are __(#) wells present on the project site and the locations are shown and labeled. (Check all of the following that apply)
      - __ The wells are not in use and have been properly abandoned.

___   The wells are not in use and will be properly abandoned.
___   The wells are in use and comply with 30 TAC §238.
___   There are no wells or test holes of any kind known to exist on the project site.

11.   Geologic or manmade features which are on the site:

___   All **sensitive and possibly sensitive** geologic or manmade features identified in the Geologic Assessment are shown and labeled.
___   No **sensitive and possibly sensitive** geologic or manmade features were identified in the Geologic Assessment.
___   **ATTACHMENT C - Exception to the Geologic Assessment.** An exception to the Geologic Assessment requirement is requested and explained in **ATTACHMENT C.** Geologic or manmade features were found and are shown and labeled.
___   **ATTACHMENT C - Exception to the Geologic Assessment.** An exception to the Geologic Assessment requirement is requested and explained in **ATTACHMENT C.** No geologic or manmade features were found.

12.   ___   The drainage patterns and approximate slopes anticipated after major grading activities.

13.   ___   Areas of soil disturbance and areas which will not be disturbed.

14.   ___   Locations of major structural and nonstructural controls. These are the Temporary Best Management Practices.

15.   ___   Locations where soil stabilization practices are expected to occur.

16.   ___   Surface waters (including wetlands).

17.   ___   Locations where stormwater discharges to surface water or sensitive features.
      ___   There will be no discharges to surface water or sensitive features.

BEST MANAGEMENT PRACTICES

18.   Any spills must be directed to a point convenient for collection and recovery. Spills from storage tank facilities must be removed from the controlled drainage area for disposal within 24 hours of the spill.

___   In the event of a spill, any spillage will be removed from the containment structure within 24 hours of the spill and disposed of properly.
___   In the event of a spill, any spillage will be drained from the containment structure through a drain and valve within 24 hours of the spill and disposed of properly. The drain and valve system are shown in detail on the scaled drawing.

19.   ___   All stormwater accumulating inside the containment structure will be disposed of through an authorized waste disposal contractor.
      ___   Containment area will be covered by a roof.
      ___   Containment area will not be covered by a roof.
___   A description of the alternate method of stormwater disposal is submitted for the executive director's review and approval and is provided directly behind this page.

20.   **ATTACHMENT D - Spill and Overfill Control.** A description of the methods to be used at the facility for spill and overfill control are provided as **ATTACHMENT D**. Methods can include the proper transfer of fuels or chemicals from tanks into motor vehicles, and having a person present during fuel or chemical transfers.

21.   **ATTACHMENT E - Response Actions to Spills.** A description of the planned response actions to spills that will take place at the facility is provided as **ATTACHMENT E**.

ADMINISTRATIVE INFORMATION

22.   A Water Pollution Abatement Plan (WPAP) is required for construction of any associated commercial, industrial or residential project located on the Recharge Zone.

   __   The WPAP application for this project was approved by letter dated _____. A copy of the approval letter is attached at the end of this application.

   __   The WPAP application for this project was submitted to the TCEQ on _____, but has not been approved.

   __   A WPAP application is required for an associated project, but it has not been submitted.

   __   There will be no building or structure associated with this project. In the event a building or structure is needed in the future, the required WPAP will be submitted to the TCEQ.

   __   The proposed AST is located on the Transition Zone and a WPAP is not required.

23.   __   This facility is subject to the requirements for the reporting and cleanup of surface spills and overfills pursuant to 30 TAC 334 Subchapter D relating to Release Reporting and Corrective Action.

24.   __   One (1) original and three (3) copies of the completed application has been provided.

25.   Any modification of this AST Facility Plan application will require executive director approval, prior to construction, and may require submission of a revised application, with appropriate fees.

To the best of my knowledge, the responses to this form accurately reflect all information requested concerning the proposed regulated activities and methods to protect the Edwards Aquifer. This **ABOVEGROUND STORAGE TANK FACILITY PLAN APPLICATION** is hereby submitted for TCEQ review and executive director approval. The application was prepared by:

_____
Print Name of Customer/Agent

_____   _____
Signature of Customer/Agent   Date

Form 11 (Montana)

Sample Objection to Application

Form No. 611 R5/07

OBJECTION TO APPLICATION

INSTRUCTIONS

Use this form when objecting to an application for a water use permit, change authorization or reservation of water. Use one form for each application.

A person has standing to file an objection if his or her property, water rights, or interests would be adversely affected by the proposed appropriation. Individual water right owners must file separate objections.

A CORRECT AND COMPLETE OBJECTION FORM MUST BE RECEIVED OR POSTMARKED ON OR BEFORE THE DEADLINE SPECIFIED IN THE PUBLIC NOTICE.

FILING FEE: $25.00

Permit	Change
Obj # _____	
☐ Valid	☐ WQ
Reviewed by: _____	
Date: _____	

FOR DEPARTMENT USE ONLY

Postmarked Date _____
Date Received _____
Rec'd By _____
Fee Rec'd _____
Check No. _____
Refund _____

---

1. NAME OF OBJECTOR _____

   Mailing Address _____

   City _____ State _____ Zip _____

   Home Phone _____ Other Phone _____

2. APPLICATION BEING OBJECTED TO: Number _____

   Applicant Name: _____

3. STATE THE FACTUAL BASIS OF YOUR OBJECTION

   a) OBJECTION TO PERMIT APPLICATION must provide facts tending to show one or more of the criteria in Section 85-2-311, MCA are not met.

   b) OBJECTION TO CHANGE APPLICATION must provide facts tending to show one or more of the criteria in Section 85-2-402, MCA are not met.

   NOTE: Water quality objections must contain substantial credible information establishing to the satisfaction of the department that the water quality criteria cannot be met by the applicant.

   _____
   _____
   _____
   _____
   _____
   _____
   _____
   _____
   _____
   _____
   _____
   _____
   _____
   _____

MONTANA DEPARTMENT OF NATURAL RESOURCES AND CONSERVATION
1424 9TH AVENUE, P.O. BOX 201601 HELENA, MT 59620-1601 444-6610
web site: http://www.dnrc.mt.gov/wrd

4. STATE THE BASIS OF YOUR WATER RIGHT, if you are claiming your water right will be affected.

☐ (W) Statement of Claim No. _____

☐ (P) Permit to Appropriate Water No. _____

☐ (C) Certificate of Water Right No. _____

☐ (D) Final Decree No. _____

☐ (M/R) Reservation of Water No. _____

☐ (E) Exempt Existing Water Right (no claim filed; complete items below)

THIS INFORMATION ONLY REQUIRED FOR EXEMPT RIGHTS.

 Date of First Use: _____

 Name of Appropriator: _____

 Type of Use: Stock ☐   Domestic ☐

 Amount Used: Flow Rate_____ Gallons Per Minute; Volume_____Acre-Feet

 Point of Diversion:

 ____1/4 ____1/4 ____1/4 Section ____, Twp____ N/S, Rge____ E/W,_____ County

 Lot_____ Block _____Tract No. _____ Subdivision Name _____

5. STATE ANY CONDITIONS OR MODIFICATIONS UNDER WHICH YOU WOULD AGREE TO THE ISSUANCE OF THE PERMIT OR AUTHORIZATION TO CHANGE.

_____

_____

_____

_____

_____

6. ARE YOU REPRESENTED BY COUNSEL? YES ☐ NO ☐ 7. PERSON PREPARING THIS FORM, if different from objector

Name _____ Name _____

Mailing Address _____ Mailing Address _____

City, State, Zip _____ City, State, Zip _____

Phone _____ Phone _____

8. OBJECTOR'S SIGNATURE _____ **DATE** _____

WATER RESOURCES REGIONAL OFFICES

Billings
Airport Business Park
1371 Rimtop Drive
Billings, MT 59105-1978
Phone: 406-247-4415
Fax: 406-247-4416
Serving: Big Horn, Carbon, Carter, Custer, Fallon, Powder River, Prairie, Rosebud, Stillwater, Sweet Grass, Treasure, and Yellowstone Counties

Bozeman
2273 Boot Hill Court, Suite 110
Bozeman, MT 59715
Phone: 406-586-3136
Fax: 406-587-9726
Serving: Gallatin, Madison, and Park Counties

Glasgow
222 6th Street South
P.O. Box 1269
Glasgow, MT 59230-1269
Phone: 406-228-2561
Fax: 406-228-8706
Serving: Daniels, Dawson, Garfield, McCone, Phillips, Richland, Roosevelt, Sheridan, Valley, and Wibaux Counties

Havre
210 6th Avenue
P.O. Box 1828
Havre, MT 59501-1828
Phone: 406-265-5516
Fax: 406-265-2225
Serving: Blaine, Chouteau, Glacier, Hill, Liberty, Pondera, Teton, and Toole Counties

Helena
1424 9th Avenue
P.O. Box 201601
Helena, MT 59620-1601
Phone: 406-444-6999
Fax: 406-444-9317
Serving: Beaverhead, Broadwater, Deer Lodge, Jefferson, Lewis and Clark, Powell, and Silver Bow Counties

Kalispell
109 Cooperative Way, Suite 110
Kalispell, MT 59901-2387
Phone: 406-752-2288
Fax: 406-752-2843
Serving: Flathead, Lake, Lincoln, and Sanders Counties

Lewistown
613 NE Main Street, Suite E
Lewistown, MT 59457-2020
Phone: 406-538-7459
Fax: 406-538-7089
Serving: Cascade, Fergus, Golden Valley, Judith Basin, Meagher, Musselshell, Petroleum, and Wheatland Counties

Missoula
1610 South 3rd Street West, Suite 103
P.O. Box 5004
Missoula, MT 59806-5004
Phone: 406-721-4284
Fax: 406-542-1496
Serving: Granite, Mineral, Missoula, and Ravalli Counties

For Mailing, Use Post Office Box Number.

Form 12 (Alaska)

Sample Application for Temporary Use

DIVISION OF MINING, LAND AND WATER
WATER RESOURCES SECTION
www.dnr.state.ak.us/mlw/water/index.htm

Alaska Department of
**NATURAL
RESOURCES**

Anchorage Office 550 West 7ᵗʰ Avenue, Suite 1020 Anchorage, AK 99501-3562 (907) 269-8600 Fax: (907) 269-8947	Juneau Office PO Box 111020 400 Willoughby Avenue Juneau, AK 99811-1020 (907) 465-3400 Fax: (907) 586-2954	Fairbanks Office 3700 Airport Way Fairbanks, AK 99709-4699 (907) 451-2790 Fax: (907) 451-2703	*For ADNR Use Only* *Date/Time Stamp*
For ADNR Use Only *TWUP #*	*For ADNR Use Only* *CID #*	*For ADNR Use Only* *Receipt Type* WR	

APPLICATION FOR TEMPORARY USE OF WATER

INSTRUCTIONS

1. Complete one application for each project including up to five water sources (incomplete applications will not be accepted).
2. Attach legible map that includes meridian, township, range, and section lines such as a USGS topographical quadrangle or subdivision plat. Indicate water withdrawal point(s), location(s) of water use, and point(s) of return flow or discharge (if applicable).
3. Attach sketch, photos, plans of water system, or project description (if applicable).
4. Attach driller's well log for drilled wells (if available).
5. Attach copy of ADNR fish habitat permit (if applicable).
6. Attach completed Coastal Project Questionnaire (if applicable - see page 4).
7. Submit non-refundable fee (see page 4).

APPLICANT INFORMATION

Project Name

Organization Name (if applicable)   Agent or Consultant Name (if applicable)

Individual Name (if applicable)   Individual Co-applicant Name (if applicable)

Mailing Address   City   State   Zip Code

Daytime Phone Number   Alternate Phone Number (optional)

Fax Number (if available)   E-Mail Address (optional)

102-4048 (Rev. 2/06)
Page 1 of 4

PROPERTY DESCRIPTIONS

Location of Water Use

Project Area (e.g. milepost range, place name, survey number)	Meridian	Township	Range	Section	Quarter Sections	
					¼	¼
					¼	¼

Location of Water Source

Geographic Name of Water Body or Well Depth	Meridian	Township	Range	Section	Quarter Sections	
					¼	¼
					¼	¼
					¼	¼
					¼	¼
					¼	¼

Location of Water Return Flow or Discharge (if applicable)

Geographic Name of Water Body or Well Depth	Meridian	Township	Range	Section	Quarter Sections	
					¼	¼
					¼	¼

METHOD OF TAKING WATER

Pump Pump Intake _____ Inches　　Hours Working _____ Hours/Day

Pump Output _____ GPM　　Length of Pipe _____ Feet (from pump to point of use)

Gravity Pipe Diameter _____ Inches　　Length of Pipe _____ Feet (take point to point of use)

Head _____ Feet

Ditch L _____ H _____ W _____ Feet　　Diversion Rate _____ ☐ GPM or ☐ CFS

Reservoir L _____ H _____ W _____ Feet　　Water Storage _____ Acre-feet

Dam L _____ H _____ W _____ Feet　　Water Storage _____ Acre-feet

AMOUNT OF WATER

Purpose of Water Use	Quantity of Water			Season of Use	
	Maximum Withdrawal Rate	Total Daily Amount	Total Seasonal Amount	Date Work Will Start	Date Work Will be Completed

Project Totals   Total years needed: _____

PROJECT DESCRIPTION

What alternative water sources are available to your project should a portion of your requested diversion be excluded because of water shortage or public interest concerns?

Are there any surface water bodies or water wells at or near your site(s) that could be affected by the proposed activity? If yes, list any ground water monitoring programs going on at or near the sites, any water shortages or water quality problems in the area, and any information about the water table, if known.

Briefly describe the type and size of equipment used to withdraw and transport water, including the amount of water the equipment uses or holds.

Briefly describe what changes at the project site and surrounding area will occur or are likely to occur because of construction or operation of your project (e.g. public access, streambed alteration, trenching, grading, excavation).

Briefly describe land use around the water take, use, and return flow points (e.g. national park, recreational site, residential).

Will project be worked in phases? State reason for completion date.

Briefly describe your entire project:

_____

_____

_____
(Attach extra page if needed.)

11 AAC 93.220 sets out the required information on the application and authorizes the department to consider any other information needed to process an application for a temporary use of water. This information is made a part of the state public water records and becomes public information under AS 40.25.110 and 40.25.120. Public information is open to inspection by you or any member of the public. A person who is the subject of the information may challenge its accuracy or completeness under AS 44.99.310, by giving a written description of the challenged information, the changes needed to correct it, and a name and address where the person can be reached. False statements made in an application for a benefit are punishable under AS 11.56.210.

SIGNATURE

The information presented in this application is true and correct to the best of my knowledge. I understand that no water right or priority is established per 11 AAC 93.210-220, that the water used remains subject to appropriation by others, and that a temporary water use authorization may be revoked if necessary to protect the water rights of other persons or the public interest.

Signature _____   Date _____

Name (please print) _____   Title (if applicable) _____

REFERENCES

Measurement Units
GPD = gallons per day
CFS = cubic feet per second
GPM = gallons per minute
AF = acre-feet
AFY = acre-feet per year (325,851 gallons/year)
AFD = acre-feet per day (325,851 gallons/day)
MGD = million gallons per day

Conversion Table

5,000 GPD=	30,000 GPD=	100,000 GPD=	500,000 GPD=	1,000,000 GPD=
0.01 CFS	0.05 CFS	0.2 CFS	0.8 CFS	1.5 CFS
3.47 GPM	20.83 GPM	69.4 GPM	347.2 GPM	694.4 GPM
5.60 AFY	33.60 AFY	112.0 AFY	560.1 AFY	1120.1 AFY
0.2 AFD	0.09 AFD	0.3 AFD	1.5 AFD	3.1 AFD
0.01 MGD	0.03 MGD	0.1 MGD	0.5 MGD	1.0 MGD

Fee required by regulation 11 AAC 05.010(a)(8)
- **$350** for all uses of water from up to five water sources

Make checks payable to "Department of Natural Resources".

Coastal Zone
If this appropriation is within the Coastal Zone, and you are planning to use more than 1,000 GPD from a surface water source or 5,000 GPD from a subsurface water source, you need to submit a completed Coastal Project Questionnaire with this application. For more information on the Coastal Zone, contact the Office of Project Management and Permitting; Anchorage 269-7470, Juneau 465-3562, www.dnr.state.ak.us/acmp/.

Form 13 (North Dakota)

Sample Application to Construct

**APPLICATION/NOTIFICATION TO CONSTRUCT
OR MODIFY A DAM, DIKE, RING DIKE OR OTHER
WATER RESOURCE FACILITY**

Office of the State Engineer
900 East Boulevard -- Bismarck, ND 58505-0850
SFN 51695 (11/03)

SWC USE ONLY

I, the undersigned, do hereby submit the following information to the Office of the State Engineer for determination and use as a filing of information required under North Dakota Century Code §61-04-02 or as an application to construct or modify a facility under North Dakota Century Code §61-16.1-38.

(SWC USE ONLY) **No.** _____

A. GENERAL INFORMATION:

(1) This Application/Notification must include a map from an actual survey, aerial photo or topographic map. The size of the map shall be 8½ by 11 inches. The map shall have a north arrow and approximate scale. If, in the opinion of the State Engineer, the map does not contain information to properly evaluate the project, it will be returned.

(2) The proposed facility is a:

☐ Dam (Complete Sections A, C & F)   ☐ Pond, Lagoon, or Dugout (Complete Sections A, B & F)

☐ Dike (Complete Sections A, D & F)   ☐ Diversion Ditch (Complete Sections A, B & F)

☐ Ring Dike (Complete Sections A, D & F)   ☐ Other (Complete Sections A, B & F)

☐ Wetland Restoration (Complete Sections A, C, E & F)

(3) Is this Application/Notification for modification of an existing structure? ☐ Yes ☐ No

If so, what year was existing structure constructed? _____ By whom? _____

(4) Project will be located in the _____ Water Resource District

(5) Legal description to the nearest forty-acre tract: _____ ¼ _____ ¼ Section _____ Township _____ Range _____

(Optional)   Latitude _____ Longitude _____

(6) Waterway on which project will be located: _____

(7) A tributary to: _____

(8) Will the project, including any area inundated as a result of the project, be located entirely on land owned by the applicant?

☐ Yes ☐ No   If any portion of the project will be constructed on land not owned in fee title by the applicant, written authorization to construct the project must be obtained from the landowner of record and a copy of the authorization provided to this office. If the project will impound water on land not owned in fee title by the applicant, a flowage easement must be obtained by the applicant and a copy of the easement provided to this office. If any portion of the project will be constructed within the right-of-way of a section line, roadway, or railroad, or if the project will impound water within the right-of-way of a section line, roadway, or railroad, written authorization to do so must be obtained from the appropriate authority and a copy provided to this office.

(9) Project sponsor (Water Resource District/City/US Fish & Wildlife Service, etc.) if applicable _____

(10) Contractor, if known _____

(11) Anticipated construction start date _____ Completion date _____

(12) Who will be responsible for the operation and maintenance of this project? _____

B. POND, LAGOON, DUGOUT, DIVERSION DITCH, OR OTHER WATER RESOURCE FACILITY:

(1) Design Data:

a. Pond, Lagoon, or Dugout (complete below and diagram next   b. Diversion Ditch
page for each pond or cell, photocopy if necessary)

1. Surface area:	top of structure	_____ acres	1. Length	_____ feet
	service level	_____ acres	2. Bottom width	_____ feet
2. Storage:	top of structure	_____ acre-feet	3. Side slopes	_____ feet
	service level	_____ acre-feet	4. Maximum cut	_____ feet
3. Maximum depth of water		_____ feet	5. Gradient	_____ foot/foot
4. Maximum embankment height		_____ feet		

(2) Description of project, if not a Pond, Lagoon, Dugout, or Diversion Ditch: _____

B. OTHER WATER RESOURCE FACILITY (continued):

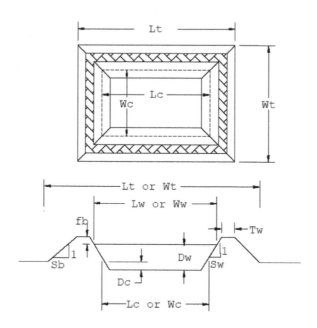

DESCRIPTION	ABBREVIATION	DIMENSION (feet)
Total length of pond (includes banks)	Lt	
Total width of pond (includes banks)	Wt	
Length of water surface at full service level	Lw	
Width of water surface at full service level	Ww	
Length of cut into the soil surface	Lc	
Width of cut into the soil surface	Wc	
Depth of cut into soil surface	Dc	
Depth of water in the pond at the full service level	Dw	
Freeboard (the distance between the full service level and the top of the structure that is used to manage wave action, usually 2-3 feet)	fb	
Top width of embankment surrounding the pond	Tw	
Outside bank sideslope ratio (usually 4:1, which is 4 horizontal feet for every 1 foot of rise)	Sb	
Inside bank sideslope ratio (will vary between 4:1 and 6:1, depending on the soil type)	Sw	

C. DAMS

(1) Drainage area above dam _____ square miles or _____ acres

(2) Purpose: _____

(3) Geometric description of dam:

    a. Maximum height (H) _____ feet, elevation _____ feet msl

    b. Top width (T) _____ feet

    c. Side slopes: upstream (S_1) _____:1

                  downstream (S_2) _____:1

    d. Type of embankment protection _____

    e. Emergency spillway: type _____

        If earthen: width _____ ft, side slopes _____:1, level section length _____ ft

        Dimensions if other than earthen _____

    f. Principal spillway:

        Outlet pipe: type _____ diameter _____ length _____ ft

        Riser: type _____ diameter _____

        Control gate: type _____ dimensions _____

    g. Drawdown Pipe: type _____ diameter _____

(4) Distance to nearest downstream occupied dwelling(s) _____

	ELEVATION (feet) Indicate datum: ☐ local ☐ NGVD 29 ☐ NAVD 88	RESERVOIR SURFACE AREA (acres)	RESERVOIR CAPACITY (acre-feet)
Top of Dam			
Emergency Spillway			
Principal Spillway			
Drawdown Pipe			
Streambed at Dam			

D. DIKE

(1) Is this application/notification for the construction of a ring dike? ☐ Yes ☐ No

    If so, will the ring dike tie into existing? ☐ dike ☐ roadway ☐ high ground ☐ other _____

(2) Purpose: _____

_____

(3) Area of land to be protected by dike _____ acres

(4) Description of Dike:

    a. Dike length _____ feet

    b. Dike design:

        1. Top width (T) _____ feet

        2. Side slopes: interior (S_1) _____:1

                   exterior (S_2) _____:1

        3. Maximum height (H) _____ feet, elevation _____ feet msl

           Minimum height (H) _____ feet, elevation _____ feet msl

        4. Embankment erosion protection: _____

(5) Will the dike flood or adversely affect adjacent, upstream or downstream land? ☐ Yes ☐ No

If yes, attach flowage easements. Easements must include a description of provisions, and names and signatures of grantors.

E. WETLAND RESTORATION

(1) The proposed wetlands are: ☐ Temporary ☐ Permanent

(2) Drainage area above dam _____ square miles or _____ acres

(3) Is this project mitigation for another project? ☐ Yes ☐ No
If yes, please describe: _____

(4) Describe the proposed operation plan for the wetland: _____

	OVERFLOW ELEVATION (feet) Indicate datum: ☐ local ☐ NGVD 29 ☐ NAVD 88	CAPACITY (acre-feet)	SURFACE AREA (acres)
Existing			
Natural			
Proposed			
Top of Structure			

F. ADDITIONAL INFORMATION, AFFIDAVIT OF DESIGN ENGINEER, AND SIGNATURE

(1) Additional information and comments: _____
_____
_____
_____

(2) A complete set of plans and specifications prepared by a professional engineer registered in the State of North Dakota must be submitted with and made part of this Application/Notification if the proposed structure will be capable of retaining, obstructing, or diverting more than 50 acre-feet of water, or if the structure is a medium or high hazard dam, as determined by the State Engineer, capable of retaining more than 25 acre-feet of water. Low hazard dams, as determined by the State Engineer, less than 10 feet in height are exempt from the requirement for professional engineering services. If plans and specifications are required, the following affidavit must be completed:

I, _____ (name), _____ (PE license number), a Professional Engineer registered in the State of North Dakota, designed and/or personally supervised the design of the project as described in this application and on any attached sheets, and construction will be inspected in accordance with North Dakota Administrative Code §89-08-03-01. Date: _____

(3) The filing of this Application/Notification in no way relieves the applicant or landowner from any responsibility or liability resulting from the construction, operation or failure of the project.

Land Owner (Print): _____
Address: _____
_____
Phone: _____
Signature: _____ Date: _____

Sponsoring Agency: _____
Address: _____
_____
Phone: _____
Signature: _____ Date: _____

Form 14 (Wyoming)

Sample Notice of Abandonment

Abandonment Notice

State Engineer's Office
Attn: Ground Water Division
Herschler Bldg., 4E
Cheyenne, WY 82002

Date: _____

Permit No. U.W. _____

Well Name _____

We have abandoned the above-mentioned water well for the following reason:
_____
_____
_____
_____

Date well was abandoned:
_____

How was well abandoned (Describe details of plugging below):
_____
_____
_____
_____
_____
_____

Signature

_____
Name

_____
Address

_____
City, State Zip Code

Form 15 (Oklahoma)

Sample Groundwater Lease Form

GROUNDWATER LEASE
(LANDOWNER PERMISSION)

I, _____, am the surface owner of _____ acres
         (name of surface land owner)

of land located in the _____,_____
                 (legal description)   (county)

County, and I hereby authorize _____ or a
                           (name of applicant for groundwater permit)

duly authorized representative to apply for a permit, locate wells on the land and withdraw

groundwater from the land described above. This lease is for: (complete applicable option)

    1. a period not to exceed _____ from the date listed below, <u>or</u>
                        (days, months, or years)

    2. _____ an indefinite term.
      (√ if applies)

_____   _____
           Signature   Street address or P.O.Box

_____   _____
   Surface Owner Name (print)   City, State, Zip

_____   _____
            Date   Telephone number

State of Oklahoma   )
)   ss.
County of _____)

Subscribed and sworn to before me this ____day of _____, 20___.

_____
Notary Public

My Commission Expires:_____
    (Seal)

Form 16 (Kansas)

Sample Minimum Desirable Streamflow Form

_____
(Date)

Kansas Department of Agriculture
Division of Water Resources
David L. Pope, Chief Engineer
109 SW 9th Street, 2nd Floor
Topeka, Kansas 66612-1283

Re: Application
File No. _____

Minimum Desirable Streamflow

Dear Sir:

    I understand that a Minimum Desirable Streamflow requirement has been established by the legislature for the source of supply to which the above referenced application applies.

    I understand that diversion of water pursuant to this application will be subject to regulation any time Minimum Desirable Streamflow requirements are not being met.

    I also understand that if this application is approved, there could be times, as determined by the Division of Water Resources, when I would not be allowed to divert water. I realize that this could affect the economics of my decision to appropriate water.

    I am aware of the above factors, and with the knowledge thereof, request that the Division of Water Resources proceed with processing and approval, if possible, of the above referenced application.

_____
Signature of Applicant

State of Kansas   )
) ss
County of _____)

_____
(Print Applicant's Name)

    I hereby certify that the foregoing instrument was signed in my presence and sworn to before me this _____ day of _____, 20___.

_____
Notary Public

My Commission Expires:

DWR 1-100.171   (Rev. 03/10/2000)

**MINIMUM DESIRABLE STREAMFLOW FORM TO BE USED WHEN
APPLICABLE WHEN FILING AN APPLICATION FOR PERMIT
TO APPROPRIATE WATER FOR BENEFICIAL USE**

The Kansas Legislature has established minimum desirable streamflows for the streams listed below. If your proposed diversion of water is going to be from one of these watercourses or adjacent alluvial aquifers, please complete the back side of this page and submit it along with your application for permit to appropriate water.

Arkansas River	Ninnescah River
Big Blue River	North Fork Ninnescah River
Chapman Creek	Rattlesnake Creek
Chikaskia River	Republican River
Cottonwood River	Saline River
Delaware River	Smoky Hill River
Little Arkansas River	Solomon River
Little Blue River	South Fork Ninnescah
Marais des Cygnes River	Spring River
Medicine Lodge River	Walnut River
Mill Creek (Wabaunsee Co. area)	Whitewater River
Neosho River	

Form 17 (Mississippi)

Sample Withdrawal Application

APPLICATION FOR PERMIT TO DIVERT OR WITHDRAW
FOR BENEFICIAL USE THE PUBLIC WATERS OF THE STATE OF MISSISSIPPI

DEPARTMENT OF ENVIRONMENTAL QUALITY, OFFICE OF LAND AND WATER RESOURCES
P.O. BOX 10631, JACKSON, MS 39289-0631; (601) 961-5202

This box is for office use only   FORM OLWR-AP-3 (REV 6/06)

Issued:	Expires:	Fee Paid:	Permit No.
Lat.	Long.	Elev.	USGS No.
Quad	ASCS Farm No.	STAC.	MSDOH No.
Aquifer:	Tract No.		Basin No.
Remarks:			Dam Inv. No.

THIS APPLICATION IS FOR (Circle one): NEW PERMIT   RENEWAL – PERMIT NO. _____

THIS APPLICATION IS FOR (Circle one):   GROUNDWATER – COMPLETE SECTIONS A, B, D

                                         SURFACE WATER – COMPLETE SECTIONS A, C, D

BENEFICIAL USE (Circle one or more):   1) Public Water System   2) Irrigation   3) Industrial   4) Fish Culture
5) Recreation   6) Institutional (e.g. Church, School)   7) Commercial (e.g. Hotel, Casino, Restaurant)   8) Fire Protection
9) Livestock   10) Flood Protection   11) Wildlife Management
12) Other: _____

SECTION A (to be completed by ALL APPLICANTS)

LANDOWNER: _____   _____
                 (Name)   (E-mail address)

               _____
               (Address)

               _____   _____
               (City)   (State & Zip)   (Telephone No.)   (Fax No.)

APPLICANT, AGENT OR LESSEE (if different from Landowner):

               _____   _____
               (Name)   (E-mail address)

               _____
               (Address)

               _____   _____
               (City)   (State & Zip)   (Telephone No.)   (Fax No.)

LOCATION of diversion/withdrawal point (A suitable map with location marked MUST accompany this application):

_____ ¼ of the _____ ¼ of Section _____ , Township _____ , Range _____ , County _____

Does the land to which this application pertains have any source(s) of water other than that for which you are now applying (circle one)? YES NO

If yes, describe the nature and amount of any additional supply and, if applicable, list permit number._____

SECTION B (to be completed for GROUNDWATER SOURCE)

1. **AQUIFER:** _____ MISSISSIPPI DEPARTMENT OF HEALTH NO.: _____

2. **DESCRIPTION** of proposed or completed well:

    (a) Driller: _____

    (b) Proposed work will begin on (date) _____ and will be completed by _____ .

        If well has already been drilled, when was well completed? _____

        Under whose name was well originally drilled (if known)?_____

    (c) Depth of well:_____ feet

    (d) Surface Casing: Length _____ feet; Diameter _____ inches; Type_____
                                                       (PVC, steel, stainless, black iron, other)

    (e) Screen: Length _____ feet; Diameter _____ inches; Type_____
                                                       (PVC, steel, stainless, open hole, other)

    (f) Pump: Type _____ ; Capacity _____ ; Setting depth _____ feet
                    (submersible, turbine, jet, flowing, other)   (gallons per minute)

    (g) Power unit: Type_____ ; Size_____horsepower
                      (electric, tractor, diesel, gasoline, butane, other)

4. **PERMITTED VOLUME:**

    (a)_____ acre-feet per year at a maximum of _____ gallons per minute

    (b) _____ million gallons per day at a maximum rate of _____ gallons per minute

(CONTINUED ON BACK)

SECTION C (to be completed for SURFACE WATER SOURCE)

1. **SOURCE** of water is from _____ which drains into _____

   which drains into _____ .
   (major stream or river)

2. **DESCRIPTION** of pump/diversion works:

   Pump: Type _____; Diameter _____ inches; Capacity _____ gallons per minute
   (turbine, jet, other)

   Power Unit: Type _____ ; Size _____ horsepower; Lift _____ feet
   (electric, tractor, diesel, gasoline, butane, other)

3. **PERMITTED VOLUME**

   _____ acre-feet per year at a maximum rate of _____ gallons per minute

SECTION D WATER USE DATA (ALL APPLICATIONS – complete section related to beneficial use)

1 **IRRIGATION:** List the number of acres of each crop to be irrigated: Rice _____; Cotton _____ ; Oats _____ ;

   Corn _____ ; Soybeans _____ ; Pasture _____ ; Truck _____ ; Wheat _____ ; Grain Sorghum _____ ;

   Wildlife _____ ; Other (specify) _____ total acres

   (a) Method of Irrigation (circle one) –   Center Pivot   Flood   Furrow

   (b) Land Condition (circle one) -   Precision Land Formed   Smoothed

   (c) ASCS Farm No. _____   Tract No. _____

2. **FISH CULTURE:** Explain how water will be used: _____

   How often will reservoir(s) be emptied and refilled? _____ ; Reservoir acreage _____

3. **PUBLIC WATER SYSTEM**

   Choose "a" or "b". (a) The number of people served is _____ or (b) The number of connections is _____ .

   What is the estimated average daily consumption during periods of maximum use at the end of each five-year period during the

   next twenty (20) years? _____   _____ ;   _____   _____ ;   _____   _____ ;   _____   _____
   (volume)   (year)   (volume)   (year)   (volume)   (year)   (volume)   (year)

4. **INDUSTRIAL:** If the water is released into a watercourse, indicate the amount released each year _____ (MGD)

   Rate of release _____ (MGD); NPDES Permit No. _____

   Explain any changes in quality of water to be released: _____

   Explain how water will be used: _____

   How much groundwater will be used for once-through non-contact cooling? _____

5. **RECREATION:** Explain how water will be used: _____
   _____

6. **OTHER USE:** Explain in detail (if needed, attach another page): _____
   _____

7. **REMARKS:** _____
   _____

List below the person to be contacted for additional information if required.

_____
(Name)

_____
(Address)

_____
(City, State, ZIP)

_____   _____
(Telephone)   (Fax)

_____
(E-mail address)

The **ACCOMPANYING MAP** is hereby declared a part of this application. For irrigation and fish culture in the Delta, an ASCS photograph is required. For all others, an appropriate map is required. The **TEN DOLLAR ($10.00) permit fee** is enclosed herewith.

_____
(Signature)

Subscribed and sworn to before me this _____ day of _____, _____ at _____ County of _____

My commission expires _____ , _____ Notary Public.

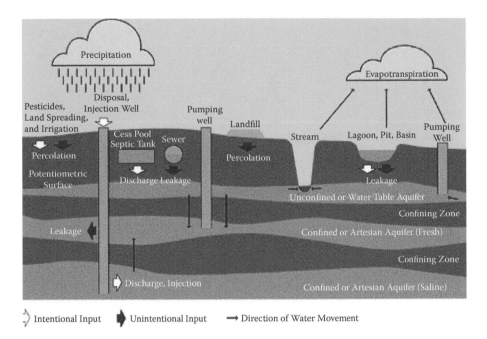

Intentional Input   Unintentional Input   → Direction of Water Movement

FIGURE 1   Typical routes of ground water contamination. (Adapted from US EPA Office of Water Supply and Solid Waste Management Programs, Waste Disposal Practices and Their Effects on Groundwater, Washington, DC, 1977.)

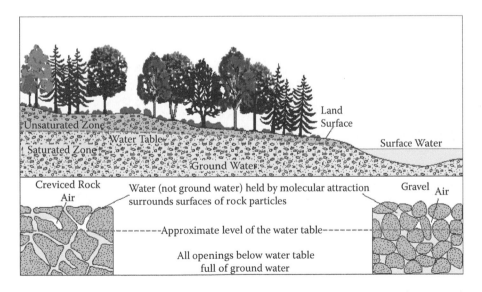

FIGURE 2   Ground saturation. (Adapted from USGS website, available at http://ga.water. usgs.gov/edu/earthgwaquifer.html, Feb. 2007.)

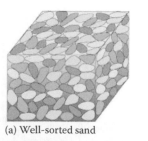 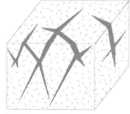 

(a) Well-sorted sand   (b) Fractures in granite   (c) Caverns in limestone

Modified from Heath (1998)

FIGURE 3   Ground water storage. (Available at http://pubs.usgs.gov/circ/2003/circ1262/#heading 156057192, September 2007.)

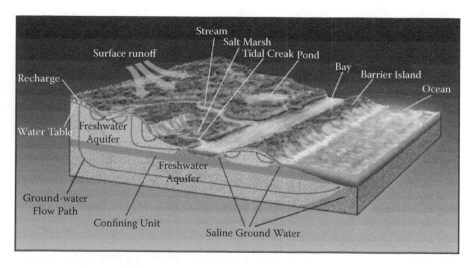

FIGURE 4   Ground water flow paths in ideal watershed. (Available at http://pubs.usgs.gov/circ/2003/circ1262/#heading156057192, September 2007.)

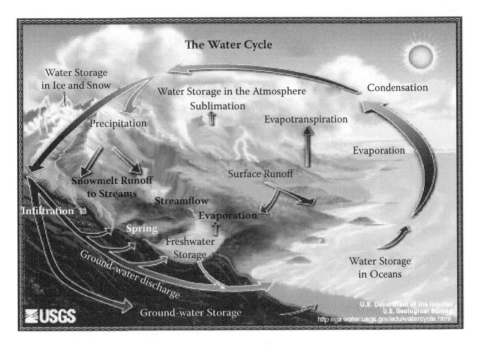

FIGURE 5   The water cycle. (Available at http://groundwater.sdsu.edu, September 2007.)

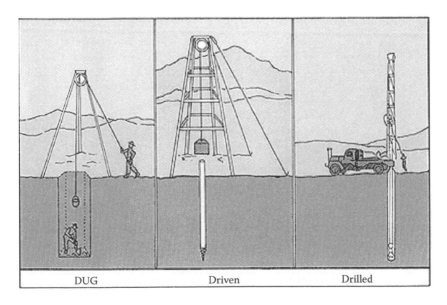

FIGURE 6   Well types. (Adapted from USGS website, available at http://ga.water.usgs.gov/edu/earthgwwells.html, February, 2007.)

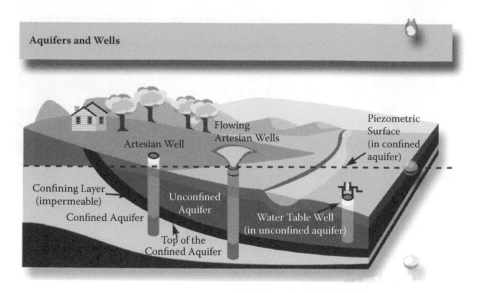

FIGURE 7   Aquifers and wells. (Adapted from USGS website, available at http://ga.water. usgs.gov/edu/earthgwwells.html, February, 2007.)

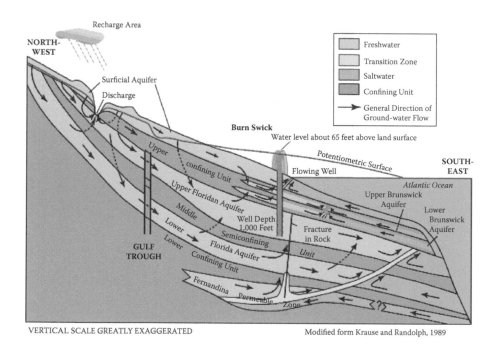

FIGURE 8   Artesian well. (Adapted from USGS website, available at http://ga.water.usgs.gov/ edu/gwartesian.html, February, 2007.)

Monday, February 19, 2007 03: 06ET

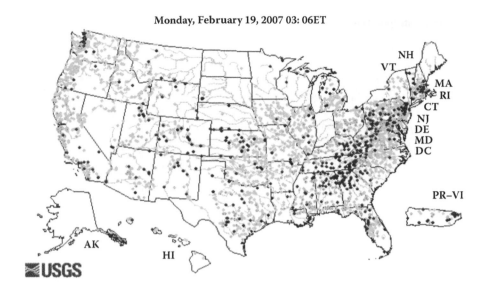

		Explanation–Percentile Classes				
●	●	○	○	○	●	●
Low	<10	10–24	25–75	76–90	>90	High
	Much below normal	Below normal	Normal	Above normal	Much below normal	

FIGURE 9 Real-time stream flows to historic stream flow comparison. The map shows the stream flow cites monitored by USGS and the flow comparison for February 19, 2007. Current flow listings can be obtained from http://water.usgs.gov/waterwatch, Feb. 2007.

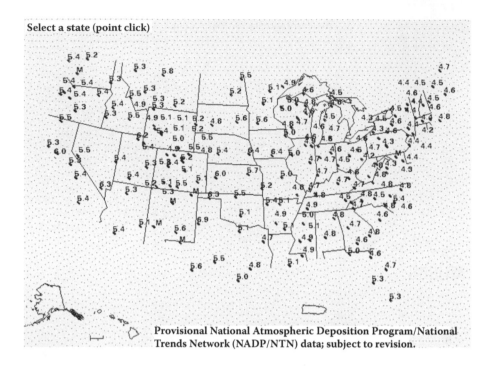

Select a state (point click)

Provisional National Atmospheric Deposition Program/National
Trends Network (NADP/NTN) data; subject to revision.

FIGURE 10   pH of precipitation for November 26 through December 23, 2001. Current pH data shown on the map are precipitation-weighted means calculated from preliminary laboratory results provided by the NADP/NTN Central Analytical Laboratory at the Illinois State Water Survey and are subject to change. The 190 points shown on this map represent all sites that were in operation during the reporting period. A notation of D instead of a pH value at a site indicates that there was no precipitation; a notation of M indicates that data for the site did not meet preliminary screening criteria for this provisional report. (Available at http://water.usgs.gov/nwc/NWC/pH/html/ph.html, March 2007.)

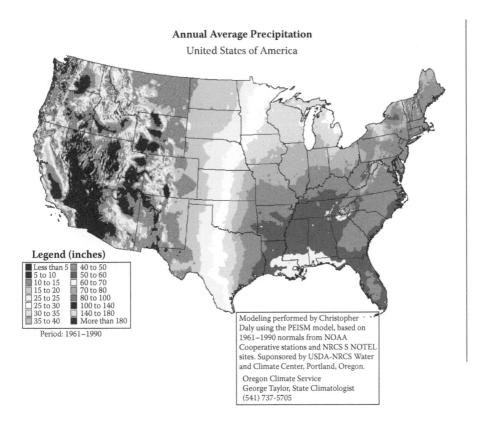

FIGURE 11 Average annual precipitation in the United States. (Available at http://www-das. uwyo.edu/~geerts/cwx/notes/chap17/rain_usa.html, March 2007.)

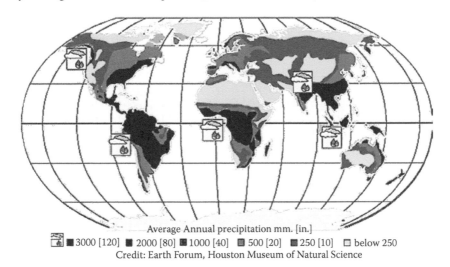

FIGURE 12 Average annual world rainfall. (Modified from figure available at http://ga.water. usgs.gov/edu/watercycleprecipitation.html, March 2007.)

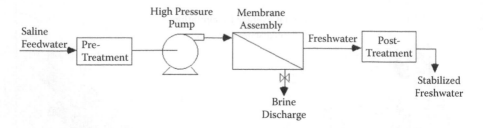

FIGURE 13   Reverse osmosis. (Available at http://www.coastal.ca.gov/desalrpt/dchap1.html, March 2007.)

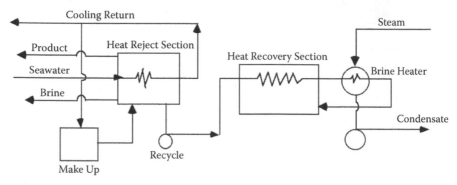

Multi-State Flash (Recycle)

FIGURE 14   Distillation.   (Available at   http://www.coastal.ca.gov/desalrpt/dchap1.html, March 2007.)

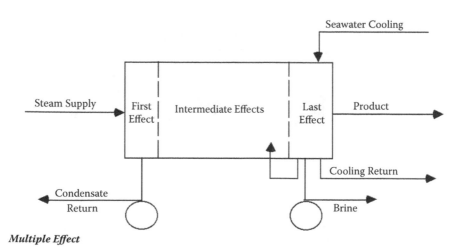

Multiple Effect

FIGURE 15   Distillation.   (Available at   http://www.coastal.ca.gov/desalrpt/dchap1.html, March 2007.)

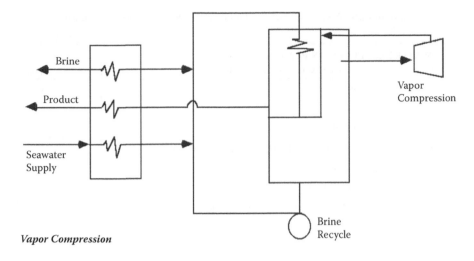

Vapor Compression

FIGURE 16 Distillation. (Available at http://www.coastal.ca.gov/desalrpt/dchap1.html, March 2007.)

Estimates for electricity use requirements for various technologies for seawater desalination are:

Multistage Flash (MSF)	3,500–7,000 kWh/AF
Multiple Effect Distillation (MED)	2,500–5,000 kWh/AF
Vapor Compression (VC)	10,000–15,000 kWh/AF
Reverse Osmosis (RO)–single pass	5,800–11,000 kWh/AF
Reverse Osmosis (RO)–double pass	6,500–12,000 kWh/AF

FIGURE 17 Energy required for desalination. (Available at http://www.coastal.ca.gov/desalrpt/dchap1.html, March 2007.)

Costs of water from Desalination Plants & Other Sources (1992 Cost Basis)

	$ Cost (per AF)
SEAWATER DESALINATION PLANTS	
• Chevron Gaviota Oil and Gas Processing Plant	4,000
• City of Morro Bay	1,750
• City of Santa Barbara**	1,900
• Marin Municipal Water District*	1, 600–1,700
• Metropolitan Water District (MWD) of Southern California*	700
• Monterey Bay Aquarium*	1,800
• PG&E Diablo Canyon Power Plant	2,000
• San Diego County Water Authority (South Bay Desalination Plant)*	1,100–1,300
• SCE, Santa Catalina Island	2,000
• U.S. Navy, San Nicolas Island	6,000
OTHER WATER SOURCES	
• City of Santa Barbara	
Lake Cachuma—existing source	35
Groundwater—existing	200
Groundwater wells in mountains—new source	600–700
Expanding reservoir—new	950
Tying into State Water Project	1,300
Temporary State Water Project deliveries Via MWD	2,300
• Metropolitan Water District (MWD) of Southern California	
Colorado River—existing	27
California Water Project—existing	195
Imperial Irrigation District—new	130
Water storage project—new (no water now)	90
• San Diego County	
MWD—existing	270
New water projects—new	600–700

Notes:

   * Cost estimate for a proposed plant.

  ** Cost amortized over 5 years.

   Unless otherwise noted, cost estimates for desalination plants are costs to produce the water. Cost estimates listed under other sources are the costs to the city, county, or water district.

FIGURE 18   Estimated desalinization costs. (Available at http://www.coastal.ca.gov/desalrpt/dchap1.html, March 2007.)

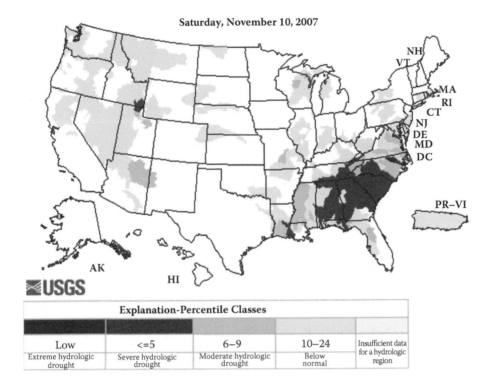

FIGURE 19   Drought watch by USGS. (Available at http://water.usgs.gov/waterwatch/?m=dryw&w=map&r=us, November11, 2007.)

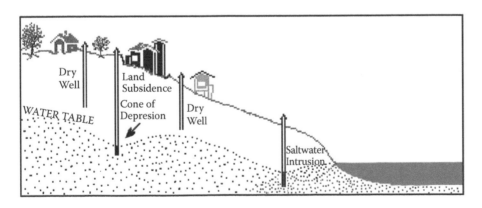

FIGURE 20   Impacts of groundwater depletion. (Available at http://groundwater.sdsu.edu, September 2007.)

FIGURE 21 Land subsidence in United States. (From United States Geological Service, available at http://water.usgs.gov/ogw/pubs/fs00165, September 2007.)

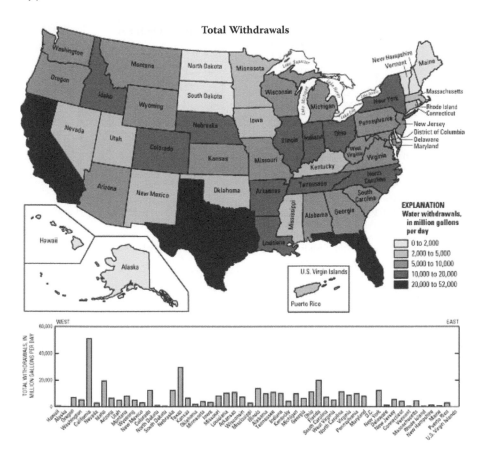

FIGURE 22   Total water use in United States. (Available at http://ga.water.usgs.gov/edu/maptotal.html, March 2007.)

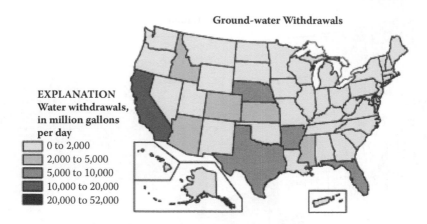

FIGURE 23   Groundwater withdrawals in United States. (Available at http://ga.water.usgs.gov/edu/maptotalgw.html, March 2007.)

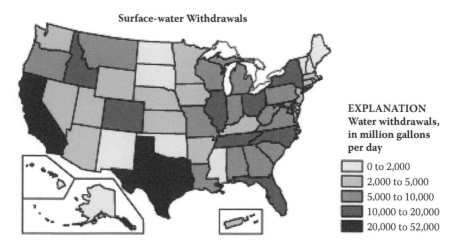

FIGURE 24   Total surface withdrawals in United States. (Available at http://ga.water.usgs.gov/edu/maptotalsw.html, March 2007.)

FIGURE 25 Bureau of Reclamation water operations. (Available at http://www.usbr.gov/main/water, March 2007.)

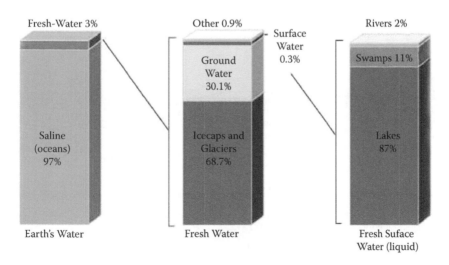

FIGURE 26 Distribution of Earth's water. (Available at http://groundwater.sdsu.edu/, September 2007.)

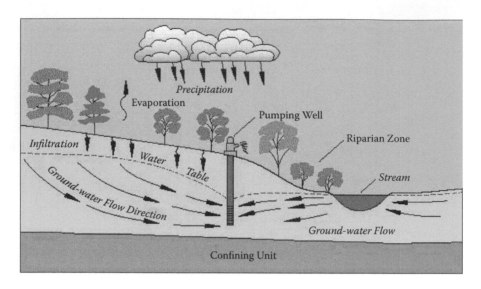

FIGURE 27   Influent to partially effluent stream due to pumping. (Available at http://groundwater. sdsu.edu, November 11, 2007.)

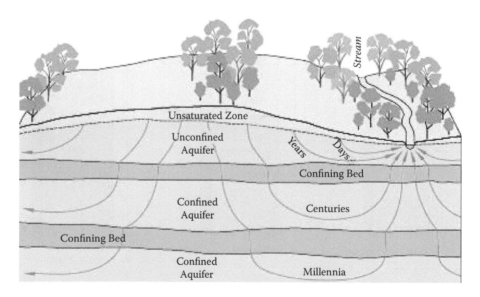

FIGURE 28   Age of groundwater. (Available at http://groundwater.sdsu.edu, September 2007.)

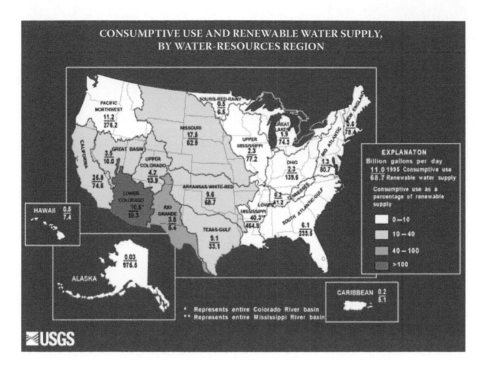

FIGURE 29 Consumptive use to renewable supply. (Available at http://water.usgs.gov/watuse/misc/consuse-renewable.html, November, 2007.)

Index